Applied Industrial Catalysis
Volume 2

Contents of Other Volumes

Applied Industrial Catalysis

Volume 2

Edited by

BRUCE E. LEACH

Conoco Inc.
Research and Development
Ponca City, Oklahoma

1983

ACADEMIC PRESS

A Subsidiary of Harcourt Brace Jovanovich, Publishers

New York   London
Paris   San Diego   San Francisco   São Paulo   Sydney   Tokyo   Toronto

ACADEMIC PRESS, INC.
111 Fifth Avenue, New York, New York 10003

United Kingdom Edition published by
ACADEMIC PRESS, INC. (LONDON) LTD.
24/28 Oval Road, London NW1 7DX

Library of Congress Cataloging in Publication Data

Main entry under title:

Applied industrial catalysis.

   Includes index.
   1. Catalysis. 1. Leach, Bruce E.
TP156.C35A66 1983   660.2'995   82-22751
ISBN 0-12-440202-X (v. 2)

PRINTED IN THE UNITED STATES OF AMERICA

83 84 85 86   9 8 7 6 5 4 3 2 1

Contents

1 Catalyst Design and Selection

ARTHUR W. SLEIGHT AND UMA CHOWDHRY

2 Uses and Properties of Select Heterogenous and Homogenous Catalysts

FRANK S. WAGNER

3 Hydrogenations—General and Selective

A. B. STILES

4 Dehydrogenation and Oxidative Dehydrogenation
A. B. STILES

5 The Sasol Fischer–Tropsch Processes
MARK E. DRY

6 Methanol Synthesis
F. MARSCHNER AND F. W. MOELLER

7 Oxidation Catalysts for Sulfuric Acid Production
J. R. DONOVAN, R. D. STOLK, AND M. L. UNLAND

Contributors

Numbers in parentheses indicate the pages on which the authors' contributions begin.

UMA CHOWDHRY (1), Central Research and Development Department, E. I. du Pont de Nemours and Company, Experimental Station, Wilmington, Delaware 19898

J. R. DONOVAN (245), Monsanto Enviro-Chem Systems, Inc., St. Louis, Missouri 63131

MARK E. DRY (167), Sasol Technology (Pty) Limited, Sasolburg, South Africa, 9570

F. MARSCHNER (215), Mineraloltechnik GmbH, D-6000 Frankfurt am Main 11, Federal Republic of Germany

F. W. MOELLER (215), Lurgi Kohle und Mineraloltechnik GmbH, D-6000 Frankfurt am Main 11, Federal Republic of Germany

ARTHUR W. SLEIGHT (1), Central Research and Development Department, E. I. du Pont de Nemours and Company, Experimental Station, Wilmington, Delaware 19898

A. B. STILES (109, 137), University of Delaware, Center for Catalytic Science and Technology, Colburn Laboratory, Newark, Delaware 19711

R. D. STOLK (245), Monsanto Enviro-Chem Systems, Inc., St. Louis, Missouri 63131

M. L. UNLAND (245), Monsanto Enviro-Chem Systems, Inc., St. Louis, Missouri 63131

FRANK S. WAGNER (27), Strem Chemicals Inc., Newburyport, Massachusetts 10950

Preface

Catalysts play a major role in the production of chemical intermediates from oil, natural gas, coal, and natural feedstocks. Catalysis represents an active research field with increasing cooperation between academic and industrial scientists. The objective of these volumes is a practical description of catalysis as practiced by industry to explain important aspects of commercial operation with the use of specific examples.

The reader should recognize that the processes discussed are examples of current industrial practice but obviously do not disclose proprietary information. Competitive processes are available for the synthesis of most chemical intermediates. The emphasis is on commercial practice and not on theory, kinetics, and reaction mechanisms. Excellent reference works already exist for these important aspects of catalysis.

This second of three volumes begins with helpful guidance in catalyst selection by A. W. Sleight and U. Chowdhry. Forty catalysts most often cited in the literature are reviewed by F. S. Wagner. Catalytic uses of these forty compounds are reviewed using only the most recent references.

Hydrogenation and dehydrogenation reactions produce a wide variety of chemical intermediates. A. B. Stiles reviews catalyst and reactor selections for these important synthetic reactions.

The synthesis of hydrocarbons from carbon monoxide and hydrogen has recently received renewed attention. M. E. Dry reviews the history and development of Fischer–Tropsch plant technology from the perspective of commercial experience.

The synthesis of methanol from carbon monoxide and hydrogen ranks third in volume behind ammonia and ethylene. The history and development of methanol synthesis catalysts are described by F. Marschner and F. W. Moeller.

Sulfuric acid is the largest volume chemical commodity produced. J. R. Donovan, R. D. Stolk, and M. L. Unland describe the catalysts, reaction equilibria and kinetics, and reaction variables for oxidation of sulfur dioxide to sulfur trioxide.

The editor acknowledges Drs. D. P. Higley and J. B. Winder, who assisted in reviewing the chapters. Dr. C. M. Starks, who encouraged me to edit this work, and Sherry Head, who assisted with communication to authors and typed manuscript revisions, deserve special recognition.

It is the editor's hope that this book will contribute to an understanding of industrial catalysis and industrial processes and further scientific cooperation between academic and industrial scientists.

CHAPTER 1

Catalyst Design and Selection

ARTHUR W. SLEIGHT   UMA CHOWDHRY

Central Research and Development Department
E. I. du Pont de Nemours and Company
Experimental Station
Wilmington, Delaware

I. Introduction

For a catalyst to be useful in a commercial process, it must usually meet certain requirements of activity, selectivity, and lifetime. There are situations in which selectivity might be discounted, for example, in a catalyst for the complete combustion of hydrocarbons to carbon dioxide and water. However, there has generally been increased emphasis on catalyst selectivity in recent years even to the point where activity and lifetime are sometimes sacrificed. Important factors have been the increasing cost of feedstocks, energy, and disposal of undesirable by-products.

There are often compromises to be made with respect to catalyst activity, selectivity, and lifetime. A catalyst with very high activity due to a very high surface area may possess intrinsic problems involving lifetime and even selectivity. In general, the most active catalysts are amorphous or are very poorly crystalline materials with considerable disorder. Most highly selective catalysts on the other hand tend to be crystalline and present a more uniform distribution of active sites.

In principle, one can quickly screen large numbers of catalysts for activity and selectivity. Evaluation of lifetime is by its very nature time-consuming. Many publications (including patents) present interesting examples of catalyst activity and selectivity with little or no information on catalyst life. In view of the difficulty of evaluating the lifetime for every catalyst with attractive selectivity and activity, an understanding of catalyst design and selection with regard to catalyst life becomes especially desirable.

II. Principles of Design and Selection

A. PROPOSED REACTION

Before selection of a catalyst, one must carefully consider the conditions to which the catalyst will be subjected in a reactor. Starting with thermodynamic considerations, appropriate ranges of temperature, pressure, and feed composition must be defined for the desired reactants and products. Competing side reactions and product degradation reactions must also be considered. Usually, there are further practical considerations. In oxidation reactions, explosive mixtures of reactants are generally, but not always, avoided. Corrosion problems in a reactor can influence the selection of reactor conditions.

There are many economic factors to be considered in defining reasonable reactor conditions. Optimum energy utilization is important as well as the cost of the feed material and the reactor itself. Frequently it is desirable that the feed to the reactor contain an amount of the least expensive reactant in excess of that required for a stoichiometric reaction. Desirable reactor conditions may also be influenced by considerations of the value of various by-products or the difficulty of separating the various reaction products.

B. POSSIBLE CATALYSTS

After deciding upon appropriate reaction conditions, the next phase in catalyst selection might be consideration of what types of materials are likely to survive under what in many cases may be regarded as very hostile conditions. For hydrogenation reactions, stable materials are generally metals or oxides not reduced by hydrogen, e.g., Cr_2O_3. For oxidation reactions, stable materials are normally oxides or noble metals.

There is a considerable amount of misleading literature concerning the types of materials considered for various reactions. Frequently, oxide catalysts are reduced to the metal or a metal oxide–metal mixture when used. Indeed, there may be some indication of this in that a pretreatment or activation step is recommended for the catalyst. An oxide catalyst may become a sulfide when used as a hydrodesulfurization catalyst, or it may become a fluoride when used as a fluorination catalyst. In fact, there are situations in which a given oxide catalyst finds application in several different reactions. In one case it may in truth be a sulfide; in another case it may be a mixture of a metal and a metal oxide. If one is interested in the rational design of a catalyst based on the underlying chemistry, it is essential to know the nature of the real working catalyst which in some cases bears little resemblance to the catalyst a manufacturer supplies.

Many catalysts contain metals known to exhibit several different oxidation states. For example, in an oxide catalyst containing iron there may be Fe^{2+}, Fe^{3+}, or both. In fact, there is good reason to believe that this mixed or variable oxidation state directly influences catalytic properties. However, when considering a variable oxidation state catalyst, there should be some appreciation of the plausible oxidation state under a given set of conditions. For example, molybdenum oxides are in a low oxidation state ($+4$ or lower) under highly reducing conditions but are close to the maximum oxidation state of $+6$ when used as oxidation catalysts with excess oxygen. Although one might inquire about the catalytic properties of MoO_2 versus MoO_3 for a particular reaction, only one of these two compounds can be maintained under a given set of reactor conditions.

A complication in considering the stability of a catalyst in a catalytic reactor is that the conditions inside the reactor usually vary dramatically from one region of the reactor to another. A successful catalyst must then survive these various conditions. In fluid bed or riser reactors, catalyst particles rapidly move through various parts of the reactor where different conditions exist. In a fixed bed reactor, the conditions at the inlet generally are very different that at the outlet. The temperature may be significantly different, and the gas phase composition is definitely different. For example, in an oxidation reaction operated under lean oxygen conditions to avoid the explosive region, there may be no oxygen at the outlet. In this case the catalyst is exposed to significant oxygen pressure at the beginning of the bed but to essentially no oxygen at the end of the bed. If the catalyst is a metal oxide, the oxidation state of the metal, and its catalytic properties, might well differ in various regions of the bed.

C. PROBABLE CATALYSTS

Having decided which types of materials are likely to be stable under reaction conditions, one must then decide which of these materials might actually be a catalyst for the desired reaction. First, one considers past experience with similar reactions. For example, molybdates and vanadates are good candidates for a wide variety of selective oxidation reactions, and palladium is a good catalyst for a wide variety of hydrogenation reactions.

One can also consider known chemistry: solution chemistry, noncatalytic reaction chemistry, homogeneous catalysis for possible heterogeneous analogs, and surface chemistry. For example, if one is interested in an acid-catalyzed reaction, the literature could be evaluated for solid compounds of high acidity. One should also consider whether or not such candidates will be stable under the reactor conditions. Then candidates should be obtained and screened in a laboratory reactor.

Zeolites present special opportunities for rational catalyst design and selection in view of their unique selectivity capabilities. The pores in zeolites are well-defined and of molecular dimensions. Therefore, one can design catalyst selectivity based on dimensions of products, reactants, or transition state intermediates. Also, one has considerable control over the acidity which can be very high.

In choosing probable catalyst candidates, one must also consider the undesirable side reactions that might be promoted by various catalysts. One must as well examine the known chemistry of the product. Is the product likely to survive reactor conditions in the presence of the candidate catalyst? Some catalysts are more impressive for what they do not catalyze than for what they do catalyze. For example, a good selective oxidation catalyst gives high conversions to the desired product, yet there is very little overoxidation to carbon dioxide and water even in the presence of a great excess of oxygen.

Consideration should be given at this point in catalyst design and selection to whether or not the catalyst will be supported. If a support is desirable for a reason such as dispersion or strength, the support should be carefully chosen. Candidate supports must first be considered from the point of view of stability. Because alumina dissolves in acid, it cannot, for example, be used as a support in a slurry bed reactor run at low pH. Likewise, silica cannot be used as a support for a fluorination reaction since SiF_4 is volatile. Another aspect of choosing the proper support is the catalytic activity of the support itself. Silica is generally regarded as more inert than alumina; however, the catalytic activity of alumina is sometimes advantageous, for example, in reforming reactions.

Some consideration should also be given to the physical requirements of the catalyst. For example, what are the physical strength requirements? If the catalyst will be used as pellets, the crush strength may well be important. If the catalyst is to be used in a fluid bed or slurry bed reactor, attrition resistance will be important. The physical form of catalysts used in laboratory reactors should, at least in the later stages of evaluation, closely resemble that anticipated in the application.

D.   CATALYST SCOUTING

Having decided upon what types of catalysts to obtain or prepare, how should the catalytic properties be evaluated? Two different strategies can be implemented. One is to design a laboratory reactor that can most closely match the envisioned application. However, with such a reactor it may be essentially impossible to obtain intrinsic rate and kinetic data. Usually, it is preferable to start with a laboratory minireactor designed not to simulate a plant reactor but rather to produce intrinsic kinetic and mechanistic information. This affords the possibility of obtaining a better understanding of the reaction at the same time one is scouting for different catalysts.

Frequently in a catalyst scouting program, each catalyst is examined under very few conditions, and the only information obtained is reactant conversion and selectivity at this conversion. There are several problems with this approach. One is that a good catalyst can easily be overlooked, since conditions appropriate for the catalyst may not have been employed. Another problem is that little useful information is obtained that will be of value for the next round of catalyst design and selection. One will have obtained an indication that some catalysts are good and others are not, but there will be little understanding of *why* some are good and others are not. The increasing automation of laboratory reactors for catalyst scouting has led to a situation where the examination of each catalyst under many conditions is feasible. It is, of course, essential that the pattern of conditions used be such that the catalyst is not adversely affected by inappropriate conditions before it has been evaluated under its optimum working conditions. For example, the temperature is normally increased rather than decreased during a set of runs. As a check on changes occurring in the catalyst while in the reactor, the final reactor conditions should be the same as certain prior reactor conditions.

E.   MECHANISMS

Once leads to a good catalyst for a given reaction have been found, it is desirable to obtain some mechanistic information. Such information can be very useful in optimizing the catalyst and can also influence the optimization of reactor operating conditions. Even in the case where a catalyst acceptable for a plant is found through a semiempirical scouting procedure and no further optimization appears worthwhile, one can be confident that there will eventually be problems in the plant reactor. The more one already knows about the mechanism, the more likely one is to find a quick solution.

Mechanistic studies can start simply with rate data obtained for various temperatures and reactant mixtures under differential conditions where heat and mass transfer are under control. One might also consider isotopic labeling and kinetic isotope experiments. A reasonably sound picture of certain critical steps may emerge from such a study. The rate-limiting step may become apparent. When one is attempting to increase the activity of a selective but relatively inactive catalyst, it can be very useful to know what is limiting the rate of the reaction. A comprehensive mechanistic understanding of a reaction carried out over a heterogeneous catalyst is only rarely achieved. However, instrumentation advances in recent years make such a goal much less elusive today than 10 yr ago. Substantial understanding of some aspects of the mechanism may be obtained by combining sophisticated reactor studies with thorough catalyst characterization, chemisorption measurements, temperature-programmed desorption (TPD), and surface spectroscopic studies that examine the catalyst surface and reaction intermediates on the catalyst surface.

III. Surface Sites

Surface sites are clearly of paramount importance in the design and selection of a catalyst. Frequently, the focus is only on this aspect of a catalyst, and this can be shortsighted. However, it is difficult to overemphasize the requirement that a selective catalyst must have just the right type of surface sites and that it cannot have too many of the wrong type.

Acid and base sites on the surface of a catalyst are important for a wide variety of reactions. They may be either on the Brönsted or the Lewis type. Frequently, dissociative chemisorption requires an adjacent pair of acid and base sites. For example, the dissociative adsorption of methanol on an oxide surface to form methoxy and hydroxyl groups can take place at a Brönsted base site (B) next to a Lewis acid (L) site:

$$CH_3OH \rightarrow CH_3O-L + H-B$$

There are many different ways to measure the acid or base properties of a surface [1] and not surprisingly these methods tend to give somewhat different results.

Traditionally, types of surface sites are assessed by a variety of chemisorption experiments. This method becomes even more powerful if it is followed by temperature-programmed desorption or decomposition of the chemisorbed species. Additional assessment of the nature of surface sites is obtained if the chemisorbed species are observed directly by a technique such as infrared spectroscopy.

Chemisorption measurements of the actual reactants and products of the reaction under consideration are desirable. This is sometimes, but not always, possible. If the rate-limiting step in the reaction occurs strictly on the catalyst surface, then chemisorption measurements of a reactant may well yield the number of active sites. However, if the rate-limiting step in the reaction is the chemisorption step itself, and this is often the case, meaningful chemisorption measurements of this reactant may not be possible. In such cases, chemisorption measurements of other molecules may be useful in probing surface sites, and this may ultimately yield the number of active sites.

On oxide catalyst surfaces, the most common chemisorption measurements are those directed at an assessment of the acid – base properties of the surface. On metal catalyst surfaces, the most common chemisorption measurements are of CO and hydrogen. In many situations dissociative hydrogen chemisorption and nondissociative CO chemisorption tend to give a reasonable picture of metal dispersion on a support. If this is coupled with another technique such as TPD, the chemisorption data may give a reliable indication of the number of active sites for a given reaction.

IV. Bulk Properties

The bulk properties of a catalyst are sometimes neglected because their importance is less clear than that of surface properties. Indeed, there are probably many cases where the bulk properties of a catalyst play no significant role. However, for very fast and very exothermic reactions, the thermal conductivity of the bulk can be important. If the surface is left in a thermally excited state for too long a time, reactions other than those desired may occur.

Oxidation reactions frequently operate through redox mechanisms whereby the catalyst is continually reduced by hydrocarbon molecules at one set of surface sites and continuously reoxidized by molecular oxygen at another set of surface sites. For such a process to function successfully, there must be good communication between the two sites. Electrons must flow from site to site — thus the electronic structure of the bulk is relevant. Also, oxygen mobility in the bulk as well as at the surface is important in replacing oxygen at the surface site when it becomes depleted through reaction. Thus, in a typical selective oxidation catalyst, one must consider the proper set of surface sites, the electronic properties of the bulk, and the oxygen mobility in the bulk.

V.   Metal – Support Interactions

Frequently catalyst supports are considered inert materials having the simple role of dispersing the catalyst. In fact, supports very often significantly alter catalytic selectivity. There are many different types of interactions between catalysts and their supports. The effect of these interactions on catalysis is not always well understood.

The reducibility of metals depends on the support used. The best documented case is iron. Although iron is readily reduced completely to the metal on silica and carbon supports, hydrogen treatment generally fails to reduce all iron past the divalent state when it is supported on alumina. Thus, if one desires metallic iron for catalysis, an alumina support might best be avoided. On the other hand, if one is interested in catalysis by divalent iron under highly reducing conditions, alumina might be the preferred support. An analogous metal–support interaction has been observed for other metals, for example, rhodium.

Another type of chemical interaction between a metal and its support has been observed for platinum on alumina under high-temperature reduction with hydrogen. Some alumina is taken into solid solution with platinum. At present it is not clear how widespread or important this effect is. Certainly one might expect that an impurity in a support, such as iron in alumina, would become alloyed with the supported metal. Such alloying could have significant effects on catalytic properties.

The size, shape, and distribution of metal particles can also vary considerably depending on the support used. One critical factor is that metals tend to wet some supports and not others. For example, metal particles that tend toward cubic or octahedral morphology might form rafts instead if the metal tended to wet the support. The size of metal particles can depend on the pore structure of the support and also on the mobility of very small metal particles on the support surface. The actual pattern of metal particles on the support surface may also be governed by the surface and pore structure of the support.

Electronic interactions between a metal and its support can be significant, especially if the metal particles are very small. From work function considerations, one expects that for metals supported on oxides there will generally be a net electron flow from the metal to the oxide. Thus, metal particles tend to take on a small positive charge when supported on an oxide such as silica. However, for a reduced metal oxide such as TiO_{2-x}, the reverse electron flow occurs, since the work function of the support is less than that of the metal [2]. Although the direction of electron flow has been confirmed by various experiments, the magnitude of the charge on the metals has not been established. Presumably, the magnitude of the metal charge depends on the size of the metal particles as well as on the particular metal and support. Surface electronic states, essentially what most chemists refer to as Lewis acid or base sites, are also expected to transfer electron charge to or from small metal particles. The relative impact of the bulk and surface electronic structure of a support on the electronic state of supported metals needs considerable further investigation.

Spillover is yet another type of metal–support interaction. The most common and best established example of this is hydrogen spillover. Hydrogen is adsorbed and dissociated on metal particles. Then it may spill over onto the support surface where it reacts with other molecules adsorbed there.

It is also possible that a special situation for catalysis exists at the interface between a metal and its support. For example, a molecule might chemisorb on a support close enough to the metal to interact with it directly. This type of metal–support interaction tends to be important only for highly dispersed metals.

The pore structure of the support can also have considerable influence on catalysis by the metal. This can be a dramatic effect when a metal is supported within the pores of a zeolite structure. The pore structure of amorphous supports can also influence catalytic properties [3].

VI. Other Multiphase Catalysts

Many catalysts other than supported metals are multiphase systems. For example, catalysts for selective oxidation frequently contain 10 or more elements forming several phases, some crystalline and some not [4]. It can be very difficult to understand the roles of the various phases in such catalysts that were generally arrived at by largely empirical methods. A common mistake made is to assume that all ingredients of such catalysts have some direct catalytic function. This is not generally true. Usually, some components are present for reasons of ease of synthesis and promotion of desired microstructure, catalyst strength, and catalyst life.

The most selective catalyst for the oxidation of methanol to formaldehyde is one based on iron, molybdenum, and oxygen. The catalyst used commercially is simply a physical mixture of ferric molybdate and molybdenum trioxide. These phases are highly crystalline. There is no support, and usually there are no promoters. There has been considerable controversy on the question of why both phases apparently need to be present in a commercially viable catalyst. A typical question involves which phase is really the catalyst. We now know that both ferric molybdate and molybdenum trioxide are excellent catalysts for the oxidation of methanol to formaldehyde [5]. In fact, from the point of view of intrinsic catalytic properties, there are practically no differences between the two phases. However, neither phase by itself can be a practical catalyst. Molybdenum trioxide is too volatile under reactor conditions, and it very readily loses surface area. Iron tends to stabilize molybdenum oxide against volatilization and loss of surface area. However, the surface of ferric molybdate gradually loses some molybdenum through volatilization, and then a loss in selectivity occurs. In the presence of excess molybdenum this surface depletion does not occur. Therefore, the mixing of ferric molybdate and molybdenum trioxide results in a catalyst with a lifetime far in excess of that possessed by either component alone.

VII.   Catalyst Microstructure

An understanding of the fundamental processes that govern the evolution of catalyst microstructure during catalyst processing is essential to rational catalyst design. The phrase "catalyst processing" is chosen to include catalyst synthesis as well as catalyst fabrication (Fig. 1). Some of the critical microstructural features of a catalyst are surface area, pore volume, pore size and distribution, particle size and distribution, arrangement of phases, particle morphology, and degree of agglomeration. These features can affect catalyst activity, selectivity, and lifetime. The important message to be conveyed in this section is that the microstructure of the catalyst should be examined at each step along the catalyst processing route. Control over the evolution of this microstructure can then be systematically attempted. The kinetic processes of nucleation and growth of particles during synthesis, particle morphology change, matter transport, and pore coarsening or shrinkage during fabrication are generally not well understood or controlled. A data bank of carefully documented observations of microstructural evolution during catalyst processing and use would be an invaluable asset in improving our understanding of catalytic phenomena.

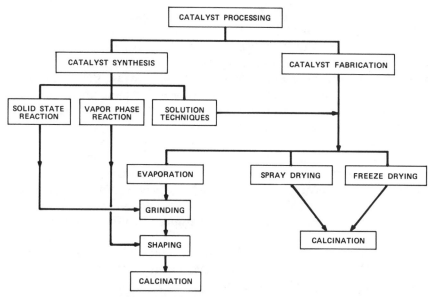

Fig. 1.   Steps involved in catalyst processing.

Although catalyst microstructural requirements are dictated by the type of catalytic reaction as well as reactor configuration, it is generally true that catalyst activity increases with surface area. Section VII.A describes some nontraditional techniques for making oxide catalysts with high surface areas. To increase the surface area of the catalytically active phase, some oxides and most metals are supported on porous, refractory supports. A high-surface-area support may, however, be detrimental if undesirable side reactions occur in the pores of the support or on its surface. Techniques must then be sought to obtain a high dispersion of the active phase on a low-surface-area inert support, and this is difficult to achieve in practice. In the final analysis, it is catalyst performance that one is seeking to optimize. Thus, all studies relating to microstructural control must aim at preservation of the microstructural design under catalytic conditions for a reasonable period of time.

A.   SYNTHESIS ROUTES

The range of techniques used for preparing mixed oxide catalysts or supports includes solid state reactions, vapor phase processes, and solution techniques. Traditional solid state approaches of milling metals or oxides together and firing at high temperatures to effect formation of the desired alloy or compound result in low-surface-area materials generally not suitable for catalytic applications.

Vapor phase techniques are becoming increasingly popular, for they lead to very fine unagglomerated powders that often have spherical morphology. Control in multicomponent systems is difficult because the components have different vapor pressures. A disadvantage of scaling up vapor phase processes is that large volumes of gas are usually trapped in the powder, hence very large collection systems are necessary.

Condensation of the desired vapor phase species can be used to obtain a fine powder. In general, MoO_3 crystallizes with a platey habit, but submicrometer spheres may be produced by controlled condensation of MoO_3 vapors. Evaporation of oxides by electron beam heating can also be used [6]. Plasmas of up to 20,000 K can now be generated, and highly refractory oxides, carbides, and nitrides of submicrometer particle size have been prepared [7, 8].

Vapor phase decomposition of organometallics [9] has been shown to yield high-purity, finely divided materials, but this technique tends to be limited to preparing single-component systems. The commercial production of TiO_2 and carbon blacks is an example of vapor phase reactions involving halides and hydrocarbons. Laser processing is an emerging technique in which a high-intensity laser beam is used to initiate gas phase reactions. Silicon powder has been synthesized by the laser-driven decomposition of SiH_4 (gas), with a resultant surface area of 55 m^2/g [10].

The most widely used techniques for catalyst preparation start with a solution of the desired metal salts. This is the optimum way to effect atomic mixing of the desired components. The solid catalyst precursor is then formed by direct evaporation of solvent, spray- or freeze-drying, coprecipitation, or gel formation. The physical methods of evaporation and spray- and freeze-drying will be discussed in Section VII.B. A brief overview of the factors governing precipitate morphology and size is appropriate at this juncture.

The morphology of precipitates is determined by a number of factors, the most important being the relative rates of nucleation and growth. These rates in turn depend to a large degree on the supersaturation of the system. The final particle size depends on the number of nucleation sites and the rate at which material is precipitated. Sparingly soluble materials, e.g., carbonates, usually precipitate as very small particles. With slow, controlled growth of sparingly soluble salts, it is often feasible to produce monodisperse precipitates. At high degrees of supersaturation, the primary critical nucleus may be smaller than a unit cell and starts to grow without having an ordered crystal structure. Amorphous or partially crystalline precipitates can form in this manner [11]. At low degrees of supersaturation, precipitate crystals are well formed, the shape depending on crystal structure and on the interfacial processes prevalent during growth. The precipitate morphology is profoundly affected by the rate of crystal growth. For slow growth rates, the crystals take on a compact shape related to the crystal structure. Ions in solution in the vicinity of the crystal–liquid interface play an important role in habit modification. At high degrees of supersaturation, large, dendritic precipitates are often observed. At even higher levels of supersaturation, very small particles form, which may agglomerate or form sols.

Mosts of the reported data on precipitation kinetics indicate that the process is too slow to be diffusion-controlled, and usually interfacial kinetics play the dominant role [11]. Important factors affecting precipitation are pH, solution concentration, temperature, order of mixing, stirring rate, and impurity ions.

The order of mixing of reagents is found to be important in many catalyst preparations. If a solution containing metal salts is added slowly to a stirred bath containing the precipitating reagent, the solubility product of all the cations will be exceeded simultaneously and a more homogeneous precipitate will result. If the reverse operation is carried out, the cations are likely to be precipitated sequentially and inhomogeneities will result [12]. For superior catalytic properties homogeneity is not necessarily desirable. In many cases, controlled heterogeneity as described for the propylene ammoxidation catalyst may prove superior [13].

Rapid mixing of reagents increases the number of nuclei formed and results in finely divided precipitates. A rapid stirring rate can affect particle size and diminish agglomeration. The presence of extraneous ions affects the surface chemistry of precipitates. After precipitation, a high concentration of electrolyte is present that can destroy the electric double layer surrounding particles and cause them to flocculate. If the excess electrolyte is washed off, the particles may form a stable colloid that is difficult to filter. On a laboratory scale, centrifuging of fine suspensions can be used to separate the solid component from solution. Very finely divided materials can be obtained in this manner.

Precipitates can be simple compounds, solid solutions, or mixtures of phases. Even simple compounds are not always easy to prepare in a homogeneous fashion with the desired stoichiometry. The preparation of ferric molybdate, which is a good methanol oxidation catalyst, poses some problems. The addition of a ferric nitrate solution to an ammonium molybdate solution causes precipitation to occur. However, if pH, order of addition, rate of mixing, and presence of other ions are not controlled, an inhomogeneous precipitate forms. When this is subsequently calcined, either Fe_2O_3 or MoO_3 is often found to coexist with $Fe_2(MoO_4)_3$. To avoid these problems, it is advantageous to start with amorphous mixtures containing a homogeneous distribution of all the necessary cations. Such an amorphous precursor can be obtained by rapid dehydration of solutions containing metallic salts and an organic polyfunctional acid such as citric, tartaric, or lactic acid [14]. Such acids possess at least one hydroxy and one carboxylic function. An amorphous, transparent, uniformly colored material is obtained when rapid dehydration of a solution containing ferric nitrate, ammonium molybdate, and citric acid is carried out. Upon pyrolysis at 400°C, a pale-green powder of crystalline stoichiometric $Fe_2(MoO_4)_3$ is obtained.

The hydrolysis of organometallic compounds is used effectively to make high-purity, high-surface-area materials. In this method, a water-sensitive organometallic compound, generally an alkoxide, is decomposed by hydrolysis [15]. High-purity aluminum oxyhydroxide made in this manner is used for catalyst manufacture.

While the kinetics of precipitation are interface-controlled, the processes of precipitate dissolution and Ostwald ripening are most likely diffusion-controlled [11]. A system containing dispersed precipitate particles of varying sizes tends to lower the interfacial free energy by agglomeration or Ostwald ripening. Small particles are in equilibrium with a certain supersaturation S_1, and larger particles with S_2, where $S_1 > S_2$. The system adjusts itself to a supersaturation between S_1 and S_2, with the result that small particles dissolve and large ones grow. Agglomeration is another mechanism for lowering the total interfacial energy. Its rate is a function of the number of particles and their surface charge. The rate of agglomeration is negligible if the system is dilute, with less than 10^6 particles/ml, even if the particles have no charge. For systems in which a strong surface charge is established, particles can be stabilized without a detectable rate of agglomeration, as in colloidal silica.

The area of colloidal chemistry has tremendous potential for fundamental research in catalyst design. The ability to control particle size, surface charge, and particle morphology in some systems [16] will allow studies on the effects of these variables on catalytic performance to be conducted in a systematic manner.

Sol–gel techniques for catalyst synthesis have become increasingly popular in recent years. These methods involve preparing a sol that generally contains particles less than 200 Å in size held in suspension in a liquid medium. The solids concentration of these sols is often low, but colloidal silica containing 40 wt% SiO_2 is commercially available. By controlling the pH of the sol, hence the charge on the particle surface, gelation can be obtained [17]. This causes the particles to flocculate and form a continuous rigid network. Extremely uniform mixing of components can be achieved. Various solutions of cations can be incorporated into the pores of the gel skeleton, akin to impregnating a support. The manner of removal of water from the gel is critically important because it can alter the microstructure of the product. This will be discussed further in Section VII.B.

The above-mentioned techniques generally describe the preparation of unsupported catalysts. Supported oxide catalysts are often synthesized by coprecipitating various metal salts in the presence of colloidal silica, for instance. The propylene ammoxidation catalyst supported on silica is often made in this manner. Supported metal catalyst preparation involves the introduction of a solution, aqueous or nonaqueous, containing the desired salt onto a porous support.

Impregnation can be performed in several ways [18]. On a laboratory scale, an excess of the impregnating solution is often used. The maximum concentration of the active component then depends on the solution concentration. If all the solvent evaporated instantaneously, the solute would be uniformly deposited. However, because of capillary forces and a distribution of pore sizes, solvent evaporation is slow and a nonuniform solute deposit results. To obtain high concentrations of the active component, successive impregnations and drying operations can be performed. The incipient wetness technique is utilized when it is desirable to have just enough liquid to fill the pores of the support. The method used is to evacuate the support and, while stirring it, spray the desired solution onto it. The volume of solution must not be greater than the support's absorption capacity. A high catalyst concentration at the exterior of support particles can be obtained by impregnating the support with a salt solution and subsequently precipitating a hydroxide at the mouth of the pores in the support. Gas phase impregnation, with metal carbonyls, for example, can also be carried out. Ion-exchange techniques give a more even distribution of catalyst particles when compared to impregnation methods. In addition, the catalyst particles tend to have smaller average diameters [19].

Fusion and leaching techniques have been employed to make porous metal catalysts [20]. The traditional catalyst for ammonia synthesis is made by the fusion of magnetite, Fe_3O_4. Textural promoters such as alumina, magnesia, and silica and chemical promoters such as potassium salts are added in small quantities to dry magnetite powder. Fusion is effected by induction heating or arc melting. The melt is then cast and ground to a suitable size. The catalyst is typically reduced to metallic iron in flowing hydrogen at about 500°C. The promoters are thought to form a phase between magnetite grains and help to confer porosity and prevent agglomeration during reduction to metallic iron. Raney nickel is the best known of the catalysts prepared by leaching [20]. A nickel–aluminum alloy is used that usually contains 50 mol% Ni. The aluminum is leached out using an aqueous alkali solution. Some residual aluminum and hydrated aluminum oxide remain and improve the catalyst's thermal stability.

B. CATALYST FABRICATION

The manner of solvent removal from the precipitate, gel, slurry, or supported catalyst is important, for it can affect the surface area and pore structure of the material. The particular method chosen also depends on the form in which the catalyst will be used. Direct evaporation is commonly used but can cause segregation of components. The rate of drying also affects the microstructure and should be controlled. In the case of silica gels containing a large percentage of water, freezing the gels produces interesting results. Upon freezing, polymerization of the silica renders it insoluble in water, and upon thawing the silica is found to have adopted the morphology of the ice crystals. Thus if dendritic growth of ice crystals is initiated, silica fibers can result [21]. Exceedingly high surface areas on the order of 1000 m^2/g have been obtained using the freezing method. Replacing the water in the gel with alcohol and subjecting it to critical conditions in an autoclave have resulted in products with high surface areas as well as very large pores [22]. The use of liquids having a lower surface tension than water, such as acetone, in washing the gel before drying prevents collapse of narrow pores due to capillary forces. One of the disadvantages of the gel technique for making high-surface-area solids is the low solids concentration often employed. Large quantities of solvent need to be removed; thus, there is an economic penalty. The powder that results is extremely friable and difficult to handle. It usually needs to be compacted prior to further processing.

Spray-drying is a technique widely used for making fluidizable catalysts. A solution or slurry containing the desired metal ions is sprayed through a nozzle into a heated drying chamber. The rapid drying of the solution droplets prevents excessive segregation. The particle size of the resulting spherical beads depends on the concentration and viscosity of the solution or slurry to be spray-dried, the size of the nozzle, the atomizing air pressure, and other factors. Spherical particles ranging from submicrometer to 200 μm in size can be prepared. Thorough drying of these particles in the chamber is essential because subsequent calcination can cause the liquid remaining in the spheres to force its way out, causing blow holes in the spheres. Binders may be introduced into the solution prior to spray-drying to increase the "green" strength of the product, but their removal can cause similar problems. For catalytic reactions in which mass and heat transfer limitations are problematic, donut-shaped spray-dried particles can in fact be useful. These "amphora" catalysts have a higher surface/volume ratio than solid spheres and pack efficiently in a fixed bed reactor. The critical parameters for making amphorae are slurry rheology and the rate of drying.

Amphora catalysts have been reported to be superior to tablets and extrudates for several oxidation reactions and for catalytic reforming [23].

During calcination of spray-dried powders, segregation of components can occur. To circumvent this problem, a technique known as evaporative decomposition of solutions (EDS) can be utilized [24]. The solution is atomized, and the fine mist so obtained is pumped into a vertical tube furnace with a long heated zone. Decomposition of aluminum nitrate solutions in the range 750–950°C results in amorphous alumina. The powders produced in this manner are generally amorphous or poorly crystalline but readily crystallize at temperatures well below normal sintering temperatures.

Freeze-drying is a technique employed to obtain free-flowing powders with high surface areas. A solution or slurry is sprayed into liquid nitrogen or into a chilled organic liquid with which water is immiscible. The solution droplets freeze as they come into contact with the chilled, rapidly stirred bath. Compositional gradients within the particle are minimized by the freezing process. The frozen material is then transferred to a precooled chamber which is evacuable. The water in the material is removed by pumping, and a high-surface-area friable powder results, which is usually amorphous.

Grinding to effect size reduction is often necessary prior to shaping the catalyst in the form of pellets. A slurry is also sometimes milled to reduce the particle size prior to spray-drying so that the resulting spherical agglomerate has a denser packing of crystallites. Milling can also be used to mix catalyst powders uniformly with binders that aid in pelleting or extrusion processes. The equipment conventionally used is a ball mill in which the powder to be ground is placed along with grinding media. The mill is placed on rollers and allowed to rotate for a period of time. For particles below 10 μm in size, grinding processes are extremely inefficient [25]. This is due primarily to the tendency of small particles to reagglomerate under grinding pressure. When particles are small, it is difficult to concentrate the crushing energy on a small number of particles at a time. The wear of the grinding media introduces unwanted impurities into the catalyst. Vibratory milling is more efficient than ball-milling. An attrition mill offers some advantages over both ball and vibratory mills [26]. It consists of a movable cage of vertical bars rotated in a slurry containing the solid to be ground. The slurry must consist of a coarse grinding medium and a fine material that is to be ground. Remarkable size reductions have been reported with this technique, with an order of magnitude of improved efficiency over vibratory milling. The grinding results in very high stresses being generated at particle surfaces. Under this stress, plastic flow occurs and highly strained particles form.

X-ray patterns of such powders reveal lattice distortions, amorphous layers and, in some instances, structural transformations. The highly energetic surfaces so formed also tend to sinter rapidly, thus relieving their strain energy.

Catalyst pellets are used in many fixed bed applications, and their size, density, and pore structure affect catalytic performance. The physical characteristics of the powder determine the manner in which it will fill the die and compact under pressure. Free-flowing powders are necessary for uniform filling of the die cavity. This can be achieved by spray-drying to obtain spherical morphology or by the addition of agents, such as Cabosil fumed silica, that help to keep the powder dry and prevent caking. Friction with the die wall causes density gradients among the pellets [28]. This can be minimized by lubricating the die prior to pelleting. The compaction pressure used should be monitored and kept constant if uniformly dense pellets are desired. Biaxial pressure during compaction and low aspect ratios for the pellets generally give more uniform results. When a powder is placed in a rubber mold and hydrostatic pressure is applied, the forming process is known as isostatic pressing and usually results in more uniform packing and a narrow pore size distribution.

Extrudates are also widely used in the catalyst industry. In the extrusion process pastelike material is pushed through an orifice that has the size and shape of the cross section of the object. The flow pattern of the material through the extruder governs the quality of the product. Thus, it is important to optimize the catalyst material/binder ratio as well as the type of binder and solvent beforehand. Particle and binder morphology determines the pore structure of the extrudate.

The task of preserving the high surface area of the green body during calcination is a difficult one. There always exists a thermodynamic driving force that decreases the surface free energy of a solid. Thus, when high-surface-area solids are used, there is always a tendency toward sintering. The rate of sintering is determined by temperature, atmosphere, and the physical and chemical nature of the solid. At sufficiently low temperatures, the rate of sintering is insignificant. Initial sintering generally occurs by surface diffusion. When the surface temperature reaches one-third of the bulk melting point, this process becomes significant. As the temperature is increased, a transition in rate-controlling mechanisms generally occurs. Bulk lattice or grain boundary diffusion gains control of the sintering rate, and large changes in particle arrangement and pore structure are observed. At even higher temperatures, evaporation–recondensation becomes important.

The important variables that affect microstructural evolution during calcination are (1) particle size and shape, (2) heating rate, (3) final temperature, and (4) atmosphere. Herring's scaling laws [29] show that, for two particles with radii r_1 and r_2, where $r_2 = \lambda r_2$, the total time required for equivalent geometric changes between pairs of particles is related to particle size by $\Delta t_2 = \lambda^n \Delta t_1$, where t_1 and t_2 are the times necessary to accomplish a given fractional shrinkage in two respective sizes of particles. The value of the exponent n gives information regarding the mechanism of matter transport. For $n = 1$, Herring showed that geometric changes occurred by viscous flow, whereas $n = 3$ for a diffusional mechanism. In general n is much larger than unity, indicating that fine particles sinter more rapidly than large particles. Several processes may operate simultaneously to cause the centers of two contacting grains to approach each other. Either viscous flow or dislocation movement can cause shrinkage. Alternatively, matter can be transported to the neck between particles by a diffusional process. These processes are to be avoided if sintering is undesirable. On the other hand, evaporation and recondensation or surface diffusion cause neck growth and increase adhesion or strength without causing shrinkage. These latter two processes cause pore shape change but do not cause pore shrinkage. The rounding of pores that results from these two processes reduces the curvature at the neck between particles, and this in turn decreases the driving force for sintering. In most cases, surface diffusion has a much lower activation energy than bulk lattice or grain boundary diffusion which causes sintering and loss of surface area. Surface diffusion is generally significant at low temperatures, and it affects catalyst synthesis as well as catalyst aging.

Before calcination, the residual water content of the material should be controlled. The presence of water vapor significantly affects the surface mobilities of ions, and in many instances it is observed that catalyst sintering during calcination can be minimized by thorough driving of the powder. This should be carried out at low temperatures under a vacuum or in the presence of a flowing, dry gas.

To preserve a high surface area, it is desirable to choose the lowest possible calcination temperature. This should, however, be well above the temperature at which the catalyst will be used. The heating rate during calcination can have varied effects depending on the mechanism of sintering that is operative. If a liquid phase forms during firing, a fast heating rate may avoid rapid sintering. The use of a slow heating rate allows the liquid to coat the particles and promote densification. On the other hand, a slow heating rate can decrease the sintering rate in cases where a solid state diffusional mechanism is operative. At low temperatures, slow heating rates allow surface diffusional processes to reduce the curvature at the neck between particles. This in turn decreases the driving force for sintering at temperatures where bulk diffusion controls the rate. Thus, faster heating rates may either decrease or increase sintering depending on the prevailing circumstances. This points out the need to understand the physiochemical nature of the catalyst under investigation.

It is sometimes observed that the surface area increases upon firing, contrary to expectation. Several explanations are possible. Particle shape influences surface area as well as particle size, with spheres having the minimum surface/volume ratio. If a spherical precursor particle were to transform into a faceted platey structure, the surface area would increase. If a binder was used in the initial preparation, its removal during firing would leave a greater pore volume, thereby increasing the solid vapor surface area. In some cases, for supported metals, redispersion is known to occur [30]. It has been reported that redispersion occurs on heating Pt supported on γ-Al_2O_3 in air between 500 and 800°C. This behavior is not well understood but may be associated with the formation of platinum oxide which interacts with the support. The oxide may wet the support and spread on it. On reduction this thin oxide layer may break up into small metal particles.

A high surface area is not always desirable, especially for the support. Except in bifunctional catalysis where the support plays a direct and useful catalytic role, supports are generally expected to be inactive. However, the widely used supports, such as silica and alumina, often possess some catalytic activity which may be undesirable. This activity can be reduced by selective poisoning of the unwanted sites, but this may be difficult to achieve in practice. In an effort to lower the surface area of the support without affecting the dispersion of the catalyst, the support can be presintered, its surface etched, and the catalyst impregnated on the surface thereafter. Various deposition techniques including vapor phase reactions, ion plating, and rf sputtering can be used to coat dense particles, but obtaining high dispersions of the catalyst using these techniques is a challenge. Often, a compromise has to be sought, and some selectivity loss due to the support surface has to be tolerated. Supports are sometimes entirely avoided because of unacceptable selectivity losses.

The pore structure of a catalyst can have an important effect on its performance. The optimum pore structure depends, for instance, on the exothermicity of a reaction and the molecular dimensions of the reactants. Both activity and selectivity can be affected by pore structure. As much as a 10% loss in selectivity is observed to occur in certain selective oxidation reactions as a result of improper pore structure. This may be due to the occurrence of undesirable homogeneous gas phase reactions in large pores.

Some methods used to control pore structure have already been mentioned. Other methods include calcining at high temperatures to effect pore closure and steam treatments to increase pore diameter. A systematic variation in pore size can be achieved by adding varying amounts of a sacrifical binder to a catalyst prior to calcination. Removal of the binder during firing leaves behind pores. It is essential to know the surface area and pore size distribution of a catalyst in order to understand its catalytic behavior.

C.   CHARACTERIZATION OF MICROSTRUCTURE

The importance of careful catalyst characterization cannot be overstated. It is only through an understanding of microstructural evolution during catalyst synthesis and fabrication that we can hope to build a credible scientific base for the preparation of heterogeneous catalysts.

Many techniques are available for the characterization of catalyst microstructure. The pitfalls associated with interpretation of the results obtained using various sophisticated tools are many, and caution must be exercised. Ideally, it is desirable to use several different techniques to assess the same parameter. For instance, to characterize particle size and size distribution, various tools are employed. For large particles greater than $20-30\ \mu m$, optical microscopy, conventional sieve analysis, and sedimentation techniques are easily employed and the results should be in agreement. In the $1-100\ \mu m$ size range, an x-ray sedigraph, a Coulter counter, a centrifuge, or a laser diffraction unit can be employed [31]. The results can depend on powder density and state of agglomeration. Combining these results with optical microscopy and scanning electron microscopy (SEM) is most useful. Doubts regarding sampling problems when employing microscopy techniques can be alleviated to some extent in this manner. The greatest difficulties arise in trying to analyze particles below $1\ \mu m$ in size. For 0.1- to 1-μm particle diameters, Coulter counters and laser diffraction units are employed, but the problems associated with dispersing the powders make interpretation of the data ambiguous. For particles in the range $100-1000\ \text{Å}$, the SEM is a versatile and indispensable tool. It can give a quick picture of the size and morphology of particles with relative ease. Light-scattering techniques can be usefully employed for the characterization of sols which often contain particles in the $100-200\ \text{Å}$ range. Estimates of an average crystallite size in the $100\ \text{Å}-1\ \mu m$ range can also be obtained from x-ray line-broadening measurements. This measurement does not depend on the state of agglomeration of the crystallites and may give a crystallite size much smaller than that obtained with other techniques that measure particle size.

For many mixed oxide or metal catalysts dispersed on a support, crystallite sizes are often less than 100 Å. In such cases x-ray diffraction techniques may not be useful if the extent of line broadening makes interpretation difficult. A transmission electron microscope (TEM) can be used to resolve particles down to a few angstroms in size. In multiphase systems particles with different morphologies can readily be detected. With the use of a TEM, selected area diffraction of regions 1 μm in size can reveal structural information. Microdiffraction techniques can be used to obtain electron diffraction patterns from areas several tens of angstroms in size. Much information regarding size, morphology, nature of defects, and crystal structure can be obtained but is dependent on the skill of the operator. A modern-day scanning transmission electron microscope (STEM), combines the features of a SEM and a TEM to yield a very powerful instrument [32]. In addition to high-resolution and diffraction capabilities, it can be used to analyze chemically particles down to 50 Å in size using x-ray fluorescence and electron energy loss spectroscopy. A STEM can be used to detect poisons or promoters present in very small amounts on a catalyts sample. In multicomponent oxide catalysts, the number and distribution of various phases can be observed. Component segregation to surfaces or grain boundaries can be detected. Composition profiles across zeolite particles have been measured [33], and this information can provide insight into the nucleation and growth processes that govern particle formation. With the use of such information, rational catalyst synthesis and design can be contemplated. This makes the STEM an invaluable tool for the study of catalyst microstructure evolution.

The distribution of elements and component segregation on a 1-μm scale can be readily detected by an electron microprobe. This instrument can be useful for examining catalyst pellets and large catalyst particles. Energy-dispersive x-ray analysis often made in conjunction with scanning electron microscopy is useful in the same regard.

The surface area of a catalyst is routinely measured using the nitrogen B.E.T. physisorption technique. Based on assumptions about particle shape and a knowledge of particle density, an average particle diameter can be calculated. If the powder is agglomerated, this technique for measuring particle size will yield results that do not in general agree with those obtained from x-ray line broadening measurements. In the case of supported metals and multicomponent oxide systems, the total surface area of the sample is not a very useful value. Selective chemisorption techniques are commonly employed in the case of some supported metals. Assumptions regarding the lack of adsorption on the support surface have to be made, and careful interpretation of the data is warranted. When the data are used in combination with electron microscopy and x-ray line broadening measurements, the results can be very useful. Techniques for selective chemisorption of gases in the case of multiphase supported oxide systems are in general not available.

Pore structure is a very important property of catalysts. Its characterization is generally accomplished with gas physical adsorption – desorption data using mercury porosimetry. For pores in the size range 20 – 500 Å, the nitrogen adsorption technique is reliable and very useful. Detailed interpretations regarding pore shape and size based on the adsorption – desorption hysteresis loops are available [34]. For large pores in the range 100 – 150 μm, mercury porosimetry is often used. Since the contact angle between mercury and nonwetted solids exceeds 90°, mercury can enter a pore only if a positive pressure is applied. If the volume of mercury that has been forced into the pore space of a catalyst as a function of applied pressure is known, the pore size distribution can be estimated. Several assumptions regarding pore geometry have to be made, and it is also assumed that the pores are open to the surface and are not interconnected. Micropores less than 20 Å in diameter are not reliably measured by either technique. Molecular probes of various sizes and shapes are recommended for the study of such micropores. Thus, although it is extremely important to understand pore structure, reliable measurement can be challenging.

The surface area and pore structure of a catalyst are interrelated and, in turn, the mechanical properties of a catalyst are strongly influenced by its porosity. In general catalyst strength varies inversely with pore volume, but some optimum pore volume is essential for mass transfer during catalysis.

VIII.   Catalyst Strength

The resistance of a catalyst pellet to crushing in a fixed bed or to attrition resistance in a fluidized bed is an important practical consideration. The creation of fines in a catalyst bed can lead to undesirably high pressure drops across the reactor or to catalyst being carried out of the reactor. Chemical attack and thermal cycling can also lead to mechanical property degradation during catalyst use. The tensile strength of a pellet and its length/diameter ratio are important parameters that should be optimized. A long, cylindrical pellet is weaker than a short, thick pellet. It should be borne in mind that maximum strength is achieved with minimal porosity, but that significant porosity is essential for optimum catalytic properties. A compromise between these two opposing needs should be sought.

To measure tensile strength, a pellet is generally subjected to a uniform uniaxial stress and the stress increased until failure occurs. Tensile strength is defined as $\sigma = P/A$, where P is the load and A the cross-sectional area. The observed strength can vary from 100 psi for some highly porous materials to 10^6 psi for whiskers of highly crystalline ceramics [35]. Surface flaws can drastically reduce the strength of materials. The cleanliness of surfaces should not be overlooked, as fractures can begin to propagate from impurity particles bonded to clean surfaces. Stresses set up during the cooling of powders and pellets after calcination can lead to microcracks which eventually propagate under reactor conditions. Thermal quenching and cycling should be avoided if possible. Microcracks also form during grinding, as mentioned before. In the case of metals that are ductile, plastic deformation prevents the propagation of cracks. For polycrystalline ceramics, however, there is no equivalent energy-absorbing process and the crack continues until failure occurs. Pores can help in preventing crack propagation, and again optimum porosity is desirable.

IX. Catalyst Life

All catalysts sooner or later deactivate under reactor conditions. Extending catalyst life is a very desirable goal from an economic standpoint, and there are many mechanisms of deactivation to consider. Sintering and consequent reduction in surface area can cause a loss of catalyst activity. The growth of catalyst crystallite size by sintering can occur, and if this is a major cause of deactivation, regeneration of the catalyst may be difficult to achieve. Other causes of deactivation may be coking, poisoning, phase separation, dealloying, and volatilization of active species. Several of these may occur simultaneously, making the problem very complex.

A major cause of catalyst deactivation is coking or poisoning. Regeneration is sometimes achieved by steam or oxidizing atmosphere treatments at high temperatures. Under such conditions, sintering may be accelerated and the initial activity of the catalyst may never be restored. Coke formation is sometimes found to be depressed by the addition of small amounts of potassium or magnesia to the catalyst [36].

Phase transformations can occur under catalytic conditions. When transitional aluminas are used as supports, phase changes are not uncommon, accompanied by sintering and loss of activity. For some hydrogenation reactions, it is necessary that an alloy noble metal catalyst be used. If separation of the alloying components occurs, the catalyst may deactivate. Component segregation under catalytic conditions is known to occur frequently. Often for electronic reasons, the proximity of two elements is critical for selectivity. During long-term use, this segregation may be unavoidable, but if properly understood, the regeneration procedure can ensure replenishment of the desired species.

Lifetime requirements depend on catalyst replacement costs and the possibility of regeneration. Some useful catalysts last for many years. More commonly, a catalyst has an effective lifetime of seconds to a few years and regeneration procedures are necessary. Preferably a catalyst is regenerated without removal from the reactor. A cracking catalyst, for example, is regenerated every few seconds as part of a continuous cycle. The design and selection of a catalyst for long life are challenging and are best accomplished if one understands how the catalyst functions and how it ages. Successful regeneration schemes can also be envisioned if such understanding is at hand.

X.   Future of Catalyst Design and Selection

The actual rational design of a particular catalyst for a particular reaction is left as an exercise for the reader. The objective of this chapter has not been to provide easy procedures for catalyst design and selection but merely to discuss some of the more significant factors involved in this process. Traditionally, this method has been highly empirical, but steady progress is being made toward a rational approach. The merger of different scientific disciplines and the rapid growth of physical techniques for studying catalysts and catalysis are having the desired impact.

References

1. K. Tanabe, "Solid Acids and Bases," p. 5–43. Academic Press, New York, 1970.
2. A. W. Sleight, *Science,* 1980, **208**, 895.
3. T. Inui, T. Sezume, K. Miyaji, and Y. Takegami, *J. Chem. Soc. Chem. Commun.,* 1979, **20** 873.
4. B. C. Gates, J. R. Katzer, and G. C. A. Schuit, "Chemistry of Catalytic Processes," p. 349. McGraw-Hill, New York, 1979.
5. C. J. Machiels and A. W. Sleight, *Proc. Fourth Int. Conf. Chem. Uses Molybdenum,* August, 1982.
6. J. D. F. Ramsey and R. G. Avery, *J. Mater. Sci.,* 1974, **9**, 1681.
7. J. Canteloup and A. Mocellin, *J. Mater. Sci.,* 1976, **11**, 2352.
8. I. G. Sayce and B. Selton, *Spec. Ceram.,* 1975, **5**.
9. K. S. Mazdiyasni, C. T. Lynch, and J. S. Smith, *J. Am. Ceram. Soc.,* 1965, **48**(7), 372.
10. J. S. Haggerty and W. R. Cannon, "Annual Report on Contract" (No. N00014–77–C–0581, M.I.T. Report No. MIT EL79–047), 1980.

11. A. G. Walton, "The Formation and Properties of Precipitates." Wiley (Interscience), New York, 1967.
12. D. W. Johnson, Jr., and P. K. Gallagher, "Ceramic Processing before Firing," (G. Y. Onoda, Jr. and L. L. Hench, eds.). Wiley, New York, 1978.
13. I. Matsuura and M. J. W. Wolfs, *J. Catal.*, 1975, **37**, 174.
14. C. Marcilly, P. Courty, and B. Delmon, *J. Am. Ceram. Soc.*, 1970, **53**(1), 56.
15. D. F. Saunders and A. Packter, *Chem. Ind. London*, 1970, **18**, 594.
16. E. Matijevic, *Acc. Chem. Res.*, 1981, **14**, 22.
17. R. K. Iler, "The Chemistry of Silica." Wiley, New York, 1979.
18. F. Traina and N. Pernicone, *Chim. Ind.*, 1970, **52**(1), 1.
19. T. A. Dorling, B. W. J. Lynch, and R. L. Moss, *J. Catal.*, 1971, **20**, 190.
20. J. R. Anderson, "Structure of Metallic Catalysts." Academic Press, New York, 1975.
21. W. Mahler and M. Bechtold, *Nature*, 1980, **285**(5759), 27.
22. S. J. Teichner, G. A. Nicolaon, M. A. Vicarini, and G. E. E. Gardes, *Adv. Coll. Interf. Sci.* **5**, 245.
23. D. R. Herrington, *Chemtech*, 1982, **1**, 42.
24. D. M. Roy, R. R. Neurgaonkar, T. P. O'Holleran, and R. Roy, *J. Am. Ceram. Soc.*, 1977, **56**, 11.
25. J. L. Pentecost, "Treatise on Materials Science and Technology," (F. Y. Wang, ed.), Vol. 9, Academic Press, New York, 1976.
26. L. Y. Sadler, D. A. Stankey, and D. R. Brooks, "Powder Technology," Vol. 12, p. 19, 1975.
27. A. R. Cooper, Jr. and W. H. Goodnow, *Bull. Am. Ceram. Soc.*, 1962, **41**(11), 760.
28. R. L. Coble and J. E. Burke, *Prog. Ceram. Sci. B*, 1964,
29. C. Herring, *J. Appl. Phys.*, 1950, **21**, 301.
30. S. A. Khassan, E. I. Emel'yanova, and B. D. Polkovnikov, (1961). *Dokl. Akad. Nauk SSSR*, 1961, **139**, 1101.
31. T. Allen, "Particle Size Measurement." Wiley, New York, 1977.
32. J. B. Vandersande and E. L. Hall, *J. Am. Ceram. Soc.*, 1979, **62**(5), 246.
33. C. E. Lyman, P. W. Betteridge, and E. F. Moran, *Inorg. Chem.*, in press.
34. G. D. Parfitt and K. S. W. Sing, "Characterization of Powder Surfaces." Academic Press, London, 1976.
35. W. D. Kingery, H. K. Bowen, and D. R. Uhlmann, "Introduction to Ceramics." Wiley, New York, 1976.
36. D. A. Dowden, *Chem. E. Symp. Sci.*, 1968, **27**, 18.

CHAPTER 2

Uses and Properties of Select Heterogenous and Homogenous Catalysts

FRANK S. WAGNER

Strem Chemicals Inc.
Newburyport, Massachusetts

I. Introduction

Catalysts play a major role in the science of chemistry. The greater portion of chemical feedstocks produced in the world are obtained through catalytic reactions on crude oil, natural gas, and coal. Suitable catalysts are crucial in the commercial-scale production of many plastics, drugs, foodstuffs, and specialty chemicals. The increasingly important field of biosynthesis is based on remarkably specific and efficient enzymatic catalysts.

Catalytic reactions are divided into the two main categories of homogenous and heterogenous. A heterogenous catalyst is a suitable chemical compound that is insoluble in the reaction medium. The catalyst may be pure, mixed with other catalysts, or dispersed on an inert support. Some common heterogenous catalysts are metals and metal oxides. The advantages in using heterogenous catalysts are their low cost, ease of recovery, and adaptibility to either batch or flow reactors. The disadvantages of these catalysts are their general lack of specificity and the energy requirements of high temperatures and pressures required in many catalytic systems.

Homogenous catalysts are usually metal complexes that are soluble in the reaction medium. The advantages of homogenous catalysts are their specificity and generally low temperature and pressure requirements. Specificity is possible because the catalytic behavior of a particular metal complex can be subtly altered by varying the ligands, coordination number of the complex, or the oxidation state of the central metal atom. The disadvantages of homogenous catalysts are the difficulty in separating the catalyst from the product, degradation of the catalyst, and the high initial cost, especially if platinum group metals are required. Despite the economic advantages of heterogenous catalysts, several homogenous catalytic systems are being used to produce chemical feedstocks in the chemical industry today.

A trend in the chemistry of catalysts is the preparation of polymer-supported homogenous catalysts that seem to combine the advantages of both heterogenous and homogenous systems.

The object of this chapter is to provide the investigator with a guideline in selecting a catalyst for a particular transformation. The available literature on catalytic reactions is enormous. Over eight journals published worldwide are dedicated to the subject of catalysis. Numerous books and review articles on catalysis appear with increasing frequency in the chemical literature. Obviously, several restrictions are needed to deal with the subject. This chapter reviews the use of forty catalysts, both heterogenous and homogenous. In the opinion of the author these forty catalysts are most often cited in the literature and in terms of usefulness, span the range of common types of catalytic reactions. Only the literature from 1979 to the end of 1981 is reviewed. Finally, the catalyst used in most of the reactions listed in this chapter is a single chemical compound. Many researchers formulate a specific catalyst for a particular reaction. The catalyst may be complex mixtures of several compounds. Catagorization of these systems is difficult and is generally avoided in this review. This problem is especially evident in the patent literature and most often occurs with heterogenous catalysts.

The chapter is divided into two main sections. Section II lists the names of forty catalysts arranged alphabetically. Directly under the name of the catalysts is information on chemical abstracts indexing, properties, and, if applicable, references on the preparation of the catalyst. The description of the catalyst is followed by a listing of chemical transformations represented in most cases by a balanced equation. The chemical transformations are grouped alphabetically according to the type of catalytic reaction, e.g., hydrogenation, reforming, etc.

Section III lists alphabetically the various types of catalytic reactions. Under each heading is a list of catalysts that can be used to effect a transformation, e.g., hydrogenation can be done using platinum, palladium, rhodium, etc. The homogenous catalysts are designated by an asterisk. The compounds followed by a reference number are additional catalysts other than the forty catalysts listed in Section II.

II.   Catalysts

1.   Alumina (Aluminum Oxide)

CA Registry	1344-28-1
Formula	Al_2O_3
Properties	white, air-stable solid, d 3.5–3.9

Alkylation

1. $m\text{-}C_6H_4(CH_3)_2 + C_6H_5CH{=}CH_2 \rightarrow C_6H_5CH(CH_3)C_6H_3(CH_3)_2$ [1]

2. $+ CH_3OH \rightarrow$ $+ H_2O$ [2]

3. $C_6H_5OH + CH_2{=}CHCH_2CH_3 \rightarrow o\text{-}CH_3CH_2CH(CH_3)C_6H_4OH$ [3]

Alkylation (Reductive)

1. $H_2NCH_2CH_2OH + H_2NCH_2CH_2NH_2 \xrightarrow{H_2} H_2NCH_2CH_2NHCH_2CH_2NH_2 + H_2O$
[4]

2. $C_6H_5OH \xrightarrow{H_2} C_6H_{12}C_6H_4OH$ [5]

Cyclization

1. $\xrightarrow{NH_3}$ HN $O + HN$ $NH + H_2O$ [6]

2. $H_2C{=}CHCH{=}CH_2 + CH_3NH_2 \rightarrow N$ $+ 3H_2$ [7]

3. $N(CH_2CH_2OH)_3 \rightarrow$ $O\quad NCH_2CH_2OH + H_2O$ [8]

Dehydration

1. $CH_3CH_2OH \rightarrow H_2C{=}CH_2 + H_2O$ [9]
2. $CH_3CH_2C(CH_3)(OH)CH_2CH_3 \rightarrow CH_3CH_2(CH_3)C{=}CHCH_3 + H_2O$ [10]

3. $HO(CH_2)_4OH \rightarrow$ $O + H_2O$ [11]

Dehydrosulfurization

1. Dehydrosulfurization of petroleum residues   [12–14]

Dimerization

1. Dimerization of alkylstyrenes   [15]

Fischer–Tropsch

1. $CH_3OH \rightarrow C_nH_{2n+1}$ [16]

Hydrocracking

1. Hydrocracking of hydrocarbons   [17–21]

Hydrolysis

1. $H_3CC(O)OCH_3 + H_2O \rightarrow CH_3OH + H_3CC(O)OH$ [22]

2. $C{=}O + H_2O \rightarrow HOCH_2CH_2OH + CO_2$ [23]

Isomerization

1. Isomerization of xylenes   [24–26]
2. Isomerization of methylcyclohexenes   [27]
3. Trimethylcyclohexenedione → Trimethylhydroquinone   [28]

Oxidation

1. $RC_6H_5 \xrightarrow{O_2} RC_6H_4OH$ [29]

Reduction

1. $NO_x + (3 + 2x)H_2 \rightarrow NH_3 + xH_2O$ [30]

2. *Bis(benzonitrile)dichloropalladium(II)*

CA Registry   14220-64-5
Formula $PdCl_2(C_6H_5CN)_2$
Properties   orange, yellow crystalline solid, m.p. is 129–130°C, air
stable, soluble in dichloromethane, insoluble in water
Preparation   [31]

Aromatization

1. 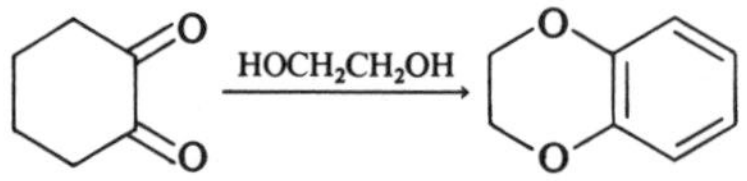   [32]

Carbonylation

1. $2C_6H_5C{\equiv}CC_6H_5 + CO \rightarrow$ 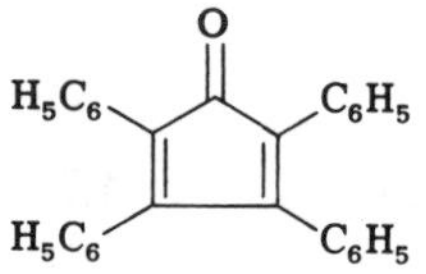   [33]

2. $HOC_6H_4R \xrightarrow{CO} RC_6H_4OC(O)OC_6H_4R$ [34, 35]

Carboxylation

1. $H_2C{=}CHCH_2OH + CO \rightarrow H_2C{=}CHCH_2C(O)OH$ [36]

Carboxylation (Oxidative)

1. $R^1HC{=}CR^2R^3C{=}CHR^4 + CO \xrightarrow{O_2, ROH} RO(O)CHR^1R^2C{=}CR^3CHR^4C(O)OR$

[37]

Dimerization

1. $2H_2C{=}CHCN \rightarrow NCCH_2HC{=}CHCH_2CN$ [38]
2. $HC{\equiv}CR + H_2C{=}CHCH_2X \rightarrow H_2C{=}CHCH_2HC{=}CRX$ [39]

Hydration

1. $RCN + H_2O \rightarrow RC(O)NH_2$ [40]

Hydrogenation

1. $C_6H_5NO_2 + 3H_2 \rightarrow C_6H_5NH_2 + 2H_2O$ [41, 42]

Isomerization

1. Isomerization of allyl acetates   [43]
2. $C_6H_5CH(OC(O)CH_3)HC=CH_2 \rightarrow C_6H_5HC=CHCH_2OC(O)CH_3$ [44]
3. 2,3-Dihydrofurans $\rightarrow$ 2,5-Dihydrofurans   [45]

4.   [46]

Transesterification

1. $CH_3CH_2C(O)OH + H_2C=CHOC(O)CH_3 \rightarrow H_2C=CHC(O)OCH_2CH_3 + CH_3C(O)OH$
[47]

3. *Bis(cyclopentadienyl)titanium Dichloride*
   Dichlorobis(h⁵-2,4-cyclopentadien-1-yl)titanium

CA Registry	1271-19-8
Formula	$(C_5H_5)_2TiCl_2$
Properties	red, crystalline solid, m.p. is 289°C, air-stable solid, sparingly soluble in nonpolar solvents, soluble in polar solvents
Preparation	[50 (p. 76)–52]

Coupling (Reductive)

1. $HOCH_2CH_2C\equiv CH \xrightarrow{R_2AlCl} CH_3CH_2HC=CHCH_2CH_2OH$ [48, 49]

Dehalogenation

1. $RX + NaBH_4 \rightarrow RH + NaX +$ boron compounds   [53]

Dehydrogenation

1. $Mg + 2C_5H_6 \rightarrow Mg(C_5H_5)_2 + H_2$ [54]

Dimerization

1. $H_2C=CH_2 + H_2C=C(CH_3)HC=CH_2 \rightarrow H_3C(CH_3)C=CHCH_2HC=CH_2$ [55]

Exchange

1. $CH_3CH_2CH_2MgBr + H_2C=CH(CH_3)C=CH_2 \rightarrow$
$[H_2CC(CH_3)CCH_2]MgBr + CH_3CH_2CH_3$ [56]

Fischer–Tropsch

1. $CO + H_2 \rightarrow C_nH_{2n+1}$ [57]

Grignard

1. $(CH_3)_2CHMgBr + (CH_3)_2CO \rightarrow (CH_3)_2CHOH + (CH_3)_2CHC(OH)(CH_3)_2$ [58]
2. $CH_3CH_2C(O)OCH_3 + CH_3CH_2CH_2MgBr \rightarrow CH_3CH_2CH(OH)CH_2CH_2CH_3$ [59]

Hydrometallation

1. $RHC{=}CH_2 + HBR''_2 \rightarrow RCH_2CH_2BR''_2$ [60]
2. $RHC{=}CHR' + HAl(NR''_2)_2 \rightarrow$
    $RCH_2CHR'Al(NR''_2)_2$ (R''—isopropyl)   [61, 62]

Isomerization

1. Bicyclo[4,3,0]nona-3,7-diene $\rightarrow$ Bicyclo[4,3,0]nona-2,9-diene   [63]

Oxidation

1. $H_2C{=}CHCH{=}CH_2 + CH_3C(O)OH \xrightarrow{O_2}$
    $CH_3C(O)OCH_2CH(OC(O)CH_3)CH{=}CH_2 + CH_3C(O)OCH_2HC{=}CHCH_2O(O)CH_3$
    [64]

Polymerization

1. $C_6H_5C{\equiv}CH \rightarrow (C_6H_5CCH)_x$ [65]

Reduction

1. $RHC{=}CH_2 \xrightarrow{\text{LiAlH}_4} RCH_2CH_3$ [66]

4. *Carbonylchlorobis(triphenylphosphine)rhodium(I)*

CA Registry	13938-94-8
Formula	$RhCl(CO)(P(C_6H_5)_3)_2$
Properties	yellow, crystalline solid, m.p. is 195°C, air stable, moderately soluble in benzene, slightly soluble in ether and hydrocarbons
Preparation	[67, 68]

Carbonylation

1. $H_2C{=}CHCH_2NH_2 + CO \rightarrow$ [69, 70]
    (ring structure: HN—C=O)

2. $RC_6H_4NO_2 + 3CO \rightarrow RC_6H_4NCO + 2CO_2$ [71]

3. (epoxide) $+ CO \rightarrow$ (cyclic carbonate)   [72]

Dimerization

1. $2RHgX \rightarrow R{-}R + HgX_2 + Hg$ [73]

Hydroesterification

1. $C_6H_5C{\equiv}CC_6H_5 + CO + CH_3CH_2OH \rightarrow$

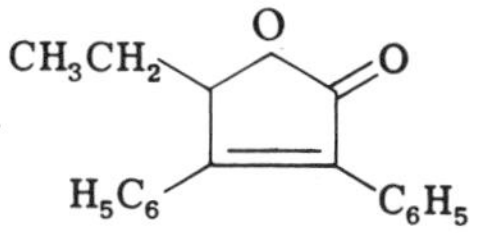

[74]

2. $C_6H_5NO_2 + 3CO + CH_3OH \rightarrow C_6H_5NHC(O)OCH_3 + 2CO_2$ [75]

Hydroformylation

1. $HC(O)H + CO + H_2 \rightarrow HOCH_2C(O)H$ [76, 77]
2. $CH_3(CH_2)_3HC{=}CH_2 + CO + H_2 \rightarrow CH_3(CH_2)_5C(O)H + CH_3(CH_2)_3CH(C(O)H)CH_3$
   [78]
3. $HC(O)H + CO + H_2 \rightarrow HOCH_2C(O)H + H_2 \rightarrow HOCH_2CH_2OH$ [79, 80]
4. $CH_3C(O)OCH{=}CH_2 + CO + H_2 \rightarrow CH_3C(O)OCH(C(O)H)CH_3$ [81]
5. Limonene + CO + H$_2 \rightarrow$ 3-(4-methyl-3-cyclohexen-1-yl)butyraldehyde   [82]
6. $CH_3CH_2CH_2HC{=}CH_2 + CO + H_2O \rightarrow$
   $$CH_3CH_2CH_2C(C(O)H)CH_3 + CH_3(CH_2)_4C(O)H \qquad [83, 84]$$

Hydrosilylation

1. $RHC{=}CHR + HSiR'_3 \rightarrow RCH_2CHR(SiR'_3)$ [85]
2. $C_6H_5C{\equiv}CH + HSiR_3 \rightarrow C_6H_5(SiR_3)C{=}CH_2$ [86]

5. *Carbonylhydrotris(triphenylphosphine)rhodium(I)*

CA Registry	17185-29-4
Formula	$RhH(CO)(P(C_6H_5)_3)_3$
Properties	yellow, microcrystalline solid, m.p. is 120°C (in air), soluble in benzene, chloroform, and dichloromethane, solutions are air sensitive
Preparation	[87]

Carbonylation

1. $RNO_2 + 3CO \rightarrow RNCO + 2CO_2$ (R—alkyl)   [88]

Cyclization

1. $H_2C{=}CHCH_2CH_2C(O)H \rightarrow$

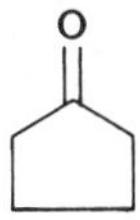

[89]

Homologation

1. $CH_3OH + CO + H_2 \rightarrow CH_3CH_2OH$ [90]

Hydroesterification

1. $C_6H_4NO_2 + 3CO + CH_3OH \rightarrow C_6H_4NHC(O)OCH_3 + 2CO_2$ [75]

Hydroformylation

1. $HOCH_2HC=CH_2 + H_2 + CO \rightarrow HOCH_2CH_2CH_2C(O)H + HOCH_2C(C(O)H)CH_3$
   [91–94]
2. $CH_3HC=CHCH_3 + CO + H_2 \rightarrow CH_3(CH_2)_3C(O)H + CH_3CH_2CH(C(O)H)CH_3$
   [95, 96]
3. $CH_3(CH_2)_5HC=CH_2 + CO + H_2 \rightarrow$
   $\quad\quad\quad\quad CH_3(CH_2)_5CH_2CH_2C(O)H + CH_3(CH_2)_5CH(C(O)H)CH_3$ [97]
4. $RHC=CH_2 + CO + H_2 \rightarrow RCH(C(O)H)CH_3 + RCH_2CH_2C(O)H$ [98–100]
5. $RR'C=CH_2 + CO + H_2 \rightarrow CRR'HCH_2C(O)H$ [101, 102]
6. $H_2C=C(CH_3)C(O)OCH_3 + CO + H_2 \rightarrow H(O)CCH_2C(CH_3)C(O)OCH_3$ [103]
7. $HC(O)H + CO + H_2 \rightarrow HOCH_2C(O)H + H_2 \rightarrow HOCH_2CH_2OH$ [79]
8. $NCC(=CH_2)CH_2CH_2CN + CO + H_2 \rightarrow NCC(CH_3)(C(O)H)CH_2CH_2CN$ [104]

Hydrosilylation

1. $C_6H_5C\equiv CH + HSiR_3 \rightarrow C_6H_5(SiR_3)C=CH_2$ [86]
2. $HC\equiv CH + 3HSiR_3 \rightarrow R_3SiHC=CH_2 + R_3SiCH_2CH_2SiR_3$ [105]

Isomerization

1. $H_2C=CHCH_2NHC(O)CH_3 \rightarrow cis\text{-}CH_3HC=CHNHC(O)CH_3$ [106]

Oxidation

1. $RHC=CH_2 \xrightarrow{O_2} RC(O)CH_3$ [107]

Polymerization

1. $H_2C=C(CH_3)C(O)OCH_3 \rightarrow (-CH_2C(CH_3)C(O)OCH_3)_x$ [108]

6. Chlorodicarbonylrhodium(I) Dimer
(Tetracarbonyldi-u-chlorodirhodium)

CA Registry	14523-22-9
Formula	$(Rh(CO)_2Cl)_2$
Properties	red, crystalline solid, m.p. is 126°C, soluble in most organic solvents, solutions are air sensitive
Preparation	[109, 110]

Carbonylation

1. $RC_6H_4NO_2 + 3CO \rightarrow RC_6H_4NCO + 2CO_2$ [71, 111–113]

2. $+ CO \rightarrow R'(NCO)C=CR''R'''$ [114]

3. $+ CO \rightarrow$ $+ (CH_3CH_2CH_2)_2CO$ [115]

Carboxylation

1. $RCH_2OH + CO \rightarrow RCH_2C(O)OH$ [116–119]

Dehydrogenation

1. $RCH_2OH \rightarrow RC(O)H + H_2$ [120]

Dimerization

1. $2RHgX \rightarrow R{-}R + HgX_2 + Hg$ [73, 121]
2. $2H_2C{=}CHCN \rightarrow NCCH_2HC{=}CHCH_2CN$ [38]

Fischer–Tropsch

1. $CO + H_2 \rightarrow C_nH_{2n+1}$ [122]

Hydroformylation

1. $C_6H_5C_6H_4CH{=}CH_2 + CO + H_2 \rightarrow$
$\qquad\qquad C_6H_5C_6H_4CH_2CH_2C(O)H + C_6H_5C_6H_4CH(C(O)H)CH_3$ [123]
2. $CH_3C(O)OCH{=}CH_2 + CO + H_2 \rightarrow CH_3C(O)OCH(C(O)H)CH_3$ [81]
3. $H_2C{=}CH_2 + CO + H_2 \rightarrow CH_3CH_2C(O)H$ [124]
4. $RHC{=}CH_2 + (CH_3)_2NH + CO + H_2 \rightarrow RCH_2CH_2CH_2N(CH_3)_2$ [125]
5. $CH_3(CH_2)_8HC{=}CH_2 + CO + H_2 \rightarrow$
$\qquad\qquad CH_3(CH_2)_8CH_2CH_2C(O)H + CH_3(CH_2)_8CH_2CH_2CH_2OH$ [126]
6. $HC(O)H + CO + H_2 \rightarrow CH_3C(O)H + CH_3CH_2OH$ [127]

Hydrosilylation

1. $HC{\equiv}CH + 3HSiR_3 \rightarrow R_3SiHC{=}CH_2 + R_3SiCH_2CH_2SiR_3$ [105]

Reduction

1. $CO_2 + H_2 \rightarrow CO + H_2O$ [129]

Reforming

1. 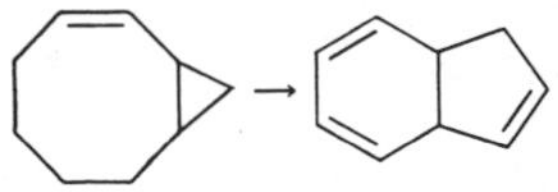 + 2H₂   [130]

2. 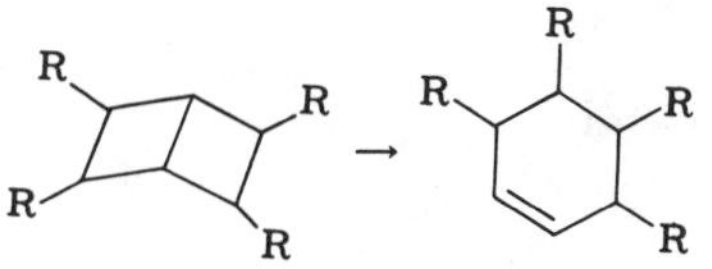   [131]

3. 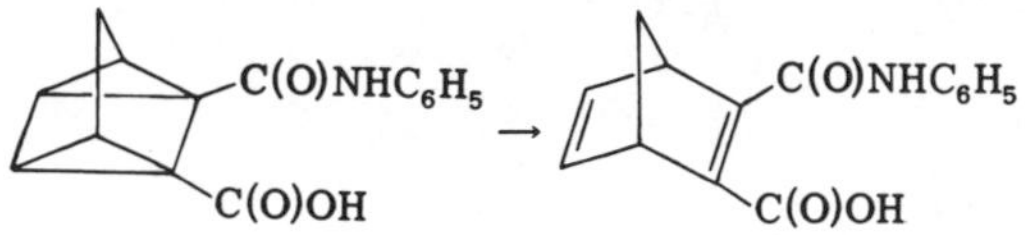   [128]

Water Gas Shift Reaction

1. $CO + H_2O \rightarrow CO_2 + H_2$ [132]

7. *Chlorotris(triphenylphosphine)rhodium(I)*

CA Registry	14694-95-2
Formula	$RhCl(P(C_6H_5)_3)_3$
Properties	red–purple solid, soluble in chloroform and dichloromethane, moderately soluble in benzene, insoluble in hydrocarbons, solutions are air sensitive
Preparation	[133, 134]

Addition

1. $RC(O)OH + H_2C{=}CHCN \rightarrow RC(O)OCH_2CH_2CN$ [135]

Alkylation

1. $CH_3(CH_2)_3NH_2 + CH_3OH \rightarrow CH_3(CH_2)_3NHCH_3 + H_2O$ [136]

Carbonylation

1. $H_2C{=}CHCH_2NH_2 + CO \rightarrow$ 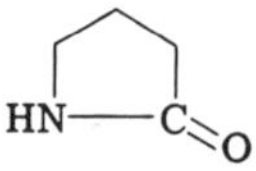   [69, 70]

2. $RHC{=}CH_2 + HSiR'_3 + CO \rightarrow RCH_2HC{=}CHOSiR'_3$ [137]

Carboxylation

1. $CH_3OH + CO \rightarrow CH_3C(O)OH$ [117, 138, 139]

Cyclization

1. $H_2C{=}CHCH_2CH_2C(O)H \rightarrow$ 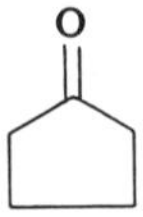   [140]

2. $(H_2C{=}CHCH_2)_2CR_2 \rightarrow$ 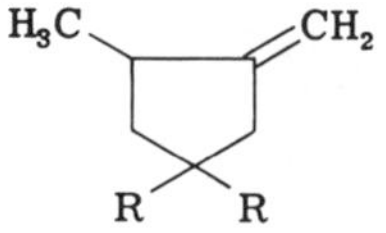   [141]

Dehydrogenation

1. $CH_3CH(OH)CH_3 \rightarrow CH_3C(O)CH_3 + H_2$ [142]

Dimerization

1. $2RHgX \rightarrow R{-}R + HgX_2 + Hg$ [73, 121]

Fischer–Tropsch

1. $CO + H_2 \rightarrow C_nH_{2n+1}$ [122]

Hydrogenation

1. Alkene hydrogenation using polymer-supported catalyst   [143 – 145]
2. $CO_2 + H_2 \rightarrow HC(O)OH$ [146, 147]
3. Hydrogenation of polymers   [148]
4. Hydrogenation of vegetable oils   [149]
5. $RC(O)R' + H_2 \rightarrow RCH(OH)R'$ [150]

Hydrogenolysis

1. $C_6H_5N{=}NC_6H_5 + 2H_2 \rightarrow 2C_6H_5NH_2$ [151]
2. $(CH_3)_2CHOH + C_6H_5NHNHC_6H_5 \rightarrow (CH_3)_2CO + 2C_6H_5NH_2$ [152]

Hydrosilylation

1. $C_6H_5C{\equiv}CH + HSiR_3 \rightarrow C_6H_5(SiR_3)C{=}CH_2$ [86]
2. $RHC{=}CH_2 + HSiR_3 \rightarrow RCH(SiR_3)CH_3$ [153, 154]
3. $RHC{=}CHR + HSiR'_3 \rightarrow RCH_2CHR(SiR'_3)$ [85]
4. $RS(CH_2)_nHC{=}CH_2 + HSiR_3 \rightarrow RS(CH_2)_nCH(SiR_3)CH_3 + RS(CH_2)_nCH_2CH_2SiR_3$
    [155]
5. $HC{\equiv}CH + HSiR_3 \rightarrow H_2C{=}CHSiR_3$ [156]

Isomerization

1. Isomerization of alkenes   [157 – 159]

Oxidation

1. $2R_3P + O_2 \rightarrow 2R_3PO$ [160]
2. $o\text{-}O_2NC_6H_4CH{=}CH_2 \xrightarrow{O_2} o\text{-}O_2NC_6H_4C(O)H$ [161]
3. $\left(\!{}^{(CH_2)_n}_{HC{=\!=}C-Si(CH_3)_3}\!\right) \xrightarrow{O_2} {}^{OC-(CH_2)_{n-1}}_{HC{=\!=}C-Si(CH_3)_3}$ [162]

Reforming

1. $H_2C{=}CHCH_2C(O)OCH_2HC{=}CH_2 \rightarrow H_2C{=}CHCH_2HC{=}CHCH_2C(O)OH$
    [163]

Review — [164]

8. Chromium Oxide

CA Registry   1308-38-9
Formula Cr_2O_3
Properties   green solid, d is 5.21, m.p. 2435°C, air-stable solid

Alkylation

1. $C_6H_5OH \xrightarrow{CH_3OH} o\text{-}C_6H_4(OH)(CH_3) + 2,6\text{-}C_6H_3(OH)(CH_3)_2$ [165, 166]

Aromatization

1. $H_2C=CHCH=CHCH=CH_2 \rightarrow$ aromatic products   [167]

Cracking

1. Decomposition of hydrazine   [168]

Cyclization

1. $H_2C=CHCH_2CH_2HC=CH_2 \rightarrow$ [169]

Dehydrogenation

1. $C_6H_5CH_2CH_3 \rightarrow C_6H_5CH=CH_2 + H_2$ [170–172]
2. Olefins $\rightarrow$ Diolefins   [173]

Fischer–Tropsch

1. $CO + H_2 \rightarrow$ aromatics   [174]

Hydrodealkylation

1. $C_6H_5CH_3 + H_2 \rightarrow C_6H_6 + CH_4$ [175, 176]

Hydrogenation

1. $H_2C=CH_2 + H_2 \rightarrow H_3CCH_3$ [177]

Oxidation

1. $2SO_2 + O_2 \rightarrow 2SO_3$ [178]
2. $2CO + O_2 \rightarrow 2CO_2$ [179]
3. $2CH_3OH + O_2 \rightarrow 2HC(O)H + 2H_2O$ [180]
4. $2H_2 + O_2 \rightarrow 2H_2O$ [181]

Reforming

1. $CH_3(CH_2)_3OH \xrightarrow{CS_2}$ [182]

Water Gas Shift Reaction

1. $CO + H_2O \rightarrow CO_2 + H_2$ [183]

Review—[184]

9. *Cobalt*

CA Registry   7440-48-4
Formula   Co
Properties   black powder, m.p. is 1495°C, *d* 8.9

Alkylation (Reductive)

1. $HOCH_2CH_2NH_2 + NH_3 \xrightarrow{H_2} H_2NCH_2CH_2NH_2 + H_2O$ [185, 186]

Aromatization

1. $H_3C(CH_2)_4CH_3 \rightarrow C_6H_6 + 4H_2$ [187]

Cracking

1. $CH_3OH \rightarrow CO + 2H_2$ [188]

Dehydrogenation

1.   [189]

Dehydrosulfurization

1. Dehydrosulfurization of naphtha   [190]
2. Dehydrosulfurization of thiophene   [191]

Fischer–Tropsch

1. $CO + H_2 \rightarrow C_nH_{2n+1}$ [192–195]

Hydrodenitrogenation

1. Hydrodenitrogenation of quinoline and acridine   [196]

Hydroformylation

1. Hydroformylation of alkenes   [197]

Hydrogenation

1. $RC_6H_4C(O)CH_3 + H_2 \rightarrow RC_6H_4CH(OH)CH_3$ [198]
2. $p\text{-}NCC_6H_4CN + 4H_2 \rightarrow p\text{-}H_2NCH_2C_6H_4CH_2NH_2$ [199]
3. $H_2NCH_2CH_2NHCH_2CH_2CN + 2H_2 \rightarrow H_2NCH_2CH_2NHCH_2CH_2CH_2NH_2$ [200]

Hydrogenolysis

1. Hydrogenolysis of 2,3-dimethylbutane   [201]
2. Hydrogenolysis of hydrocarbons   [202]

Isomerization

1. *trans* $\rightarrow$ *cis*   [203]

Oxidation

1. $CH_3C_6H_4(O)H \xrightarrow{O_2} HO(O)CC_6H_4C(O)OH$ [204]

2. $p\text{-}O_2NC_6H_4CH_3 \xrightarrow{O_2} p\text{-}O_2NC_6H_4C(O)OH$ [205]

3. $H_2C=CHCH_3 \xrightarrow{O_2,NH_3} H_2C=CHCN$ [206]

4. $H_2C=C(CH_3)C(O)H \xrightarrow{O_2} H_2C=C(CH_3)C(O)OH$ [207]

5.   [208]

6. $H_2C=CHCH=CH_2 \xrightarrow{O_2,CH_3C(O)OH} CH_3C(O)OCH_2(CH_2)_2CH_2OC(O)CH_3$ [209]

10. Cobalt Carbonyl (Di-μ-carbonylhexacarbonyldicobalt)

CA Registry   10210-68-1
Formula $Co_2(CO)_8$
Properties   orange, crystalline solid, m.p. is 50°C, d 1.73, air- and heat-sensitive solid, soluble in most organic solvents
Preparation   [50 (p. 98), 210–212]

Alkylation

1. $RNCO + R'CHO \rightarrow RN=CHR' + CO_2$ [213]

Fischer–Tropsch

1. $CO + H_2 \rightarrow C_nH_{2n+1}$ [214, 215]

Homologation

1. $CO + H_2 + CH_3OH \rightarrow CH_3CH_2OH + CH_3C(O)OCH_3 + CH_3C(O)OCH_2CH_3$ [216–218]

2. $C_6H_5CH_2OH + CO + H_2 \rightarrow C_6H_5CH_2CH_2OH$ [219]

Hydrocarboxylation

1. $RHC=CH_2 + CO + H_2O \rightarrow RCH_2CH_2C(O)OH$ [220]
2. Alkenes (C_9-C_{19}) + CO + $H_2 \rightarrow$ fatty acids $(C_{10}-C_{20})$ [221]

Hydroesterification

1. $BrCH_2Cl + CO + CH_3OH \xrightarrow{NaOCH_3} H_2C(C(O)OCH_3)_2$ [222]
2. $ClCH_2C(O)OCH(CH_3)_2 + CO + (CH_3)_2CH(OH) \rightarrow H_2C(C(O)OCH(CH_3)_2)_2$ [223]
3. $C_6H_5NH_2 + CO \xrightarrow{CH_3OH} C_6H_5NHC(O)OCH_3$ [224]
4. $H_2C=CHCN + CO + H_2 \rightarrow CH_3CH(CN)C(O)OCH_3$ [225]
5. $CH_3(CH_2)_9HC=CH_2 + CH_3C(O)NH_2 + CO + H_2 \rightarrow$
     $CH_3(CH_2)_{11}CH(C(O)OH)(NHC(O)CH_3)$ [226, 227]

Hydroformylation

1. $RC(O)H + HSiCH_3(CH_2CH_3)_2 + CO \rightarrow RCH(OSiCH_3(CH_2CH_3)_2)C(O)H$

[228, 229]

2. $H_2C=CH_2 + CO + H_2 \rightarrow CH_3CH_2C(O)H$ [230]

3. $H_2C=CHCH_2O(O)CCH_3 + CO + H_2 \rightarrow H(O)C(CH_2)_3O(O)CCH_3$ + other products

[231]

4. $H_3CHC=CH_2 + CO + H_2 \rightarrow (CH_3CH_2CH_2)_2CO$ + other products   [232]

5. $CH_3OH + CO + H_2 \rightarrow CH_3C(O)H + H_2O$ [233]

Hydrogenation

1. Hydrogenation of coal liquids   [234, 235]

2. $RC_6H_4NO_2 + 3H_2 \rightarrow RC_6H_4NH_2 + 2H_2O$ [236]

Reviews—[237, 238]

11.   Copper Chromite

CA Registry   12018-10-9
Formula $CuO—Cr_2O_3$
Properties   black solid

Alkylation

1. Alkylation of phenols   [239]

Alkylation (Reductive)

1. $RNH_2 + 2R'C(O)H + 2H_2 \rightarrow RN(CH_2R')_2 + 2H_2O$ [240]

Dehydrogenation

1. $HOC_6H_4CH_2OH \rightarrow HOC_6H_4C(O)H + H_2$ [241]

2. 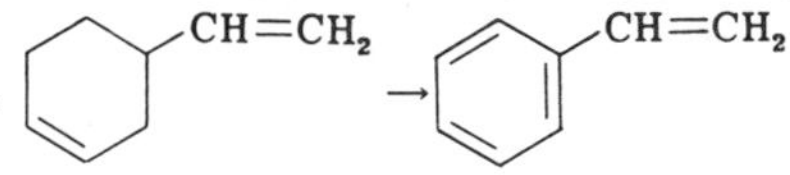 $+ 2H_2$ [242]

Fischer–Tropsch

1. $CO + H_2 \rightarrow$ aromatics   [243]

Hydrogenation

1. 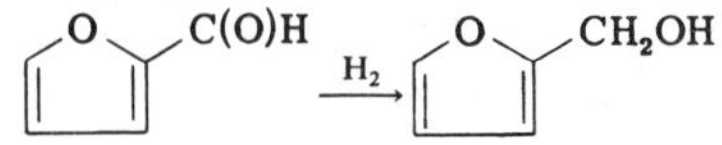   [244]

2. Unsaturated fatty acids → saturated fatty acids   [245–247]

3. $NCCH_2CH_2CH_2OH$ [248]

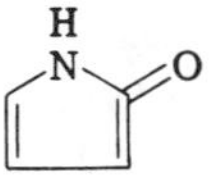

4. $RC_6H_4HC{=}CH_2 + H_2 \rightarrow RC_6H_4CH_2CH_3$ [249]

Hydrolysis

1. $RC(O)OR' + H_2O \rightarrow RC(O)OH + R'OH$ [250]

Isomerization

1. Cycloalkenols $\rightarrow$ cycloalkanones   [251]

12. Copper Oxide

CA Registry   1317-38-0
Formula   CuO
Properties   black solid, m.p. is 1326°C, d 6.3–6.49

Dehydrogenation

1. 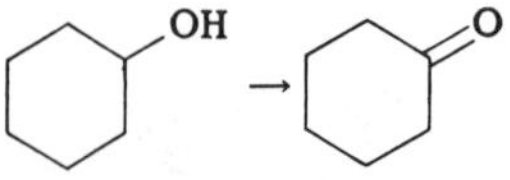 $\rightarrow$ + H_2 [252]

Dehydrohalogenation

1. Dehydrohalogenation of halogenated, unsaturated hydrocarbon polymers   [253]

Fischer–Tropsch

1. $CO + 2H_2 \rightarrow CH_3OH$ [254]

Hydrogenation

1. $HOCH_2C(O)OCH_2CH_2OH + 2H_2 \rightarrow 2HOCH_2CH_2OH$ [255]
2. $CH_3HC{=}CHC(O)H + 2H_2 \rightarrow CH_3CH_2CH_2CH_2OH$ [256]
3. $RC(O)OR' + 2H_2 \rightarrow RCH_2OH + R'OH$ [257]
4. $RNO_2 + 3H_2 \rightarrow RNH_2 + 2H_2O$ (R—aryl)   [257]
5. $RC(O)R' + H_2 \rightarrow RCH(OH)R'$ [258]

Hydrogenolysis

1. 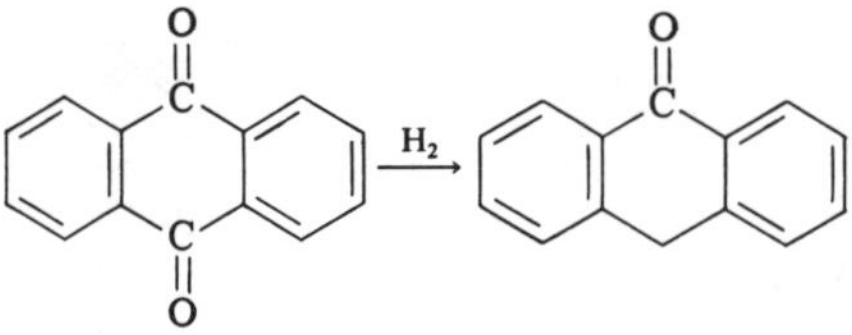 $\xrightarrow{H_2}$ [259]

Oxidation

1. $2CH_3OH + 3O_2 \rightarrow 2CO_2 + 4H_2O$ [260]

2. $H_2C{=}CHC(O)H \xrightarrow{O_2} H_2C{=}CHC(O)OH$ [261]

3. $CH_3C(O)CH_2CH_3 \xrightarrow{O_2} CH_3C(O)C(O)CH_3 + CH_3C(O)OH + CH_3C(O)H$ [262]

4. $2H_2 + O_2 \rightarrow 2H_2O$ [181]
5. $2CO + O_2 \rightarrow 2CO_2$ [263]

13. *Dichlorobis(triphenylphosphine)palladium(II)*

CA Registry	13965-03-2
Formula	$PdCl_2(P(C_6H_5)_3)_2$
Properties	yellow solid, moderately soluble in chloroform and dichloromethane, insoluble in alcohols
Preparation	[264]

Alkylation

1. $m\text{-}O_2NC_6H_4Br + H_2C{=}CHC(CH_3)_2OH \xrightarrow{CuI} m\text{-}O_2NC_6H_4CH{=}CHC(CH_3)_2OH$

[265]

Carbonylation

1. $CH_3(CH_2)_4C{\equiv}CH + HC{\equiv}CCH_2CH_2OH \xrightarrow{CO}$

$H_2C{=}C(C(O)OCH_2C{\equiv}CH)((CH_2)_4CH_3) +$ 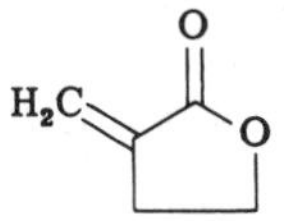 [266]

2. $I(CH_3)C{=}CH(OH) \xrightarrow{CO}$ 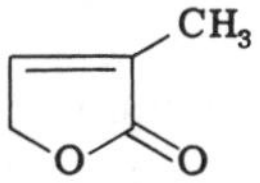 [267]

3. $RBr + CO \xrightarrow{K_2CO_3, H_2NNH_2} RC(O)OH$ [268, 269]

4. 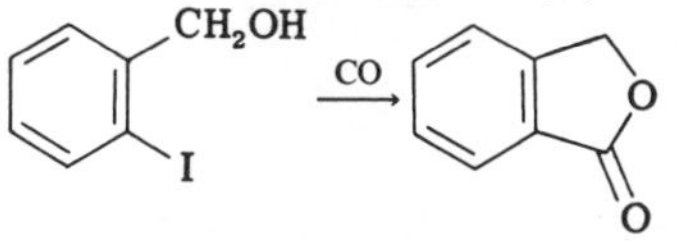 $\xrightarrow{CO}$ [270]

Coupling (Reductive)

1. $C_6H_5CH_2Br + C_6H_5C(O)Cl + Zn \rightarrow C_6H_5CH_2C(O)C_6H_5 + ZnBrCl$ [271]

Cyclization

1. $CH_3C{\equiv}CCH(OH)CH(OH)C_6H_5 \rightarrow$ 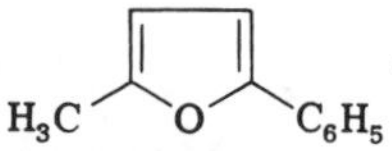 $+ H_2O$ [272]

Exchange

1. $RZnX + R'X \rightarrow R{-}R' + ZnX_2$ (R'—alkyl)   [273]
2. $(CH_3)_2CHSCH_2HC{=}CH_2 + C_6H_5MgBr \rightarrow C_6H_5CH_2HC{=}CH_2 + (CH_3)_2CHSMgBr$

[274]

Hydroesterification

1. $HC{\equiv}CH + ROH + CO \rightarrow H_2C{=}CHC(O)OR$ [266]
2. $CH_3(CH_2)_2HC{=}CH_2 + CO + CH_3CH_2OH \rightarrow CH_3(CH_2)_3CH_2C(O)OCH_2CH_3$

[275]

Hydroformylation

1. NCH=CH$_2$ + CO + H$_2$ → NCH(C(O)H)CH$_3$ [276]

Hydrogenation

1. $\xrightarrow{\text{H}_2}$ [277]

Hydrogenolysis

1. Allylic acetates → 1-alkenes   [278]

Polymerization

1. Polymerization of 2-norbornene   [279]

14. Dichlorotris(triphenylphosphine)ruthenium(II)

CA Registry   15529-49-4
Formula   RuCl$_2$(P(C$_6$H$_5$)$_3$)$_3$
Properties   black, microcrystalline solid, m.p. is 132°C, moderately
                 soluble in chloroform, acetone and benzene, insoluble
                 in water and methanol, solutions are air sensitive
Preparation   [280, 281]

Addition

1. Addition of carbon tetrachloride to polybutadiene   [282]

2. + CCl$_4$ → +   [283]

Coupling (Reductive)

1. 3R$_2$CH(OH) → R$_2$HC=CHR$_2$ + R$_2$CO + 2H$_2$O   [284]

Cyclization

1. H$_2$N(CH$_2$)$_n$NH$_2$ → + NH$_3$ [285]

Hydrogenation

1. CH$_3$CH$_2$C(O)H + H$_2$ → CH$_3$CH$_2$CH$_2$OH   [286]
2. 3(CH$_3$)$_2$CH(OH) + C$_6$H$_5$NO$_2$ → C$_6$H$_5$NH$_2$ + 3(CH$_3$)$_2$CO + 2H$_2$O   [287, 288]

3. $RC(O)R' + H_2 \rightarrow RCH(OH)R'$ [289–291]
4. $o\text{-}F_3CC_6H_4NO_2 + 3H_2 \rightarrow o\text{-}F_3CC_6H_4NH_2 + 2H_2O$ [292]

5. $\xrightarrow{H_2}$ [293]

Hydrogenolysis

1. $C_6H_5N\!=\!NC_6H_5 + 2H_2 \rightarrow 2C_6H_5NH_2$ [151]
2. $(CH_3)_2CH(OH) + C_6H_5NHNHC_6H_5 \rightarrow (CH_3)_2CO + 2C_6H_5NH_2$ [152]

3. $\xrightarrow{H_2}$ [294]

Hydrosilylation

1. $HC\!\equiv\!CH + HSiR_3 \rightarrow H_2C\!=\!CHSiR_3$ [156]
2. $R'HC\!=\!CH_2 + HSi(OR)_3 \rightarrow R'CH_2CH_2Si(OR)_3$ [295]

Oxidation

1. $2RHC\!=\!CHCH_2OH + O_2 \rightarrow 2RHC\!=\!CHC(O)H + 2H_2O$ [296]

Polymerization

1. Polymerization of 2-norbornene   [279, 297]

Reforming

1. $CH_3C(\!=\!CH_2)OCH_2HC\!=\!CH_2 \rightarrow CH_3C(\!=\!CH_2)CH_2C(CH_3)C(O)H$ [298]

15. Hexarhodium Hexadecacarbonyl (Tetra-μ_3-carbonyldodecacarbonylhexarhodium)

CA Registry   28407-51-4
Formula $Rh_6(CO)_{16}$
Properties   black, violet solid, d is 2.87, air stable, sparingly soluble
                in most solvents
Preparation   [299–302]

Carbonylation (Reductive)

1. $\xrightarrow{CO,H_2}$ + + + other products

[303]

Homologation

1. $CO + H_2 + CH_3OH \rightarrow CH_3CH_2OH + CH_3C(O)OCH_3 + CH_3C(O)OCCH_2CH_3$

[216]

Hydroesterification

1. $C_6H_5C\equiv CC_6H_5 + CO + CH_3CH_2OH \rightarrow$

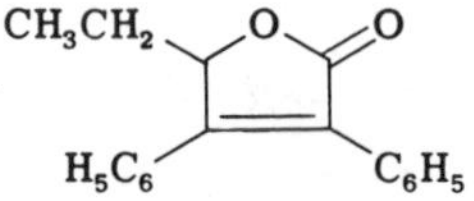

[74]

Hydroformylation

1. $RHC=CH_2 + CO + H_2 \rightarrow RCH_2CH_2C(O)H$ [304]
2. $RHC=CH_2 + 3CO + H_2O + HNR'_2 \rightarrow R(CH_2)_3NR'_2 + 2CO_2$ [305]
3. $HC(O)H + CO + H_2 \rightarrow HOCH_2C(O)H$ [76]
4. $CH_3CH_2CH_2HC=CH_2 + CO + H_2O \rightarrow$
$$CH_3CH_2CH_2C(C(O)H)CH_3 + CH_3(CH_2)_4C(O)H \qquad [84]$$
5. $RHC=CHCH_2OH + CO + H_2 \rightarrow$ 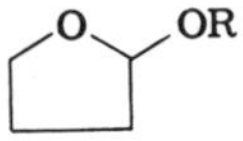   [306, 307]

Hydrogenation

1. $C_6H_5NO_2 + 2CO + H_2 \rightarrow C_6H_5NH_2 + 2CO_2$ [308, 309, 310]
2. $C_6H_5C(O)H + CO + H_2O \rightarrow C_6H_5CH_2OH + CO_2$ [311]

Oxidation

1. $2CO + O_2 \rightarrow 2CO_2$ [312, 313]

2. 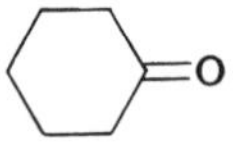 $=O \xrightarrow{O_2} HO(O)C(CH_2)_4C(O)OH$ [312]

3. $CH_3C(O)CH_3 \xrightarrow{O_2} CH_3C(O)OH$ + other products   [313]
4. $RCH_2OH + O_2 \rightarrow RC(O)OH + H_2O$ [314]

Water Gas Shift Reaction

1. $CO + H_2O \rightarrow H_2 + CO_2$ [315]

16.   Iron Dodecacarbonyl (Di-μ-carbonyldecacarbonyltriiron)

CA Registry	17685-52-8
Formula	$Fe_3(CO)_{12}$
Properties	black, green solid, d is 2.0, air-sensitive solid, sparingly soluble in most organic solvents
Preparation	[316, 317]

Fischer–Tropsch

1. $CO + H_2 \rightarrow C_nH_{2n+1}$ [318–321]

Hydroformylation

1. $RHC{=}CH_2 + 3CO + H_2O + HNR'_2 \rightarrow R(CH_2)_3NR'_2 + 2CO_2$ [305]
2. $C_6H_5NO_2 + CO + H_2 \xrightarrow{NaOCH_3}$
    $C_6H_5NHC(O)H + C_6H_5NH_2 + C_6H_5NHC(O)OCH_3 +$ other products   [322]

Hydrogenation

1. Hydrogenation of coal liquids   [234]
2. $(CH_3)_2CO + CO + H_2O \rightarrow (CH_3)_2CH(OH) + CO_2$ [323]
3. $C_6H_5C(O)H + CO + H_2O \rightarrow C_6H_5CH_2OH + CO_2$ [311]

17. Iron Oxide

CA Registry   1309-37-1
Formula Fe_2O_3
Properties   red – brown to black solid, m.p. is 1565°C, d 5.24

Alkylation

1. $C_6H_5OH \xrightarrow{CH_3OH} o\text{-}C_6H_4(OH)(CH_3) + 2,6\text{-}C_6H_3(OH)(CH_3)_2$ [165]
2. $C_6H_6 + H_3CHC{=}CH_2 \rightarrow C_6H_5CH_2CH_2CH_3$ [324]
3. $C_6H_5CH_3 + C_6H_5CH_2Cl \rightarrow (C_6H_5CH_2)C_6H_4CH_3 + HCl$ [325]

Carbonylation

1. $(CH_3)_3N + CH_3NH_2 + CO \rightarrow (CH_3)_2NC(O)H + (CH_3)_3NH^+$ [326, 327]

Dehalogenation

1. $ClCH_2CH_2Cl \rightarrow ClCH{=}CH_2 + HCl$ [328]

Dehydration

1. $CH_3CH_2OH \rightarrow H_2C{=}CH_2 + H_2O$ [329]

Dehydrogenation

1. $C_6H_5CH_2CH_3 \rightarrow C_6H_5CH{=}CH_2 + H_2$ [330–332]
2. Butenes $\rightarrow$ Butadienes   [333]

Fischer–Tropsch

1. $CO + H_2 \rightarrow C_nH_{2n+1}$ [334]

Hydrodealkylation

1. $C_6H_5CH_3 + H_2 \rightarrow C_6H_6 + CH_4$ [176]

Hydroesterification

1. $C_6H_5NO_2 + CO + CH_3CH_2OH \xrightarrow{C_6H_5NH_2} C_6H_5NHC(O)OCH_2CH_3$ [335]

Hydrogenation

1. Hydrogenation of coal   [336]

Isomerization

1. $CH_3CH_2HC=CH_2 \rightarrow (CH_3)_2C=CH_2$ [337]

Oxidation

1. $CH_3CH=CH_2 \xrightarrow{NH_3,O_2,H_2O} H_2C=CHCN$ [338]

2. $CH_3CH_2HC=CH_2 \xrightarrow{O_2}$ [339]

3. $2CO + O_2 \rightarrow 2CO_2$ [340]

4. $\xrightarrow{NH_3,O_2}$ [341]

5. $CH_3CH_3 \xrightarrow{HCl,O_2} H_2C=CH_2 + H_2C=CHCl$ [342]

Polymerization

1. $C_6H_5CH_2Cl \rightarrow (-C_6H_4CH_2-)_x$ [343]
2. $ROCH=CH_2 \rightarrow (ROCHCH_2-)_x$ [344]

Water Gas Shift Reaction

1. $CO + H_2O \rightarrow CO_2 + H_2$ [183]

Review—[184]

18. Iron Pentacarbonyl

CA Registry	13463-40-6
Formula	$Fe(CO)_5$
Properties	yellow liquid, b.p. is 103°C, m.p. −20°C, d 1.52, air-sensitive liquid, soluble in most organic solvents
Preparation	[211, 345–347]

Addition

1. $RHC=CHR' + CCl_4 \rightarrow RCH(CCl_3)CHClR'$ [348]

Carbonylation

1. $CH_3NH_2 + (CH_3)_3N + CO \rightarrow (CH_3)_2NC(O)H + (CH_3)_3NH^+$ [326, 327]
2. $RNH_2 + CO \rightarrow RC(O)NH_2$ [349, 350]

Fisher–Tropsch

1. $CO + H_2 \rightarrow C_nH_{2n+1}$ [319, 351, 352]

2. $CO + H_2 \rightarrow CH_3OCH_3 + CH_3CH_2OCH_2CH_3$ [353]

Homologation

1. $CO + H_2 + CH_3OH \rightarrow$
$$CH_3CH_2OH + CH_3C(O)OCH_3 + CH_3C(O)OCH_2CH_3 \quad [216]$$

Hydroesterification

1. $ClCH_2C(O)OCH_3 + CO \xrightarrow{CH_3OH} (CH_3O(O)C)_2CH_2$ [354]

Hydroformylation

1. $CH_3CH_2CH_2HC{=}CH_2 + CO + H_2O \rightarrow$
$$CH_3CH_2CH_2C(C(O)H)CH_3 + CH_3(CH_2)_4C(O)H \quad [84]$$

Hydrogenation

1. $C_6H_5NO_2 + 2CO + H_2 \rightarrow C_6H_5NH_2 + 2CO_2$ [308]
2. $(CH_3)_2CO + CO + H_2O \rightarrow (CH_3)_2CH(OH) + CO_2$ [323]

Isomerization

1. $H_2C{=}CHCH_2NHC(O)CH_3 \rightarrow cis\text{-}CH_3HC{=}CHNHC(O)CH_3$ [106]

2. 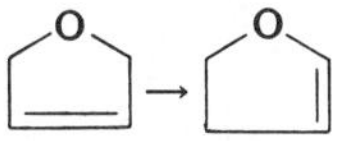 [355]

3. $CH_3CH_2CH_2HC{=}CH_2 \rightarrow CH_3HC{=}CHCH_2CH_3$ [356, 357]

Polymerization

1. $CS \rightarrow (CS)_x$ [358]

Water Gas Shift Reaction

1. $CO + H_2O \rightarrow CO_2 + H_2$ [315, 359]

Reviews — [360–363]

19. Molybdenum

CA Registry   7439-98-7
Formula   Mo
Properties   gray powder, m.p. is 2610°C, d 10.2

Alkylation

1. Low molecular weight hydrocarbons $\rightarrow$ high molecular weight hydrocarbons   [364]
2. $C_6H_6 + H_2C{=}CHCH_3 \rightarrow C_6H_5CH(CH_3)_2$ [365]

Cracking

1. $CH_3OH \rightarrow CO + 2H_2$ [188]

Dehydrosulfurization

1. Dehydrosulfurization of thiophene   [191]

Esterification

1. $RC_6H_4OH + R'C(O)OH \rightarrow RC_6H_4O(O)CR' + H_2O$ [366]

Fischer–Tropsch

1. $CO + 2H_2 \rightarrow CH_3OH$ [367, 368]
2. $CO + H_2 \xrightarrow{NH_3} CH_3CN$ [369]
3. $CO + H_2 \rightarrow C_nH_{2n+1}$ [370]

Hydrodenitrogenation

1. Hydrodenitrogenation of quinoline to acridine   [196]

Hydrogenation

1. $RHC{=}CH_2 + H_2 \rightarrow RCH_2CH_3$ [371, 372]
2. $RC_6H_5 + 3H_2 \rightarrow RC_6H_{11}$ [373]
3. $N_2 + 3H_2 \rightarrow 2NH_3$ [374]

Hydrolysis

1. $H_2C{\overset{\displaystyle\diagdown}{\underset{O}{\diagup}}}CH_2 + H_2O \rightarrow HOCH_2CH_2OH$ [375]

Metathesis

1. $2H_2C{=}CHCH_3 \rightarrow CH_3HC{=}CHCH_3 + H_2C{=}CH_2$ [367]

Oxidation

1. $H_2C{=}CHCH_3 \xrightarrow{NH_3,O_2} H_2C{=}CHCN$ [206]
2. $CH_3OH \xrightarrow{NH_3,O_2} HCN + H_2O$ [376]
3. $CH_3(CH_2)_5HC{=}CH_2 \xrightarrow{O_2} CH_3(CH_2)_5C(O)CH_3$ [377]

20. *Molybdenum Carbonyl*

CA Registry	13939-06-5
Formula	$Mo(CO)_6$
Properties	white solid, m.p. is 150°C, *d* 1.96, air stable, soluble in ethers and chlorinated hydrocarbons, sparingly soluble in alcohols
Preparation	[211, 378–382]

Cyclization

1. 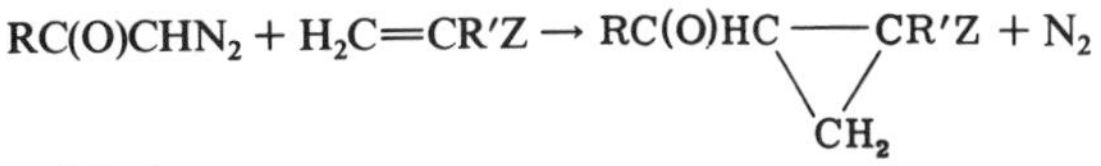 $RC(O)CHN_2 + H_2C{=}CR'Z \rightarrow RC(O)HC{-\!\!-}CR'Z + N_2$ [383]

Epoxidation

1. 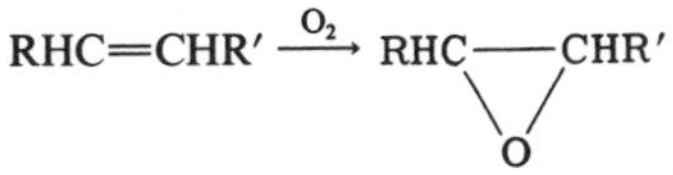 $RHC{=}CHR' \xrightarrow{O_2} RHC{-\!\!-}CHR'$ [384, 385]

2. 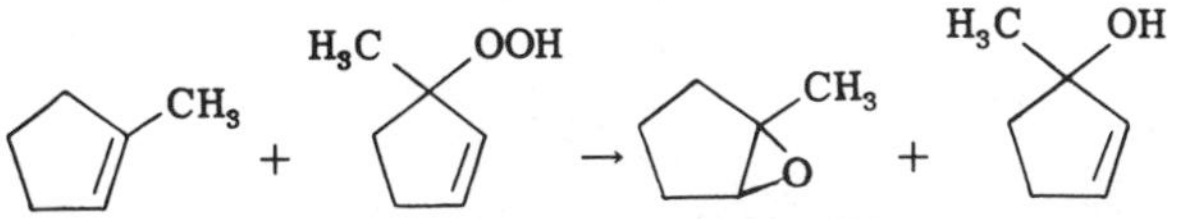   [386]

3. 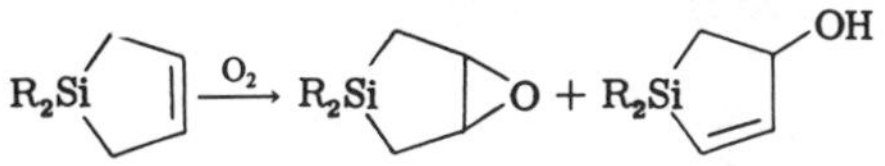   [387]

Fischer–Tropsch

1. $CO + 3H_2 \rightarrow CH_4 + H_2O$ [388]

Hydroesterification

1. $C_6H_5NH_2 + CO \xrightarrow{CH_3OH} C_6H_5NHC(O)OCH_3$ [224]

Hydrolysis

1. 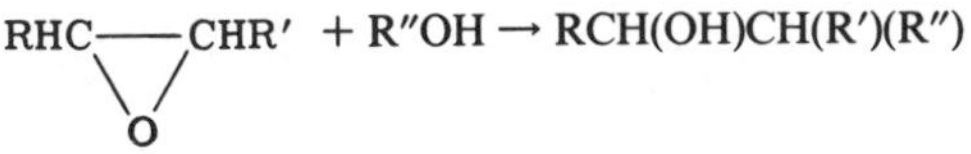 $RHC{-\!\!-}CHR' + R''OH \rightarrow RCH(OH)CH(R')(R'')$ [389]

Isomerization

1. 1-octene → octene isomers   [390]

Metathesis

1. $2C_6H_5C{\equiv}CC_6H_4CH_3 \rightarrow C_6H_5C{\equiv}CC_6H_5 + CH_3C_6H_4C{\equiv}CC_6H_4CH_3$ [391]
2. $2RC{\equiv}CR' \rightarrow RC{\equiv}CR + R'C{\equiv}CR'$ [392]

Water Gas Shift Reaction

1. $CO + H_2O \rightarrow CO_2 + H_2$ [393, 394]

Reviews—[361]

21. Molybdenum Oxide

CA Registry	1313-27-5
Formula	MoO_3
Properties	colorless to light yellow solid, m.p. is 795°C, d 4.69, air stable

Carbonylation

1. $C_6H_5NO_2 + 3CO \rightarrow C_6H_5NCO + 2CO_2$ [395]

Cracking

1. Cracking of petroleum   [396, 397]

Dehydrogenation

1. $C_6H_5CH_2CH_3 \rightarrow C_6H_5HC{=}CH_2 + H_2$ [398]

2. 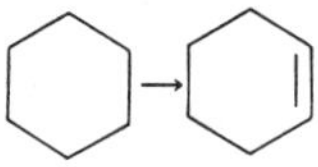 [cyclohexane] $\rightarrow$ [cyclohexene] $+ H_2$ [399]

Dehydrosulfurization

1. Dehydrosulfurization of petroleum   [400]
2. $CH_3CH_2SH + H_2 \rightarrow H_2S + CH_3CH_3$ [401]

Esterification

1. $RC(O)NH_2 + R'OH \rightarrow RC(O)OR' + NH_3$ [402]

Fischer–Tropsch

1. $CO + H_2 \rightarrow C_nH_{2n+1}$ [403–405]

Hydrodenitrogenation

1. Hydrodenitrogenation of coal oils   [406]

Hydrogenation

1. $RC(O)OR' + 2H_2 \rightarrow RCH_2OH + R'OH$ [257]

Hydrolysis

1. $H_2C{=}CH_2 + H_2O \rightarrow CH_3CH_2OH$ [407]

Isomerization

1. $CH_3(CH_2)_2HC{=}CH_2 \rightarrow CH_3CH_2HC{=}CHCH_3$ [408]

Oxidation

1. $CH_3HC{=}CH_2 \xrightarrow{NH_3,O_2} H_2C{=}CHCN$ [409, 410]

2. $H_2C{=}CHCH{=}CH_2 \xrightarrow{O_2}$ [maleic anhydride]   [411]

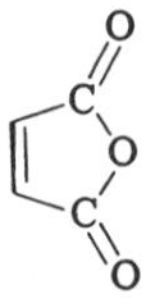

3. $CH_3OH + NH_3 + O_2 \rightarrow HCN + 3H_2O$ [412]

4. $C_6H_5CH_3 + O_2 \rightarrow C_6H_5OH + HC(O)H$ [413]

5. $(CH_3)_2C{=}CH_2 \xrightarrow{O_2} H_2C{=}C(CH_3)C(O)H + H_2C{=}C(CH_3)C(O)OH$ [414]

6. $p\text{-}NCC_6H_4C(O)H \xrightarrow{O_2,CH_3OH} p\text{-}NCC_6H_4C(O)OCH_3$ [415]

7. $N\!\!\diagup\!\!\diagdown\!\!-CH_3 \xrightarrow{NH_3,O_2} N\!\!\diagup\!\!\diagdown\!\!-CN$ [341]

Reduction

1. $RS(O)R' + H_2 \rightarrow RSR' + H_2O$ [416]

Reforming

1. Reforming of aromatics   [417]

Reviews—[418]

22. Nickel

CA Registry   7440-02-0
Formula   Ni
Properties   gray to black powder, m.p. is 1453°C, d 8.9

Alkylation

1. $C_6H_5OH \xrightarrow{CH_4} C_6H_{5-x}(OH)(CH_3)_x$ [419]
2. $C_6H_6 + H_2C{=}CHCH_3 \rightarrow C_6H_5CH(CH_3)_2$ [365]

Alkylation (Reductive)

1. $HOCH_2CH_2NH_2 + NH_3 \xrightarrow{H_2} H_2NCH_2CH_2NH_2 + H_2O$ [186]
2. $(HO(CH_2)_3)_2NH + (CH_3)_2CO + H_2 \rightarrow (HO(CH_2)_3)_2NCH(CH_3)_2 + H_2O$ [420]
3. $2C_6H_6 + 2H_2 \rightarrow C_6H_5C_6H_{11}$ [421]

Aromatization

1. $(CH_3CH_2)_2CHCH_3 \rightarrow C_6H_6 + 4H_2 + \text{other products}$ [422]
2. $CH_3(CH_2)_4CH_3 \rightarrow C_6H_6 + 4H_2$ [423]

Dehydrosulfurization

1. Dehydrosulfurization of petroleum residues   [424]

Fischer–Tropsch

1. $CO + 2H_2 \rightarrow CH_3OH$ [425, 426]
2. $CO + H_2 \rightarrow C_nH_{2n+1}$ [427–429]

Hydrogenation

1. $C_6H_6 + 3H_2 \rightarrow C_6H_{12}$ [430]

2. $RC{\equiv}CR' + H_2 \rightarrow RHC{=}CHR'$ [431]

3. $RC(O)R' + H_2 \rightarrow RCH(OH)R'$ [431]

4. $RC(O)H + H_2 \rightarrow RCH_2OH$ [421, 432]

5. $RC_6H_4C(O)OR' + 2H_2 \rightarrow RC_6H_4CH_2OH + R'OH$ [433]

6. $C_6H_5NO_2 + 3H_2 \rightarrow C_6H_5NH_2 + 3H_2O$ [434]

7. $RCN + 2H_2 \rightarrow RCH_2NH_2$ $(R{=}C_8{-}C_{24})$ [435]

8. $(NCCH_2)_2O + 4H_2 \rightarrow$ HN⎡ ⎤O $+ NH_3$ [436]

Hydrogenolysis

1. Hydrogenolysis of 2,3-dimethylbutane   [201]

2. Hydrogenolysis of isopentane   [437]

3. Hydrogenolysis of hydrocarbons   [202]

4. $C_6H_5CN + 3H_2 + C_6H_5CH_3 + NH_3$ [438]

Isomerization

1. $E{-}R(CH_3)C{=}C(CN)R' \rightarrow Z{-}R(CH_3)C{=}C(CN)R'$ [439]

2. 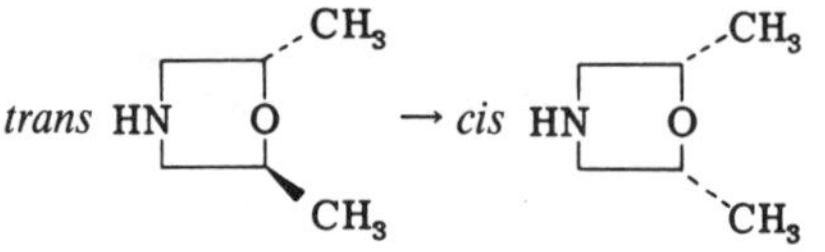 *trans* HN⎡ ⎤O·CH₃ ... $\rightarrow$ *cis* HN⎡ ⎤O·CH₃ ...   [203]

Oxidation

1. $H_2C{=}C(CH_3)C(O)H \xrightarrow{O_2} H_2C{=}C(CH_3)C(O)OH$ [207]

2. $CH_3(CH_2)_5HC{=}CH_2 \xrightarrow{O_2} CH_3(CH_2)_5C(O)CH_3$ [377]

Reviews — [440, 441]

23.   Nickel(II)Acetylacetonate(Bis(2,4-pentanedionato-0,0')nickel)

CA Registry	3264-82-2
Formula	$Ni(C_5H_7O_2)_2$
Properties	light-green solid, m.p. is 229°C, d 1.455, soluble in toluene chloroform, insoluble in ether and hydrocarbon solvents
Preparation	[442]

Alkylation

1. $CH_3C(O)CH_2C(O)CH_3 + H_2C{=}CHC(O)CH_3 \rightarrow$

$CH_3C(O)CH(CH_2CH_2C(O)CH_3)C(O)CH_3$ [443]

Coupling (Reductive)

1. $2H_2C{=}CH_2 \rightarrow H_2C{=}CHCH{=}CH_2 + H_2$ [444]

2. $2CH_3HC{=}CH_2 \rightarrow (CH_3HC{=}CH)_2 + H_2$ [445]

Cyclization

1. $2H_2C=CHCH=CH_2 \rightarrow$ ⬡   [446, 447]

Exchange

1. $C_6H_5MgBr + C_6H_5C_6H_4I \rightarrow MgBrI + C_6H_5C_6H_4C_6H_5$ [448]
2. $(CH_3)_3SiOR + C_6H_5MgBr \rightarrow C_6H_5R + BrMgOSi(CH_3)_3$ [449]

Hydrogenation

1. ⬡ $+ H_2 \rightarrow$ ⬡   [450]

2. Hydrogenation of soybean oils   [451]
3. Hydrogenation of diolefin polymers   [452]

Isomerization

1. $(CH_3)_2C=CHCH_2HC=C(CH_3)CH_2HC=CHC(O)CH_3 \rightarrow$
$(CH_3)_2C=CHCH_2CH_2(CH_3)C=CHCH=CHC(O)CH_3$ [453]

Polymerization

1. $RHC=CHR' \rightarrow (RCHCHR')_x$ [454, 455]
2. Preparation of polyurethane   [456]

24. Nickel Tetracarbonyl

CA Registry	13463-39-3
Formula	$Ni(CO)_4$
Properties	colorless liquid, b.p. is 43°C, m.p. -19.3°C, d 1.318, air-sensitive liquid, soluble in most organic solvents
Preparation	[457, 458]

Aromatization

1. $3HC\equiv CH \rightarrow C_6H_6$ [459]

Carbonylation

1. $H_2C=CHCH_2Cl \xrightarrow{CO,NaOH} H_2C=CHCH_2C(O)ONa + CH_3HC=CHC(O)ONa$
[460]

2. $RNH_2 + CO \rightarrow RC(O)NH_2$ [461]

Carboxylation

1. $CH_3OH \xrightarrow{CO,H_2} CH_3C(O)OH + CH_3C(O)OCH_3$ [462]

Hydroesterification

1. $ClCH_2C(O)OCH_3 + CO \xrightarrow{CH_3OH} (CH_3(O)OC)_2CH_2$ [354]

2. $C_6H_5NH_2 + CO \xrightarrow{CH_3OH} C_6H_5NHC(O)OCH_3$ [224]

25. Osmium Dodecacarbonyl (Dodecacarbonyltriosmium)

CA Registry   15696-40-9
Formula $Os_3(CO)_{12}$
Properties   yellow crystals, sparingly soluble in most organic sol-
               vents, air-stable solid
Preparation   [463]

Fischer–Tropsch

1. $CH_3C(O)OH \xrightarrow{CO,H_2} CH_3C(O)OCH_3 + CH_3C(O)OCH_2CH_3 + (CH_3C(O)OCH_2)_2$
                                                                    [464]
2. $CO + H_2 \rightarrow C_nH_{2n+1}$ [321, 465, 466]
3. $CO + H_2 \rightarrow CH_3OCH_3 + CH_3CH_2OCH_2CH_3$ [353]

Hydroformylation

1. $CH_3CH_2CH_2HC{=}CH_2 + CO + H_2O \rightarrow$
   $\qquad\qquad CH_3CH_2CH_2C(C(O)H)CH_3 + CH_3(CH_2)_4C(O)H$ [84]
2. $RHC{=}CH_2 + 3CO + H_2O + HNR'_2 \rightarrow R(CH_2)_3NR'_2 + 2CO_2$ [305]

Hydrogenation

1. $CH_3CH_2CH_2HC{=}CH_2 + H_2 \rightarrow CH_3(CH_2)_3CH_3$ [467]
2. $C_6H_5NO_2 + 2CO + H_2 \rightarrow C_6H_5NH_2 + 2CO_2$ [308]

Isomerization

1. $CH_3CH_2CH_2HC{=}CH_2 \rightarrow CH_3CH_2HC{=}CHCH_3$ [467]

Reforming

1. $(CH_3CH_2)_3N + (CH_3CH_2CH_2)_3N \rightarrow$
   $\qquad (CH_3CH_2)_2(CH_3CH_2CH_2)N + (CH_3CH_2)(CH_3CH_2CH_2)_2N$ [468]

26. Palladium

CA Registry   7440-05-3
Formula   Pd
Properties   gray to black powder, m.p. is 1552°C, d 11.40

Alkylation

1. Low molecular weight hydrocarbons $\rightarrow$ high molecular weight hydrocarbons   [364]

Alkylation (Reductive)

1. $C_6H_5OH + NH_3 \xrightarrow{H_2} C_6H_5NH_2 + H_2O$ [469]

Carbonylation

1. 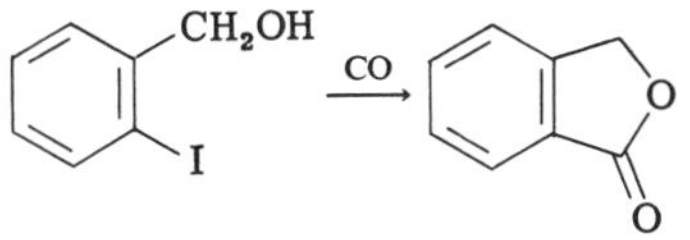 $\xrightarrow{CO}$ [270]

Carboxylation (Oxidative)

1. $R^1HC{=}CR^2CR^3{=}CHR^4 + CO \xrightarrow{O_2,ROH} RO(O)CCHR^1CR^2{=}CR^3CHR^4C(O)OR$

[470]

Decarbonylation

1. 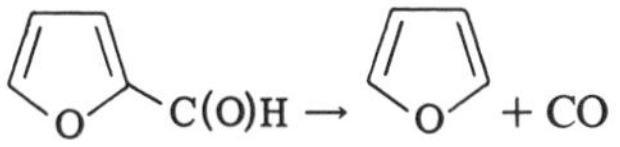 $\rightarrow$ $+ CO$ [471]

Dehydrogenation

1. $C_6H_5CH_2OH \xrightarrow{NaOH} C_6H_5C(O)ONa + 2H_2$ [472]

Fischer–Tropsch

1. $CO + H_2 \rightarrow C_nH_{2n+1}$ [473]

Hydrodealkylation

1. $RNR'_2 + H_2O \rightarrow R'_2NH + ROH$ [474]
2. $C_6H_4(OH)R + H_2 \rightarrow C_6H_5OH + RH$ [475]

Hydrogenation

1. $RC_6H_4NO_2 + 3H_2 \rightarrow RC_6H_4NH_2 + 2H_2O$ [476, 477]
2. $HC{\equiv}CH + H_2 \rightarrow H_2C{=}CH_2 + H_2 \rightarrow H_3CCH_3$ [478]
3. $RC_6H_4C(O)OR' + 2H_2 \rightarrow RC_6H_4CH_2OH + R'OH$ [433]
4. $NC(CH_2)_xRC{=}CH(CH_2)_xCN + 5H_2 \rightarrow NH_2CH_2(CH_2)_xCRHCH_2(CH_2)_xCH_2NH_2$

[479]

5. $CH_3CN \xrightarrow{H_2} (CH_3CH_2)_2NH + (CH_3CH_2)_3N$ [480]

Hydrogenolysis

1. 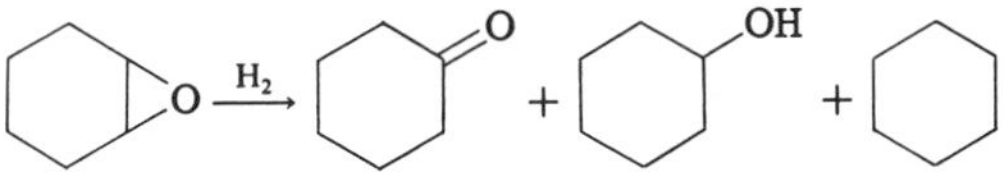 $\xrightarrow{H_2}$ $+$ $+$ [481]

Isomerization

1. $H_2C{=}CH(CH_2)_3CH(OH)HC{=}CH_2 \rightarrow H_2C{=}CH(CH_2)_4HC{=}CHOH$ [482]

2. 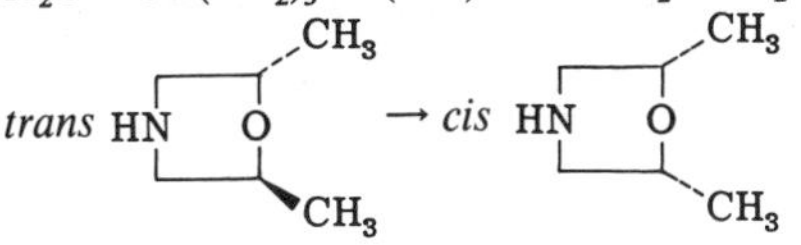 *trans* $\rightarrow$ *cis*   [203]

3. $C_6H_5OCH_2HC=CH_2 \rightarrow C_6H_5OCH=CHCH_3$ [483]

Oxidation

1. $H_2C=CH_2 \xrightarrow{O_2,H_2O} CH_3C(O)OH + H_2C=CHC(O)OCH_3$ [484]

2. $H_2C=C(CH_3)C(O)H \xrightarrow{O_2,CH_3OH} H_2C=C(CH_3)C(O)OCH_3$ [485]

3. $C_6H_5CH_3 \xrightarrow{O_2} C_6H_5C(O)OH$ [486]

4. $C_6H_5CH_2CH_3 \xrightarrow{O_2,H_2O} C_6H_5HC=CH_2$ [487]

5. $C_6H_5OCH_2CH_2OH \xrightarrow{O_2} C_6H_5OCH_2C(O)OH$ [488]

Polymerization

1. $H_2N(CH_2)_6NH_2 \rightarrow (-HN(CH_2)_6NH-)_x$ [489]

Reduction

1. $RC(O)Cl + H_2 \rightarrow RC(O)H + HCl$ [490]

Reforming

1. $RR'_2N \rightarrow R_3N + R_2R'N + RR'_2N + R'_3N$ [491]

Reviews—[492–494]

27.   Palladium Acetate (Acetic Acid, Palladium(II) salt)

CA Registry   3375-31-3
Formula $Pd(CH_3COO)_2$
Properties   golden, brown solid, soluble in most organic solvents, insoluble in water and hydrocarbons, air-stable solid
Preparation   [495]

Alkylation

1. 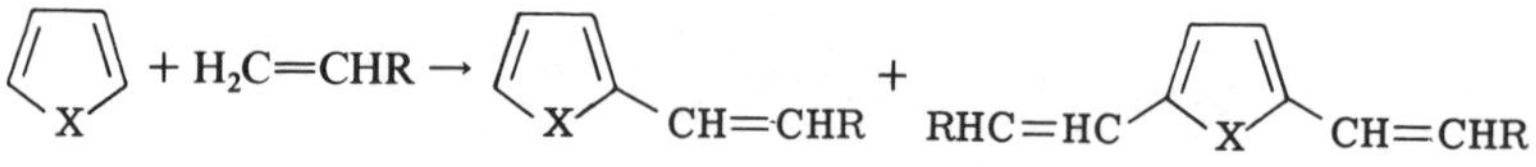 $+ H_2C=CHR \rightarrow$ $+$

                                                                    (X—O,S)   [496]

Arylation

1. $p\text{-}BrC_6H_4I + C_6H_5CH=CH_2 \xrightarrow{(CH_3CH_2)_3N} p\text{-}BrC_6H_4HC=CHC_6H_5 + HI$ [497]

Carbonylation

1. 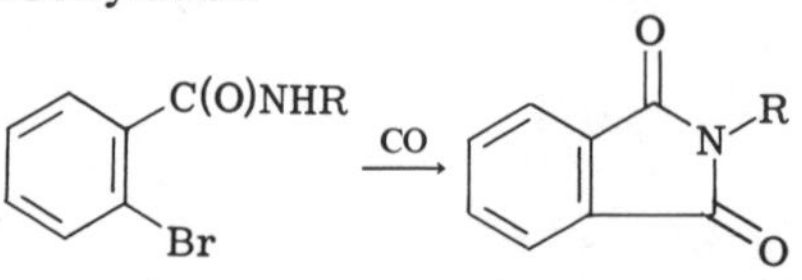 $\xrightarrow{CO}$ [498]

Carboxylation

1. $C_6H_5R \xrightarrow{CO} RC_6H_4C(O)OH$ [499]

2. $RN_2^+BF_4^- + R'C(O)ONa \xrightarrow{CO} RC(O)OC(O)R' \xrightarrow{\Delta} (RC(O))_2O + (R'C(O))_2O$

[500, 501]

3. $CH_3OH \xrightarrow{CO} CH_3OC(O)C(O)OCH_3 + (CH_3O)_2CO$ [172]

Cyclization

1. $RHC{=}CHR' + CH_2N_2 \rightarrow RHC\overset{\displaystyle \diagdown}{\underset{CH_2}{\diagup}}CHR' + N_2$ [502]

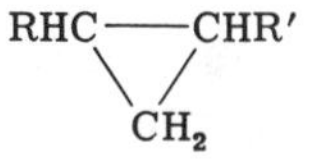

2. 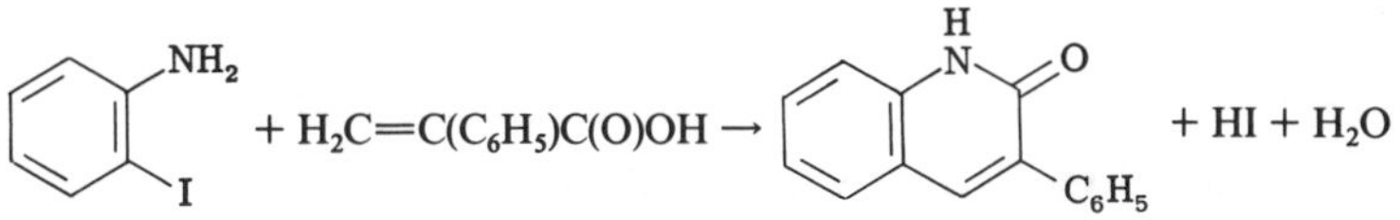 $+ H_2C{=}C(C_6H_5)C(O)OH \rightarrow$ $+ HI + H_2O$

[503]

3. $RHC{=}CHCH_2C(CH_2HC{=}CH(CH_3))(C(O)CH_3)_2 \rightarrow$

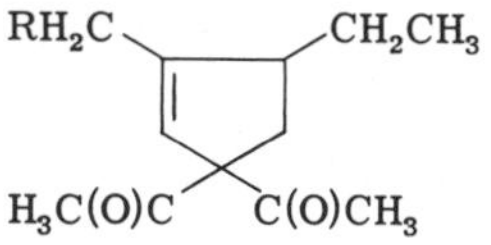

[504]

Dehalogenation

1. $o\text{-}BrC_6H_4NO_2 + (CH_3CH_2)_3NH^+HC(O)O^- \rightarrow (CH_3CH_2)_3NH^+Br^- + CO_2 + C_6H_5NO_2$

[505]

2. $Br_5C_6OC_6Br_5 \xrightarrow{NaOH} H_5C_6OC_6H_5$ [506]

Epoxidation

1. $RHC{=}CHR' \xrightarrow{O_2} RHC\overset{\displaystyle \diagdown}{\underset{O}{\diagup}}CHR'$ [507]

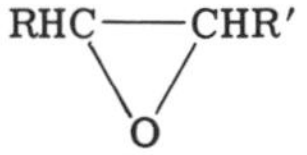

Oxidation

1. $H_2C{=}CH_2 + CH_3C(O)OH \xrightarrow{O_2} H_2C{=}CHOC(O)CH_3$ [508]

2. $RCH_2OH \xrightarrow{CCl_4, NaOH} RC(O)OH$ [509]

3. $C_6H_5CH_3 \xrightarrow{O_2} C_6H_5C(O)OH$ [486]

Polymerization

1. Polymerization of isoprene   [510]

Transesterification

1. $CH_3CH_2C(O)OH + H_2C=CHOC(O)CH_3 \rightarrow$
$$H_2C=CHC(O)OCH_2CH_3 + CH_3C(O)OH \quad [47, 511]$$

Reviews — [512, 513]

28. *Palladium Chloride*

CA Registry   7647-10-1
Formula $PdCl_2$
Properties   red, brown solid, m.p. is 678°C, insoluble in most sol-
               vents, soluble in aqueous solutions of alkali halides
Preparation   [514]

Carboxylation

1. $H_2C=CHCH_2OH + CO \rightarrow H_2C=CHCH_2C(O)OH$ [36]
2. $CH_3OH + CO \rightarrow CH_3C(O)OH$ [515]
3. 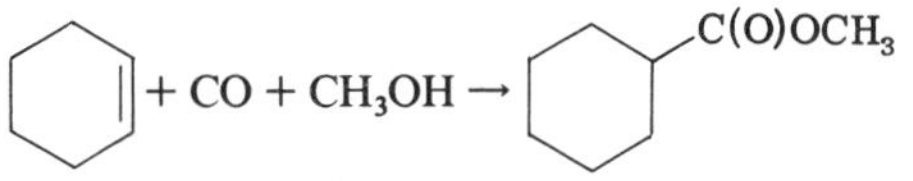 $+ CO + CH_3OH \rightarrow$ [516]

Carboxylation (Oxidative)

1. $H_2C=CHCH=CH_2 + CO \xrightarrow{O_2,CH_3OH} CH_3O(O)CCH_2HC=CHCH_2C(O)OCH_3$
$$[517]$$

2. $R^1HC=CR^2CR^3=CHR^4 + CO \xrightarrow{O_2,ROH} RO(O)CCHR^1CR^2=CR^3CHR^4C(O)OR$
$$[37, 470]$$

3. $C_6H_5C\equiv CH + CO \xrightarrow{O_2,CH_3OH} C_6H_5C\equiv CC(O)OCH_3$ [518]

4. $C_6H_5NH_2 + CO \xrightarrow{O_2,CH_3CH_2OH} C_6H_5NHC(O)OCH_2CH_3$ [519, 520]

Coupling (Reductive)

1. $C_6H_5CH_2Br + C_6H_5C(O)Cl \xrightarrow{Zn} C_6H_5CH_2C(O)C_6H_5 + ZnBrCl$ [271]

Cyclization

1. $RHC=CHCH_2C(CH_2HC=C(CH_3)H)(C(O)CH_3)_2 \rightarrow$ 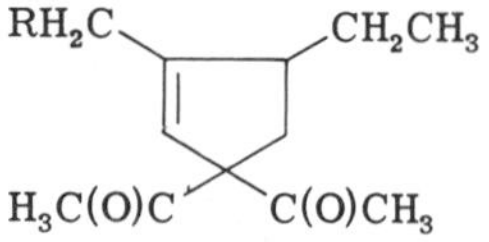
$$[504]$$

Exchange

1. $p\text{-}ClC_6H_4CF_3 + KCN \rightarrow p\text{-}NCC_6H_4CF_3 + KCl$ [521]

Hydroesterification

1. $C_6H_4(NH_2)(NO_2) + CO + CH_3OH \rightarrow C_6H_4(NO_2)(NHC(O)OCH_2CH_3)$ [522]

2. $H_3CHC=CH(CH_3) \xrightarrow{CO,CH_3OH}$
   $HC(CH_3)(OCH_3)CH(CH_3)(C(O)OCH_3) + (-CH(CH_3)C(O)OCH_3)_2$ [523, 524]

3. $CH_3C\equiv CH + CO + CH_3OH \rightarrow H_2C=C(CH_3)C(O)OCH_3$ [525]

Oxidation

1. $CH_3(CH_2)_5HC=CH_2 \xrightarrow{O_2} CH_3(CH_2)_5C(O)CH_3$ [377, 526]

2. $CH_3CH_2CH_2CH_2OH \xrightarrow{O_2} (CH_3(CH_2)_3O)_2CHCH_2CH_2CH_3 + CH_3(CH_2)_2C(O)OH$
   [527]

3. $RCH_2OH \xrightarrow{CCl_4,NaOH} RC(O)OH$ [509]

Telomerization

1. $2H_2C=C(CH_3)HC=CH_2 + (CH_3CH_2)_2NH \rightarrow$
   $H_2C=C(CH_3)(CH_2)_3C(CH_3)=CHCH_2N(CH_2CH_3)_2$ [528]

Reviews—[529]

29. Platinum

CA Registry	7440-06-4
Formula	Pt
Properties	gray to black powder, m.p. is 1769°C, d 21.45

Alkylation

1. $C_6H_6 + H_2C=CHCH_3 \rightarrow C_6H_5CH(CH_3)_2$ [365]
2. Low molecular weight hydrocarbons $\rightarrow$ high molecular weight hydrocarbons   [364]

Alkylation (Reductive)

1. $2C_6H_6 + 2H_2 \rightarrow C_6H_5C_6H_{11}$ [530]

Aromatization

1. 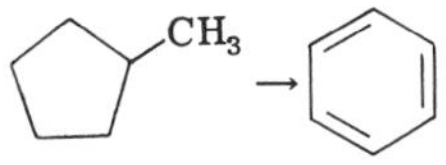 $\rightarrow$ $+ 3H_2$ [531, 532]

2. $CH_3(CH_2)_4CH_3 \rightarrow C_6H_6 + 4H_2$ [100]

Carbonylation (Reductive)

1. $NO + CO + NH_3 \xrightarrow{H_2} NH_4OCN$ [533]

Coupling (Reductive)

1. $2H_2C=CHCH=CH_2 + H_2 \rightarrow H_2C=CH(CH_2)_4HC=CH_2$ [534]

Dehydration

1. $RCH_2CH_2OH \rightarrow RHC{=}CH_2 + (RCH_2CH_2)_2O$ [535]

Dehydrogenation

1. $C_6H_{12} \rightarrow C_6H_6 + 3H_2$ [536]
2. Dehydrogenation of hydrocarbons   [537]
3. $CH_3(CH_2)_8CH_3 \rightarrow CH_3(CH_2)_7HC{=}CH_2 + H_2$ [538]

Fischer – Tropsch

1. $2CO + 4H_2 \rightarrow CH_3CH_2OH + H_2O$ [539]
2. $CO + 3H_2 \rightarrow CH_4$ [540]

Hydrogenation

1. $(C_6H_5)_3PCl_2 + H_2 \rightarrow (C_6H_5)_3P + 2HCl$ [541]

2. 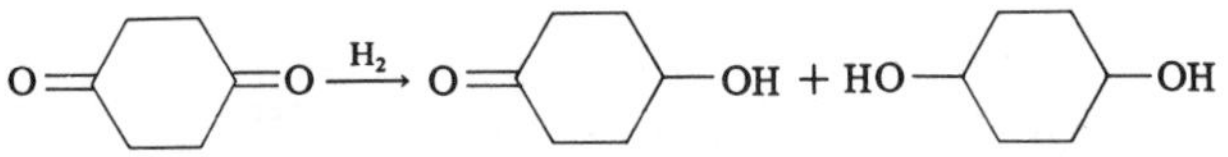   [542]

3. $RC_6H_4NO_2 + 3H_2 \rightarrow RC_6H_4NH_2 + 2H_2O$ [477]
4. $H_2C{=}CHC(O)H + H_2 \rightarrow H_2C{=}CHCH_2OH$ [543]
5. $NCCH_2CH_2C(O)NH_2 + 2H_2 \rightarrow H_2N(CH_2)_3C(O)NH_2$ [544]

Hydrogenolysis

1. Hydrogenolysis of alkanes   [545]
2. $C_6H_5CH_2CH_3 + H_2 \rightarrow C_6H_5CH_3 + CH_4$ [546]

3. 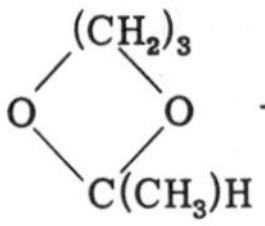 $\xrightarrow{H_2}$ $CH_3CH_2CH_2OC(O)CH_3 + CH_3CH_2O(CH_2)_3OH + $ other products

[547]

4. $C_6H_5OH + H_2 \rightarrow C_6H_6 + H_2O$ [548]

Isomerization

1. Normal paraffins $\rightarrow$ isoparaffins   [549]
2. Isomerization of hydrocarbons   [550]

Oxidation

1. $C_6H_5CH(OH)C(O)OH \xrightarrow{NaOH,Pb(II)} C_6H_5C(O)C(O)OH$ [551]

2. $CH_3CH_2OH \xrightarrow{O_2} CH_3C(O)OH$ [552]

3. $CH_3(CH_2)_2CH_2OH \xrightarrow{O_2} CH_3CH_2CH_2C(O)OH$ [553]

Reforming

1. Reforming of hydrocarbons   [554]

30. Rhodium

CA Registry   7440-16-6
Formula   Rh
Properties   black powder, m.p. is 1966°C, d 12.4

Alkylation (Reductive)

1. $2C_6H_6 + 2H_2 \rightarrow C_6H_5C_6H_{11}$ [555]

Carboxylation

1. $CH_3OH + CO \rightarrow CH_3C(O)OH$ [556]

Coupling (Reductive)

1. $2H_2C{=}CHCH{=}CH_2 + H_2 \rightarrow H_2C{=}CH(CH_2)_4HC{=}CH_2$ [534]

Cracking

1. $CH_3OH \rightarrow CO + 2H_2$ [188]

Dealkylation

1. $C_6H_5CH_3 + H_2O \rightarrow C_6H_6 + CO + 2H_2$ [557, 558]

Dehydration

1. $RCH_2CH_2OH \rightarrow RHC{=}CH_2 + (RCH_2CH_2)_2O$ [535]

Dehydrogenation

1. Dehydrogenation of hydrocarbons   [537]

Fischer – Tropsch

1. $CO + H_2 \rightarrow CH_3C(O)OH + CH_3C(O)H + CH_3CH_2OH$ [539, 559, 560]

Hydroformylation

1. $H_2C{=}CH_2 + CO + H_2 \rightarrow CH_3CH_2C(O)H$ [561]

Hydrogenation

1. $C_6H_6 + 3H_2 \rightarrow C_6H_{12}$ [562]
2. $RC_6H_4C(O)OR' + 2H_2 \rightarrow RC_6H_4CH_2OH + R'OH$ [433]
3. naphthol-OH $+ 3H_2 \rightarrow$ tetrahydronaphthol-OH   [563]

4. $NCCH_2CH_2C(O)NH_2 + 2H_2 \rightarrow H_2N(CH_2)_3C(O)NH_2$ [544]

Hydrogenolysis

1. $C_6H_5CH_2CH_3 + H_2 \rightarrow C_6H_5CH_3 + CH_4$ [546]
2. $C_6H_5CH_3 + H_2 \rightarrow C_6H_6 + CH_4$ [564]

3. 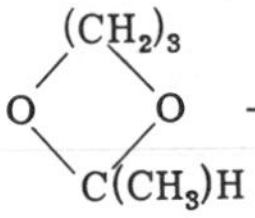 $\xrightarrow{H_2}$ $CH_3CH_2CH_2OC(O)CH_3 + CH_3CH_2O(CH_2)_3OH$ + other products

[547]

31. Rhodium Trichloride

CA Registry   10049-07-7 (anhydrous), 20765-98-4 (hydrated)
Formula $RhCl_3$
Properties   red, crystalline solid, soluble in water and alcohols
Preparation   [565, 566]

Alkylation

1. $C_6H_5NH_2 \xrightarrow{H_2C=CH_2} C_6H_5NHCH_2CH_3 +$ [567]

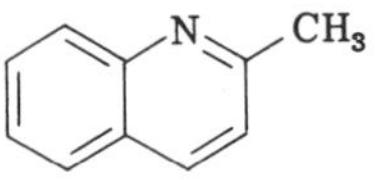

Carbonylation

1. 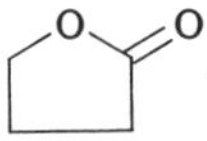 $\xrightarrow{CO}$ $HO(O)C(CH_2)_3C(O)OH$ [568]
2. $CH_3OC(O)CH_3 + CO \rightarrow (CH_3C(O))_2O$ [569, 570]
3. $CH_3CH_2OCH_2CH_3 + CO \rightarrow CH_3CH_2C(O)OCH_2CH_3$ [571]
4. $H_2C=CHCH_2NHR + CO \rightarrow$ 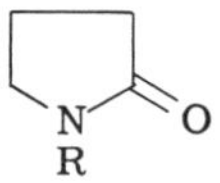   [70]

Carboxylation

1. $CH_3OH + CO \rightarrow CH_3C(O)OH$ [515, 572]

Carboxylation (Oxidative)

1. $R^1HC=CR^2CR^3=CHR^4 + CO \xrightarrow{O_2,ROH} RO(O)CCHR^1CR^2=CR^3CHR^4C(O)OR$

[37, 470]

Dimerization

1. $2RHgCl \rightarrow R-R + HgCl_2 + Hg$ [121]

Fischer – Tropsch

1. $CO + H_2 \rightarrow C_nH_{2n+1}$ [573]
2. $CO + 2H_2 \rightarrow CH_3OH$ [574]

Hydroesterification

1. $ClCH_2C(O)OCH_3 + CO \xrightarrow{CH_3OH} (CH_3O(O)C)_2CH_2$ [354]

2. $RC{\equiv}CR' + CO + R''OH \rightarrow$ [575]

Hydroformylation

1. $RHC{=}CH_2 + (CH_3)_2NH + CO + H_2 \rightarrow RCH_2CH_2CH_2N(CH_3)_2$ [125]

2. $HN\boxed{}O + CH_3(CH_2)_8HC{=}CH_2 + CO + H_2 \rightarrow O\boxed{}N(CH_2)_{11}CH_3$ [576]

Trimerization

1. $3CH_3CH_2C(O)H \rightarrow$ [577]

Water Gas Shift Reaction

1. $CO + H_2O \rightarrow CO_2 + H_2$ [578]

32. *Ruthenium*

CA Registry	7440-18-8
Formula	Ru
Properties	black powder, m.p. is 2250°C, d 12.3

Alkylation

1.

$+ CH_3(CH_2)_5HC{=}CH_2 \rightarrow$ $+ H_2$

[579]

Alkylation (Reductive)

1. $2C_6H_6 + 2H_2 \rightarrow C_6H_5C_6H_{11}$ [555]

Cracking

1. $CH_3OH \rightarrow CO + 2H_2$ [188]

Fischer–Tropsch

1. $CO + H_2 \rightarrow C_nH_{2n+1}$ [580–585]
2. $CO + 2H_2 \rightarrow CH_3OH$ [425]

Homologation

1. $C_6H_5CH_2OH + CO + H_2 \rightarrow C_6H_5CH_2CH_2OH$ [219]
2. $CH_3OH + CO + H_2 \rightarrow CH_3CH_2OH$ [586]

Hydrogenation

1. $C_6H_6 + 3H_2 \rightarrow C_6H_{12}$ [587]
2. Hydrogenation of polybutadiene   [588]
3. $RC_6H_4C(O)OR' + 2H_2 \rightarrow RC_6H_4CH_2OH + R'OH$ [433]
4. $NCCHRHC{=}CHCHR'CN + 2H_2 \rightarrow NCCHRCH_2CH_2CHR'CN$ [589]
5. $m\text{-}O_2NC_6H_4C{\equiv}CR + 3H_2 \rightarrow m\text{-}H_2NC_6H_4C{\equiv}CR + 2H_2O$ [590]

Hydrogenolysis

1. Hydrogenolysis of 2,3-dimethylbutane   [201]
2. Hydrogenolysis of hydrocarbons   [202]
3. $C_6H_5CH_2CH_3 + H_2 \rightarrow C_6H_5CH_3 + CH_4$ [546]
4.  $\triangledown + 3H_2 \rightarrow 3CH_4$ [591]

Reduction

1. $3CO + 3H_2O + N_2 \rightarrow 2NH_3 + 2CO_2$ [592]
2. $N_2 + 3H_2 \rightarrow 2NH_3$ [593, 594]

Reviews—[595–598]

33.   Ruthenium Carbonyl (Dodecacarbonyltriruthenium)

CA Registry	15243-33-1
Formula	$Ru_3(CO)_{12}$
Properties	orange crystals, d is 2.75, soluble in most organic solvents, air-stable solid
Preparation	[302, 599–601]

Fischer–Tropsch

1. $CO + 2H_2 \rightarrow CH_3OH$ [313, 318, 602–605]
2. $CH_3C(O)OH \xrightarrow{CO,H_2} CH_3C(O)OCH_3 + CH_3C(O)OCH_2CH_3 + (CH_3C(O)OCH_2)_2$
   $CH_3C(O)OCH_3 + CH_3C(O)OCH_2CH_3 + (CH_3C(O)OCH_2)_2$ [464]

Homologation

1. $RCH_2C(O)OH + CO + H_2 \rightarrow RCH_2CH_2C(O)OH$ [606, 607]

Hydroformylation

1. $RHC{=}CH_2 + 3CO + H_2O + HNR'_2 \rightarrow R(CH_2)_3NR'_2 + 2CO_2$ [305]

Hydrogenation

1. $C_6H_5C(O)H + CO + H_2O \rightarrow C_6H_5CH_2OH + CO_2$ [311]

Hydrosilylation

1. $RHC=CH_2 + R'_3SiH \rightarrow$ *trans*-$RHC=CHSiR'_3 + H_2$ [608]

Isomerization

1. $CH_3CH_2CH_2HC=CH_2 \rightarrow CH_3HC=CHCH_2CH_3$ [356]
2. $CH_3(CH_2)_3HC=CH_2 \rightarrow CH_3(CH_2)_2HC=CHCH_3 + CH_3CH_2HC=CHCH_2CH_3$
[609]

Polymerization

1. Polymerization of 2-norbornene   [279]

Reduction

1. p-$RC_6H_5NO_2 \xrightarrow{CO,NaOH} p$-$RC_6H_5NH_2$ [610]

Reforming

1. $(CH_3CH_2)_3N + (CH_3CH_2CH_2)_3N \rightarrow$
$(CH_3CH_2)_2(CH_3CH_2CH_2)N + (CH_3CH_2)(CH_3CH_2CH_2)_2N$ [468]

Water Gas Shift Reaction

1. $CO + H_2O \rightarrow H_2 + CO_2$ [315, 611]

Reviews—[612]

34. Ruthenium Dioxide

CA Registry   32740-79-7
Formula RuO_2
Properties   dark, blue solid, d is 6.97

Cracking

1. $2H_2O \rightarrow 2H_2 + O_2$ (photodecomposition)   [613, 614]

Homologation

1. $RCH_2C(O)OH + CO + H_2 \rightarrow RCH_2CH_2C(O)OH$ [606, 607, 615, 616]

Oxidation

1. $(CH_3)_3CCH(OH)C(O)ONa \xrightarrow{NaOCl} (CH_3)_3CC(O)C(O)ONa$ [617]

35. Silver

CA Registry   7440-22-4

Formula   Ag
Properties   black powder, d is 10.5, m.p. is 960.8 °C

Alkylation (Reductive)

1. $2C_6H_6 + 2H_2 \rightarrow C_6H_5C_6H_{11}$ [555]

Dehydrogenation

1. $CH_3OH \rightarrow CO + 2H_2$ [618]

Epoxidation

1. $H_2C=CH_2 \xrightarrow{O_2}$

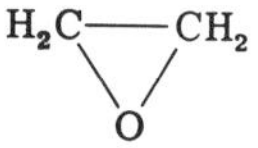

[619–621]

Hydrogenation

1. Butadiene polyperoxide $\rightarrow$ butene-3,4-diols   [622]

Oxidation

1. $HOCH_2CH_2OH \xrightarrow{O_2} H(O)CC(O)H$ [623]

2. $C_6H_5CH_2OH \xrightarrow{O_2,H_2O} C_6H_5C(O)H$ [624]

3. $CH_3OH \xrightarrow{O_2,H_2O} HC(O)H$ [625]

Reviews — [626, 627]

36.   Tetrakis(triphenylphosphine)palladium(0)

CA Registry   14221-01-3
Formula $Pd(P(C_6H_5)_3)_4$
Properties   yellow solid, air sensitive, soluble in benzene, methylene
                chloride and chloroform, insoluble in hydrocarbons,
                slightly soluble in acetone and tetrahydrofuran
Preparation   [628, 629]

Addition

1. $C_6H_5C\equiv CH + R_2(CH_3)SiSi(CH_3)R_2 \rightarrow (R_2(CH_3)Si)(C_6H_5)C=CH(Si(CH_3)R_2)$
                                                        [630–632]
2. $C_6H_5HC=CHBr + HP(O)(OCH_2CH_3)_2 \rightarrow C_6H_5HC=CHP(O)(OCH_2CH_3)_2 + HBr$
                                                        [633]

Carbonylation

1. $RBr + CO \xrightarrow{K_2CO_3,H_2NNH_2} RC(O)OH$ [269]

Coupling (Reductive)

1. $2H_2C{=}CHCH{=}CH_2 + H_2 \rightarrow H_2C{=}CH(CH_2)_4HC{=}CH_2 +$
$$H_2C{=}CH(CH_2)_3HC{=}CHCH_3 \quad [634, 635]$$

Cyclooligomerization

1. + other oligomers   [636]

Exchange

1. $RC_6H_4Br + NaCN \rightarrow RC_6H_4CN + NaBr$ [637]
2. $R_1R_2C{=}CR_3X + RLi \rightarrow R_1R_2C{=}CR_3R + LiX$ [638]
3. $o\text{-}C_6H_4Br_2 + CH_3MgBr \rightarrow o\text{-}C_6H_4(Br)CH_3 + MgBr_2$ [639]
4. $C_6H_5B(OH)_2 + RC_6H_4Br \xrightarrow{Na_2CO_3} C_6H_5C_6H_4R + BrB(OH)_2$ [640]
5. $(CH_3)_2CHSCH_2HC{=}CH_2 + C_6H_5MgBr \rightarrow C_6H_5CH_2HC{=}CH_2 + (CH_3)_2CHSMgBr$
$$[274]$$

Hydrosilylation

1. $HC{\equiv}CH + HSiR_3 \rightarrow H_2C{=}CHSiR_3$ [156]

Isomerization

1. $RHC{=}CHCH_2OC(O)CH_3 \rightarrow RCH_2HC{=}CHOC(O)CH_3$ [641]

Oxidation

1. $CH_3(CH_2)_5CH(OH)CH_3 \xrightarrow{O_2} CH_3(CH_2)_5C(O)CH_3$ [642]

Reduction

1. $RC_6H_4Br + HC(O)ONa \rightarrow RC_6H_5 + CO_2 + NaBr$ [643]
2. $RC(O)Cl + HSn(CH_2CH_2CH_2CH_3)_3 \rightarrow RC(O)H + ClSn(CH_2CH_2CH_2CH_3)_3$ [644]
3. $C_6H_5HC{=}CHCH(OC(O)CH_3)(C_6H_5) \xrightarrow{NaBH_3CN} C_6H_5HC{=}CHCH_2C_6H_5$ [645]

Reforming

1. $C_6H_5HC{=}CH(OC(O)CH_3)CN \rightarrow C_6H_5CH(OC(O)CH_3)HC{=}CHCN$ [646]

Telomerization

1. $RCH(OH)C_6H_5 + 2H_2C{=}CHCH{=}CH_2 \rightarrow$
$$RCH(C_6H_5)OCH_2HC{=}CH(CH_2)_3HC{=}CH_2 \quad [647]$$

37.   Tetrarhodium Dodecacarbonyl
(Tri-μ-carbonylnonacarbonyltetrarhodium)

CA Registry	19584-30-6
Formula	$Rh_4(CO)_{12}$
Properties	red, crystalline solid, decomposes to $Rh_6(CO)_{16}$ when heated, soluble in most organic solvents, air sensitive
Preparation	[299, 300, 648–649a]

Addition

1. $(CH_3)_3SiHC=CH_2 + HSi(OCH_2CH_3)_3 \rightarrow (CH_3)_3SiCH_2CH_2Si(OCH_2CH_3)_3$ [650]

Carbonylation

1. $C_6H_6 + C_6H_5C\equiv CC_6H_5 \xrightarrow{CO} C_6H_5HC=C(C_6H_5)_2 +$ 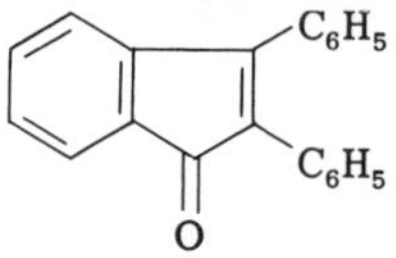   [651]

2. $H_2C=CH_2 + C_6H_6 \xrightarrow{CO} C_6H_5CH=CH_2 + CH_3CH_2C(O)CH_2CH_3$ [652]

Fischer–Tropsch

1. $CO + H_2 \rightarrow C_nH_{2n+1}$ [122, 653]
2. $CO + 2H_2 \rightarrow CH_3OH$ [574]

Homologation

1. $CH_3OH + CO + H_2 \rightarrow CH_3CH_2OH + CH_3C(O)OCH_3 + CH_3C(O)OCH_2CH_3$

[216]

Hydroesterification

1. $RC\equiv CR' + CO + R''OH \rightarrow$ 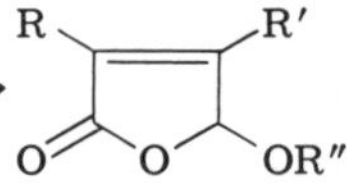   [575]

Hydroformylation

1. $RCH=CH_2 + CO + H_2 \rightarrow RCH_2CH_2C(O)H$ [304, 654, 655]
2. $HC(O)H + CO + H_2 \rightarrow HOCH_2C(O)H$ [76]

Oxidation

1. 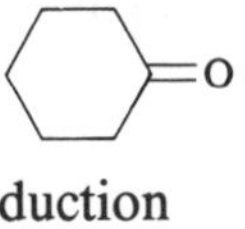 $=O \xrightarrow{O_2} HO(O)C(CH_2)_4C(O)OH$ [656]

Reduction

1. $CO_2 + H_2 \rightarrow CO + H_2O$ [129]

Water Gas Shift Reaction

1. $CO + H_2O \rightarrow CO_2 + H_2$ [132]

38. Titanium Tetrachloride

CA Registry	7550-45-0
Formula	$TiCl_4$
Properties	colorless liquid, b.p. is $136\,^\circ C$, m.p. $-24.1\,^\circ C$, moisture-sensitive liquid
Preparation	[514]

Addition

1. $RC(O)R' + CH_3OH \rightarrow RR'C(OH)(CH_2OH)$ [657]

Alkylation

1. $C_6H_6 + CH_3CH_2CH(CH_3)Cl \rightarrow C_6H_5CH(CH_3)(CH_2CH_3) + HCl$ [658]

Dehydrohalogenation

1. $\rightarrow$ $+ HCl$ [659]

Hydrometallation

1. $RHC{=}CH_2 + HAlR_2 \rightarrow RCH_2CH_2AlR_2$ [660]

Polymerization

1. Polymerization of cyclohexene [661]
2. $CH_3HC{=}CH_2 \rightarrow (-CH_2CH_2CH_2-)_x$ [662, 663]
3. Polymerization of olefins [664–666]

39. Vanadium Oxide

CA Registry	1314-62-1
Formula	V_2O_5
Properties	yellow to red solid, m.p. is $690\,^\circ C$, d 3.357

Cracking

1. Cracking of petroleum [396]

Dehydrogenation

1. $C_6H_5CH_2CH_3 \rightarrow C_6H_5CH{=}CH_2 + H_2$ [332]

Fischer – Tropsch

1. $CO + H_2 \rightarrow C_nH_{2n+1}$ [403]

Hydroesterification

1. $C_6H_5NO_2 + CO + CH_3CH_2OH \xrightarrow{C_6H_5NH_2} C_6H_5NHC(O)OCH_2CH_3$ [335]

Oxidation

1. $H_2C{=}CHCH{=}CH_2 \xrightarrow{O_2}$ 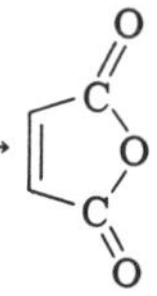   [411, 667]

2. $CH_3CH_2CH_2OH \xrightarrow{O_2} CH_3CH_2C(O)H$ [668]

3. $CH_3OH \xrightarrow{O_2} HC(O)H$ [669]

4. $2H_2 + O_2 \rightarrow 2H_2O$ [181]

5. $H_2C{=}CH_2 \xrightarrow{O_2} CH_3CH_2OH$ [407]

Reviews — [670]

40. Zinc Oxide

CA Registry   1314-13-2
Formula   ZnO
Properties   white solid, m.p. is 1975°C, d 5.606

Aromatization

1. $2(CH_3)_2C{=}CH_2 \rightarrow C_6H_4(CH_3)_2 + 3H_2$ [671, 672]

Carbonylation

1. $(CH_3)_2C{=}CH_2 + CO + 2H_2 \rightarrow (CH_3)_3C(O)CH_3$ [673]
2. $C_6H_5NO_2 + 3CO \rightarrow C_6H_5NCO + 2CO_2$ [395]

Dehydrosulfurization

1. Dehydrosulfurization of hydrocarbon streams   [674]

Fischer – Tropsch

1. $CO + H_2 \rightarrow C_nH_{2n+1}$ [675–677]
2. $CO + 2H_2 \rightarrow CH_3OH$ [254, 678]

Hydrogenation

1. Fatty acid esters → fatty alcohols   [679]
2. $HOCH_2C(O)OCH_2CH_2OH + 2H_2 \rightarrow 2HOCH_2CH_2OH$ [255]

Oxidation

1. $H_2C=CHCH_2CH_3 \xrightarrow{O_2}$ 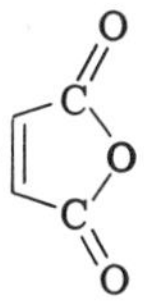 [680]

2. $(CH_3)_2C=CH_2 \xrightarrow{O_2} H_2C=C(CH_3)C(O)H$ [681]

3. $(CH_3)_2CHC(O)H \xrightarrow{O_2} (CH_3)_2CO$ [682]

4. $H_2C=CH_2 \xrightarrow{O_2, H_2O} CH_3C(O)OH + H_2C=CHC(O)OCH_3$ [484]

Water Gas Shift Reaction

1. $CO + H_2O \rightarrow CO_2 + H_2$ [683]

Review — [684]

III.   Catalytic Reactions

1.   Addition

1. Chlorotris(triphenylphosphine)rhodium(I)*
2. Dichlorotris(triphenylphosphine)ruthenium(II)*
3. Iron pentacarbonyl*
4. Tetrakis(triphenylphosphine)palladium(0)*
5. Tetrarhodium dodecacarbonyl*
6. Titanium tetrachloride*
7. Rhodium acetate [685]*

2.   Alkylation

1. Alumina
2. Chlorotris(triphenylphosphine)rhodium(I)*
3. Chromium oxide
4. Cobalt carbonyl*
5. Copper chromite
6. Dichlorobis(triphenylphosphine)palladium(II)*
7. Iron oxide
8. Molybdenum
9. Nickel
10. Nickel(II) acetylacetonate*

* Homogenous catalysts are designated by an asterisk.

11. Palladium
12. Palladium acetate*
13. Platinum
14. Rhodium trichloride*
15. Ruthenium
16. Titanium tetrachloride*
17. Nickel oxide [686]
18. Rhenium [687]

3. Alkylation (Reductive)

1. Alumina
2. Cobalt
3. Copper chromite
4. Nickel
5. Palladium
6. Platinum
7. Rhodium
8. Ruthenium
9. Silver

4. Aromatization

1. Bis(benzonitrile)dichloropalladium(II)*
2. Chromium oxide
3. Cobalt
4. Nickel
5. Nickel tetracarbonyl*
6. Platinum
7. Zinc oxide
8. Rhenium [688]

5. Arylation

1. Palladium acetate*

6. Carbonylation

1. Bis(benzonitrile)dichloropalladium(II)*
2. Carbonylchlorobis(triphenylphosphine)rhodium(I)*
3. Carbonylhydrotris(triphenylphosphine)rhodium(I)*
4. Chlorodicarbonylrhodium(I) dimer*
5. Chlorotris(triphenylphosphine)rhodium(I)*
6. Dichlorobis(triphenylphosphine)palladium(II)*

7. Iron oxide
8. Iron pentacarbonyl*
9. Molybdenum oxide
10. Nickel tetracarbonyl*
11. Palladium
12. Palladium acetate*
13. Rhodium trichloride*
14. Tetrakis(triphenylphosphine)palladium(0)*
15. Tetrarhodium dodecacarbonyl*
16. Zinc oxide
17. Dicarbonylbis(triphenylphosphine)nickel(0)* [689]
18. Manganese carbonyl* [690]

7. Carbonylation (Reductive)

1. Hexarhodium hexadecacarbonyl*
2. Platinum

8. Carboxylation

1. Bis(benzonitrile)dichloropalladium(II)*
2. Chlorodicarbonylrhodium(I) dimer*
3. Chlorotris(triphenylphosphine)rhodium(I)*
4. Nickel tetracarbonyl*
5. Palladium acetate*
6. Palladium chloride*
7. Rhodium
8. Rhodium trichloride*

9. Carboxylation (Oxidative)

1. Bis(benzonitrile)dichloropalladium(II)*
2. Palladium
3. Palladium chloride*
4. Rhodium trichloride*

10. Coupling (Reductive)

1. Bis(cyclopentadienyl)titanium dichloride*
2. Dichlorobis(triphenylphosphine)palladium(II)*
3. Dichlorotris(triphenylphosphine)ruthenium(II)*
4. Nickel(II) acetylacetonate*
5. Palladium chloride*
6. Platinum

7. Rhodium
8. Tetrakis(triphenylphosphine)palladium(0)*

11. Cracking

1. Chromium oxide
2. Cobalt
3. Molybdenum
4. Molybdenum oxide
5. Rhodium
6. Ruthenium
7. Ruthenium dioxide
8. Vanadium oxide

12. Cyclization

1. Alumina
2. Carbonylhydrotris(triphenylphosphine)rhodium(I)*
3. Chlorotris(triphenylphosphine)rhodium(I)*
4. Chromium oxide
5. Dichlorobis(triphenylphosphine)palladium(II)*
6. Dichlorotris(triphenylphosphine)ruthenium(II)*
7. Molybdenum carbonyl*
8. Nickel(II) acetylacetonate*
9. Palladium acetate*
10. Palladium chloride*
11. Cyclopentadienyldicarbonylcobalt* [691, 692]

13. Cyclooligomerization

1. Tetrakis(triphenylphosphine)palladium(0)*

14. Dealkylation

1. Rhodium

15. Decarbonylation

1. Palladium

16. Dehalogenation

1. Bis(cyclopentadienyl)titanium dichloride*
2. Iron oxide
3. Palladium acetate*

17.   Dehydration

1. Alumina
2. Iron oxide
3. Platinum
4. Rhodium

18.   Dehydrogenation

1. Bis(cyclopentadienyl)titanium dichloride*
2. Chlorodicarbonylrhodium(I) dimer
3. Chlorotris(triphenylphosphine)rhodium(I)*
4. Chromium oxide
5. Cobalt
6. Copper chromite
7. Copper oxide
8. Iron oxide
9. Molybdenum oxide
10. Palladium
11. Platinum
12. Rhodium
13. Silver
14. Vanadium oxide

19.   Dehydrohalogenation

1. Copper oxide
2. Titanium tetrachloride*

20.   Dehydrosulfurization

1. Alumina
2. Cobalt
3. Molybdenum
4. Molybdenum oxide
5. Nickel
6. Zinc oxide

21.   Dimerization

1. Alumina
2. Bis(benzonitrile)dichloropalladium(II)*
3. Bis(cyclopentadienyl)titanium dichloride*
4. Carbonylchlorobis(triphenylphosphine)rhodium(I)*

5. Chlorodicarbonylrhodium(I) dimer*
6. Chlorotris(triphenylphosphine)rhodium(I)*
7. Rhodium trichloride*
8. Bis(cyclopentadiene)nickel(0)* [693]
9. Bis(acetylacetonate)palladium(II)* [694]

22. Epoxidation

1. Molybdenum carbonyl*
2. Palladium acetate*
3. Silver
4. Tungsten carbonyl* [695]

23. Esterification

1. Molybdenum
2. Molybdenum oxide

24. Exchange

1. Bis(cyclopentadienyl)titanium dichloride*
2. Dichlorobis(triphenylphosphine)palladium(II)*
3. Nickel(II) acetylacetonate*
4. Palladium chloride*
5. Tetrakis(triphenylphosphine)palladium(0)*
6. Manganese carbonyl* [696]

25. Fischer–Tropsch

1. Alumina
2. Bis(cyclopentadienyl)titanium dichloride*
3. Chlorodicarbonylrhodium(I) dimer*
4. Chlorotris(triphenylphosphine)rhodium(I)*
5. Chromium oxide
6. Cobalt
7. Cobalt carbonyl*
8. Copper chromite
9. Copper oxide
10. Iron dodecacarbonyl*
11. Iron oxide
12. Iron pentacarbonyl*
13. Molybdenum
14. Molybdenum carbonyl*
15. Molybdenum oxide

16. Nickel
17. Osmium dodecacarbonyl*
18. Palladium
19. Platinum
20. Rhodium
21. Rhodium trichloride*
22. Ruthenium
23. Ruthenium carbonyl*
24. Tetrarhodium dodecacarbonyl*
25. Vanadium oxide
26. Zinc oxide
27. Dicarbonylacetylacetonaterhodium(I)* [697, 698]
28. Iridium dodecacarbonyl* [695]

26. Grignard

1. Bis(cyclopentadienyl)titanium dichloride*

27. Homologation

1. Carbonylhydrotris(triphenylphosphine)rhodium(I)*
2. Cobalt carbonyl*
3. Hexarhodium hexadecacarbonyl*
4. Iron pentacarbonyl*
5. Ruthenium
6. Ruthenium carbonyl*
7. Ruthenium dioxide
8. Tetrarhodium dodecacarbonyl*

28. Hydration

1. Bis(benzonitrile)dichloropalladium(II)*

29. Hydrocarboxylation

1. Cobalt carbonyl*

30. Hydrocracking

1. Alumina

31. Hydrodealkylation

1. Chromium oxide
2. Iron oxide
3. Palladium

32. *Hydrodenitrogenation*

1. Cobalt
2. Molybdenum
3. Molybdenum oxide

33. *Hydroesterification*

1. Carbonylchlorobis(triphenylphosphine)rhodium(I)*
2. Carbonylhydrotris(triphenylphosphine)rhodium(I)*
3. Cobalt carbonyl*
4. Dichlorobis(triphenylphosphine)palladium(II)*
5. Hexarhodium hexadecacarbonyl*
6. Iron oxide
7. Iron pentacarbonyl*
8. Molybdenum carbonyl*
9. Nickel tetracarbonyl*
10. Palladium chloride*
11. Rhodium trichloride*
12. Tetrarhodium dodecacarbonyl*
13. Vanadium oxide

34. *Hydroformylation*

1. Carbonylchlorobis(triphenylphosphine)rhodium(I)*
2. Carbonylhydrotris(triphenylphosphine)rhodium(I)*
3. Chlorodicarbonylrhodium(I) dimer*
4. Cobalt
5. Cobalt carbonyl*
6. Dichlorobis(triphenylphosphine)palladium(II)*
7. Hexarhodium hexadecacarbonyl*
8. Iron dodecacarbonyl*
9. Iron pentacarbonyl*
10. Osmium dodecacarbonyl*
11. Rhodium
12. Rhodium trichloride*
13. Ruthenium carbonyl*
14. Tetrarhodium dodecacarbonyl*
15. Tetracobalt dodecacarbonyl* [699]

35. *Hydrogenation*

1. Bis(benzonitrile)dichloropalladium(II)*
2. Chlorotris(triphenylphosphine)rhodium(I)*

3. Chromium oxide
4. Cobalt
5. Cobalt carbonyl*
6. Copper chromite
7. Copper oxide
8. Dichlorobis(triphenylphosphine)palladium(II)*
9. Dichlorotris(triphenylphosphine)ruthenium(II)*
10. Hexarhodium hexadecacarbonyl*
11. Iron dodecacarbonyl*
12. Iron oxide
13. Iron pentacarbonyl*
14. Molybdenum
15. Molybdenum oxide
16. Nickel
17. Nickel(II) acetylacetonate*
18. Osmium dodecacarbonyl*
19. Palladium
20. Platinum
21. Rhodium
22. Ruthenium
23. Ruthenium carbonyl*
24. Silver
25. Zinc oxide

36. Hydrogenolysis

1. Chlorotris(triphenylphosphine)rhodium(I)*
2. Cobalt
3. Copper oxide
4. Dichlorobis(triphenylphosphine)palladium(II)*
5. Dichlorotris(triphenylphosphine)ruthenium(II)*
6. Nickel
7. Palladium
8. Platinum
9. Rhodium
10. Ruthenium

37. Hydrolysis

1. Alumina
2. Copper chromite
3. Molybdenum
4. Molybdenum carbonyl*

5. Molybdenum oxide

38. *Hydrometallation*

1. Bis(cyclopentadienyl)titanium dichloride*
2. Titanium tetrachloride*

39. *Hydrosilylation*

1. Carbonylchlorobis(triphenylphosphine)rhodium(I)*
2. Carbonylhydrotris(triphenylphosphine)rhodium(I)*
3. Chlorodicarbonylrhodium(I) dimer*
4. Chlorotris(triphenylphosphine)rhodium(I)*
5. Dichlorotris(triphenylphosphine)ruthenium(II)*
6. Ruthenium carbonyl*
7. Tetrakis(triphenylphosphine)palladium(0)*
8. Dichlorobis(triphenylphosphine)platinum(II)* [700]
9. Tris(acetylacetonate)rhodium(III)* [701]

40. *Isomerization*

1. Alumina
2. Bis(benzonitrile)dichloropalladium(II)*
3. Bis(cyclopentadienyl)titanium dichloride*
4. Carbonylhydrotris(triphenylphosphine)rhodium(I)*
5. Chlorotris(triphenylphosphine)rhodium(I)*
6. Cobalt
7. Copper chromite
8. Iron oxide
9. Iron pentacarbonyl*
10. Molybdenum carbonyl*
11. Molybdenum oxide
12. Nickel
13. Nickel(II) acetylacetonate*
14. Osmium dodecacarbonyl*
15. Palladium
16. Platinum
17. Ruthenium carbonyl*
18. Tetrakis(triphenylphosphine)palladium(II)*
19. Bis(acetylacetonate)palladium(II)* [702]
20. Chromium carbonyl* [703, 704]
21. Zirconium oxide [705]

41. Metathesis

1. Molybdenum
2. Molybdenum carbonyl*

42. Oxidation

  1. Alumina
  2. Bis(cyclopentadienyl)titanium dichloride*
  3. Carbonylhydrotris(triphenylphosphine)rhodium(I)*
  4. Chlorotris(triphenylphosphine)rhodium(I)*
  5. Chromium oxide
  6. Cobalt
  7. Copper oxide
  8. Dichlorotris(triphenylphosphine)ruthenium(II)*
  9. Hexarhodium hexadecacarbonyl*
 10. Iron oxide
 11. Molybdenum
 12. Molybdenum oxide
 13. Nickel
 14. Palladium
 15. Palladium acetate*
 16. Palladium chloride*
 17. Platinum
 18. Ruthenium dioxide
 19. Silver
 20. Tetrakis(triphenylphosphine)palladium(0)*
 21. Tetrarhodium dodecacarbonyl*
 22. Vanadium oxide
 23. Zinc oxide
 24. Cobalt oxide [706]
 25. Ruthenium trichloride* [707]

43. Polymerization

  1. Bis(cyclopentadienyl)titanium dichloride*
  2. Carbonylhydrotris(triphenylphosphine)rhodium(I)*
  3. Dichlorobis(triphenylphosphine)palladium(II)*
  4. Dichlorotris(triphenylphosphine)ruthenium(II)*
  5. Iron oxide
  6. Iron pentacarbonyl*
  7. Nickel(II) acetylacetonate*
  8. Palladium

9. Palladium acetate*
10. Ruthenium carbonyl*
11. Titanium tetrachloride*
12. Bis(cyclooctadiene)nickel(0)* [708]

44. Reduction

1. Alumina
2. Bis(cyclopentadienyl)titanium dichloride*
3. Chlorodicarbonylrhodium(I) dimer*
4. Molybdenum oxide
5. Palladium
6. Ruthenium
7. Ruthenium carbonyl*
8. Tetrakis(triphenylphosphine)palladium(0)*
9. Tetrarhodium dodecacarbonyl*

45. Reforming

1. Chlorodicarbonylrhodium(I) dimer*
2. Chlorotris(triphenylphosphine)rhodium(I)*
3. Chromium oxide
4. Dichlorotris(triphenylphosphine)ruthenium(II)*
5. Molybdenum oxide
6. Osmium dodecacarbonyl*
7. Palladium
8. Platinum
9. Ruthenium carbonyl*
10. Tetrakis(triphenylphosphine)palladium(0)*
11. Tetrakis(triphenylphosphine)platinum(0)* [709]

46. Telomerization

1. Palladium chloride*
2. Tetrakis(triphenylphosphine)palladium(0)*
3. Bis(acetylacetonate)palladium(II)* [710–712]

47. Transesterification

1. Bis(benzonitrile)dichloropalladium(II)*
2. Palladium acetate*

48. Trimerization

1. Rhodium trichloride*

2. Dicarbonylbis(triphenylphosphine)nickel(0)* [713]

49. Water Gas Shift Reaction

1. Chlorodicarbonylrhodium(I) dimer*
2. Chromium oxide
3. Hexarhodium hexadecacarbonyl*
4. Iron oxide
5. Iron pentacarbonyl*
6. Molybdenum carbonyl*
7. Rhodium trichloride*
8. Ruthenium carbonyl*
9. Tetrarhodium dodecacarbonyl*
10. Zinc oxide

References

1. Yoshio Okada, Kimio Okubo, and Shoichi Kunii, Ger. Offen. Patent No. 2,921,080; *Chem. Abstr.,* 1980, **92,** 163675n.
2. John R. Dodd, U.S. Patent No. 4,187,255; *Chem. Abstr.,* 1980, **92,** 180870q.
3. Justin Herscovici *et al.,* **Rom. 64,** 821; *Chem. Abstr.,* 1979, **91,** 211069w.
4. Isao Miyanohara and Norimasa Mizui, *Jpn. Kokai Tokkyo Koho* Patent No. 80 38,329; *Chem. Abstr.,* 1980, **93,** 132066f.
5. Yasuo Yamaski and Tadashi Kawai, *Jpn. Kokai Tokkyo Koho* Patent No. 78 147,046; *Chem. Abstr.,* **90,** 151808d.
6. Fujio Nomura, Kei Furuno, and Shohei Hoshino, *Jpn. Kokai Tokkyo Koho* Patent No. 79 27,577; *Chem. Abstr.,* 1979, **91,** 20544n.
7. John Irvine Darragh, Brit. Patent No. 1,493,553; *Chem. Abstr.,* 1978, **88,** 190596m.
8. Michael E. Brennan, Philip H. Moss, and Ernest Yeakey, Ger. Offen. Patent No. 2,624,016; *Chem. Abstr.,* 1977, **86,** 140061w.
9. Helcio Valladares Barrocas, de Castro M. Da Silva, Ruy Coutinho de Assis, and Joao Baptista, Ger. Offen. Patent No. 2,834,699; *Chem. Abstr.,* 1979, **90,** 168037z.
10. Burtron H. Davis, *J. Catal.,* 1980, **61**(1), 279.
11. Herwig Hoffman, Otto H. Huchler, Herbert Mueller, and Siegfried Winderl, U.S. Patent No. 4,196,130; *Chem. Abstr.,* 1980, **93,** 186138z.
12. Glen P. Hamner, U.S. Patent No. 4,051,021; *Chem. Abstr.,* 1977, **87,** 204135d.
13. Yoshihiko Nishimoto, Naruo Yokoyama, and Tetsuya Imai, *Jpn. Kokai Tokkyo Koho* Patent No. 78 35,703; *Chem. Abstr.,* 1978, **89,** 62254g.
14. Takeshi Okano, Kanji Daigo, Jun Imai, and Tadashi Suzuki, *Jpn. Kokai Tokkyo Koho* Patent No. 79 158,392; *Chem. Abstr.,* 1980, **92,** 183521n.
15. Kazunori Takahata and Hiroshi Hasui, *Jpn. Kokai Tokkyo Koho* Patent No. 79 48,742; *Chem. Abstr.,* 1979, **91,** 74323h.
16. Bum Jong Ahn, Jean Armando, Guy Perot, Michel Guisnet, *C. R. Hebd. Seances Acad. Sci. Ser. C,* 1979, **288**(8), 245.

17. Hamid Alafandi, Dennis Stamires, U.S. Patent No. 4,198,319; *Chem. Abstr.,* 1980, **93,** 54934w.
18. Arthur Warren Chester and William Albert Stover, Ger. Offen. Patent No. 2,756,220; *Chem. Abstr.,* 1979, **90,** 124370y.
19. Theodore V. Flaherty, Jr., Richard J. Nozemack, and Hanson L. Guidry, U.S. Patent No. 4,126,579; *Chem. Abstr.,* 1979, **90,** 106726q.
20. John Herman Hansel, Robert Gordon Linton, and Charles William Stranger, Jr., Ger. Offen. Patent No. 2,825,074; *Chem. Abstr.,* 1979, **90,** 171288n.
21. Dwight Lamar McKay, *Eur. Patent Application No.* 4091; *Chem. Abstr.,* 1980, **92,** 44539m.
22. Mitsuaki Shirokami, Kazuo Tan, Toshio Uchibori, *Jpn. Kokai Tokkyo Koho* 78 31,608; *Chem. Abstr.,* 1978, **89,** 59664k.
23. Dale A. Raines and Oliver C. Ainsworth, U.S. Patent No. 4,237,324; *Chem. Abstr.,* 1981, **94,** 102824e.
24. John Mooi, U.S. Patent No. 4,118,430; *Chem. Abstr.,* 1979, **90,** 71884k.
25. Takao Nishimura, Minekazu Sueoka, Hiroshi Fujimoto, and Atsuji Saki, *Jpn. Kokai Tokkyo Koho* Patent No. 79 16,487; *Chem. Abstr.,* 1979, **91,** 140513y.
26. Yasuo Yamazaki, Atsuji Sakai, Minekazu Sueoka, Tamio Onodera, and Koji Sumitani, *Jpn. Kokai Tokkyo Koho* Patent No. 79 142,188; *Chem. Abstr.,* 1980, **92,** 180801t.
27. M. Peereboom and J. Meijering, *J. Catal.,* 1979, **59**(1), 45.
28. Michel Baudouin and Robert Perron, Ger. Offen. Patent No. 2,646,172; *Chem. Abstr.,* 1977, **87,** 134531d.
29. Cornelis Jongema, Hubertus Vonken, and Barbara Herman, Netherland Application No. 78 05663; *Chem. Abstr.,* 1980, **92,** 198095a.
30. Kohichi Kuno, Kenichi Sano, and Masayuki Aoki, *Jpn. Kokai Tokkyo Koho* Patent No. 77 09677; *Chem. Abstr.,* 1977, **87,** 89994u.
31. John R. Doyle, Phillip E. Slade, and Hans B. Jonassen, *Inorg. Synth.,* 1960, **6,** 218.
32. Enrico Mincione, Antonio Sirna, and Daniele Covini, *J. Org. Chem.,* 1981, **46**(5), 1010.
33. Elizabeth A. Kelley, G. Albert Wright, and Peter M. Maitlis, *J. Chem. Soc. Dalton Trans.,* 1979, **1,** 178.
34. John E. Hallgren, U.S. Patent No. 4,201,721; *Chem. Abstr.,* 1980, **93,** 205316z.
35. J. E. Hallgren and R. O. Matthews, *J. Organomet. Chem.,* 1979, **175**(1), 135.
36. Victor P. Kurkov, U.S. Patent No. 4,189,608; *Chem. Abstr.,* 1980, **92,** 214899y.
37. Haven S. Kesling, Jr., and Lee R. Zehner, U.S. Patent No. 4,166,913; *Chem. Abstr.,* 1980, **92,** 6086d.
38. James Robert Jennings and Dipankar Sen, Ger. Offen. Patent No. 2,752,509; *Chem. Abstr.,* 1978, **89,** 108212x.
39. Kiyotomi Kaneda *et al., J. Org. Chem.,* 1979, **44**(1), 55.
40. Antoine Gaset, Georges Constant, Philippe Kalck, and Gerard Villain, Fr. Demande Patent No. 2,419,929; *Chem. Abstr.,* 1980, **93,** 45991k.
41. Tapan Banerjee, Tapan Mondal, and Chitta R. Saha, *Chem. Ind. London,* 1979, **6,** 212.
42. Tapan K. Banerjee and Chitta R. Saha, *Indian J. Chem. Sect. A,* 1980, **19A**(10), 964.
43. Victor P. Kurkov, U.S. Patent No. 4,044,050; *Chem. Abstr.,* 1977, **87,** 167557v.
44. Larry E. Overman and Frederick M. Knoll, *Tetrahedron Lett.,* 1979, **4,** 321.
45. Tadashi Sato, Takeshi Kawashima, Osamu Maruyama, Hideo Suehiro, and Makoto Yokoyama, *Jpn. Kokai Tokkyo Koho* Patent No. 79 14,964; *Chem. Abstr.,* 1979, **91,** 20308p.
46. Durvasula V. Rao and Fred A. Stuber, *Tetrahedron Lett.,* 1981, **22**(25), 2337.
47. Fumiaki Kawamoto, Shozo Tanaka, and Yoshiro Honma, Ger. Offen. Patent No. 2,823,660; *Chem. Abstr.,* 1979, **91,** 19919g.

48. David C. Brown, Stephen A. Nichols, A. Bruce Gilpen, and David W. Thompson, *J. Org. Chem.*, 1979, **44**(20), 3457.
49. Leslie C. Smedley, Harrell E. Tweedy, Randolph A. Coleman, and David W. Thompson, *J. Org. Chem.*, 1977, **42**(25), 4147.
50. John J. Eisch and Bruce R. King, *Organomet. Synth.*, 1965, *1*.
51. Dina Nath, R. K. Sharma, and A. N. Bhat, *Inorg. Chim. Acta*, 1976, **20**(2), 109.
52. G. Wilkinson and J. M. Birmingham, *J. Am. Chem. Soc.*, 1954, **76**, 4281.
53. Bernard Meunier, *J. Organomet. Chem.*, 1980, **204**(3), 345.
54. E. A. Chernyshev, M. D. Reshetova, and I. A. Rodnikov, *Zh. Obshch. Khim.* **50**(5), 1037.
55. A. B. Amerik and V. M. Vdovin, *Izv. Akad. Nauk SSSR Ser. Khim.*, 1979, **4**, 907.
56. Fumie Sato, Hiroaki Ishikawa, and Masao Sato, *Tetrahedron Lett.*, 1980, **21**(4), 365.
57. V. V. Strelets, V. N. Tsarev, and O. N. Efimov, *Dokl. Akad. Nauk SSSR*, 1981, **259**(3), 646.
58. Fumie Sato, Takamasa Jimbo, and Masao Sato, *Tetrahedron Lett.*, 1980, **21**(22), 2171.
59. Fumie Sato, Takamasa Jimbo, and Masao Sato, *Tetrahedron Lett.*, 1980, **21**(22), 2175.
60. Kakuzo Isagawa, Hiroshi Sano, Minoru Hattori, and Yoshio Otsuji, *Chem. Lett.*, 1979, **9**, 1069.
61. E. C. Ashby and S. A. Noding, *J. Org. Chem.*, 1979, **44**(24), 4364.
62. E. C. Ashby and S. A. Noding, *J. Organomet. Chem.*, 1979, **177**(1), 117.
63. F. Turecek, H. Antropiusova, K. Mach, V. Hanus, and P. Sedmera, *Tetrahedron Lett.*, 1980, **21**(7), 637.
64. Paul R. Stapp, U.S. Patent No. 4,164,615; *Chem. Abstr.*, 1979, **91**, 210913e.
65. Amir Famili and Michael F. Farona, *Polym. Bull. Berlin*, 1980, **2**(4), 289.
66. Eugene C. Ashby and Stephen A. Noding, *J. Org. Chem.*, 1980, **45**(6), 1035.
67. D. Evans, J. A. Osborn, and G. Wilkinson, *Inorg. Synth.*, 1968, **11**, 99.
68. J. A. McCleverty and G. Wilkinson, *Inorg. Synth.*, 1966, **8**, 214.
69. John Frederick Knifton, Ger. Offen. Patent No. 2,750,250; *Chem. Abstr.*, 1978, **89**, 108199y.
70. John F. Knifton, *J. Organomet. Chem.*, 1980, **188**(2), 223.
71. Klaus Schwetlick, Klaus Unverferth, Reiner Hoentsch, and Manfred Pfeifer, Ger. (East) Patent No. 121,509; *Chem. Abstr.*, 1977, **87**, 39143v.
72. Yoshio Kamiya, Katsuhito Kawato, and Hiroyuki Ota, *Chem. Lett.*, 1980, **12**, 1549.
73. Richard Craig Larock, U.S. Patent No. 4,105,705; *Chem. Abstr.*, 1979, **90**, 71240x.
74. Pangbu Hong, Takaya Mise, and Hiroshi Yamazaki, *Chem. Lett.*, 1981, **7**, 989.
75. L. V. Gorbunova, I. L. Knyazeva, B. K. Nefedov, Kh. O. Khoshdurdyev, and V. I. Manov-Yuvenskii, *Izv. Akad. Nauk SSSR Ser. Khim.*, 1981, **7**, 1644.
76. A. Spencer, *J. Organomet. Chem.*, 1980, **194**(1), 113.
77. Alwyn Spencer, Eur. Patent Application No. 2,908; *Chem. Abstr.*, 1980, **92**, 6064v.
78. O. Richard Hughes, U.S. Patent No. 4,201,728; *Chem. Abstr.*, 1980, **93**, 149796s.
79. Richard W. Goetz, U.S. Patent No. 4,200,765; *Chem. Abstr.*, 1980, **93**, 113951z.
80. Rudolf Kummer, Kurt Schwirten, and Hans Dieter Schlinder, Can. Patent No. 1,055,512; *Chem. Abstr.*, 1979, **91**, 140340q.
81. Harold B. Tinker, Ger. Offen. Patent No. 2,623,673; *Chem. Abstr.*, 1977, **86**, 105957w.
82. Jens Hagen and Klaus Bruns, Ger. Offen. Patent No. 2,849,742; *Chem. Abstr.*, 1980, **93**, 186613a.
83. Moustafa El-Chahawi, Uwe Prange, Hermann Richtzenhain, Wilhelm Vogt, Ger. Offen. Patent No. 2,606,655; *Chem. Abstr.*, 1977, **87**, 167756j.
84. Richard M. Laine, *Ann. N.Y. Acad. Sci.*, 1980, **333**, 124.
85. Isao Koga, Yoji Terui, and Masuhito Ogushi, Ger. Offen. Patent No. 2,810,032; *Chem. Abstr.*, 1979, **90**, 23240a.

86. V. B. Pukhnarevich *et al., Zh. Obshch. Khim.,* 1980, **50**(7), 1554.
87. N. Ahmad, J. J. Levison, S. D. Robinson, and M. F. Uttley, *Inorg. Synth.,* 1974, **15,** 50.
88. L. V. Gorbunova, I. L. Knyazeva, and E. A. Davydova, *Izv. Akad. Nauk SSSR Ser. Khim.,* 1980, **5,** 1054.
89. Michael Garst and David Lukton, *J. Org. Chem.,* 1981, **46**(22), 4433.
90. H. Dumas, J. Levisallas, and H. Rudler (1979). *J. Organomet. Chem.,* 1979, **177**(1), 239.
91. N. A. De Munck, J. P. A. Notenboom, J. E. De Leur, and J. J. F. Scholten, *J. Mol. Catal.,* 1981, **11**(2–3), 233.
92. Charles U. Pittman, Jr. and William D. Honnick, *J. Org. Chem.,* 1980, **45**(11), 2132.
93. Nagahiro Saito, Shoji Arai, and Yukihiro Tsutsumi, *Jpn. Kokai Tokkyo Koho* Patent No. 80 45,646; *Chem. Abstr.,* 1980, **93,** 204061a.
94. Toshihiro Saito, Yukihiro Tsutsumi, and Shoji Arai, Ger. Offen. Patent No. 3,017,682; *Chem. Abstr.,* 1981, **95,** 42342e.
95. Peter J. Davidson and Rosemary R. Hignett, U.S. Patent No. 4,200,592; *Chem. Abstr.,* 1980, **93,** 113966h.
96. Rosemary Rena Hignett and Peter John Davidson, Brit. Patent Application No. 2,028,793; *Chem. Abstr.,* 1980, **93,** 167640r.
97. Mitsuo Matsumoto and Masuhiko Tamura, Ger. Offen. Patent No. 2,922,757; *Chem. Abstr.,* 1980, **92,** 214886s.
98. John L. Dawes, U.S. Patent No. 4,258,215; *Chem. Abstr.,* 1981, **94,** 174339y.
99. Leendert Arie Gerritsen, Hendrik Ido Ambacht, and Joseph Johannes Franciscus Scholten, Ger. Offen. Patent No. 2,802,276; *Chem. Abstr.,* 1978, **89,** 163067d.
100. L. A. Gerritsen, J. M. Herman, W. Klut, and J. J. F. Scholten, *J. Mol. Catal.,* 1980, **9**(2), 157.
101. Aaldert Johannes De Jong and Robin Thomas Gray, Ger. Offen. Patent No. 2,735,639; *Chem. Abstr.,* 1978, **88,** 169670w.
102. Robin Thomas Gray and Aaldert Johannes De Jong, Ger. Offen. Patent No. 2,753,644; *Chem. Abstr.,* 1978, **89,** 108350r.
103. Charles U. Pittman, Jr., William D. Honnick, and Jin Jun Yang, *J. Org. Chem.,* 1980, **45**(4), 684.
104. Mitsubishi Petrochemical Co., Ltd., *Jpn. Kokai Tokkyo Koho* Patent No. 81 87,549; *Chem. Abstr.,* 1981, **95,** 203367j.
105. M. G. Voronkov, V. B. Pukhnarevich, I. I. Tsykhanskaya, and Yu. S. Varshavskii, *Dokl. Akad. Nauk SSSR,* 1980, **254**(4), 887.
106. J. K. Stille and Y. Becker, *J. Org. Chem.,* 1980, **45**(11), 2139.
107. Colin W. Dudley, Gordon Read, and Peter J. C. Walker, *J. Chem. Soc. Dalton Trans.,* 1974, 1926.
108. Noriyuki Kameda, *Nippon Kagaku Kaishi,* 1976, **4,** 682.
109. Richard Cramer, *Inorg. Synth.,* 1974, **15,** 17.
110. J. A. McCleverty and G. Wilkinson, *Inorg. Synth.,* 1966, **8,** 211.
111. Mitsuo Masaki and Noboru Kakeya, *Jpn. Kokai Tokkyo Koho* Patent No. 78 40,738; *Chem. Abstr.,* 1978, **89,** 90126k.
112. K. Unverferth, R. Hoentsch, and K. Schwetlick, *J. Prakt. Chem.,* 1979, **321**(1), 86.
113. K. Unverferth, R. Hoentsch, and K. Schwetlick, *J. Prakt. Chem.,* 1979, **321**(6), 928.
114. Tsutomu Sakakibara and Howard Alper, *J. Chem. Soc. Chem. Commun.,* 1979, **10,** 458.
115. Masanobu Hidai, Masami Orisaku, and Yasuzo Uchida, *Chem. Lett.,* 1980, **6,** 753.
116. Charles Michael Bartish, Belgium Patent No. 862,828; *Chem. Abstr.,* 1978, **89,** 214906m.
117. Charles Michael Bartish, Ger. Offen. Patent No. 2,800,986; *Chem. Abstr.,* 1978, **89,** 163081d.
118. Yoshio Nakamura, *Jpn. Kokai Tokkyo Koho* Patent No. 79 44,608; *Chem. Abstr.,* 1979, **91,** 123409b.

119. Yoshio Nakamura and Yasuo Sado, *Jpn. Kokai Tokkyo Koho* Patent No. 78,119,814; *Chem. Abstr.*, 1979, **90**, 71767z.
120. Howard Alper, Khaled Hachem, Sandro Gambarotta, *Can. J. Chem.*, 1980, **58**(15), 1599.
121. Richard C. Larock and John C. Bernhardt, *J. Org. Chem.*, 1977, **42**(10), 1680.
122. A. L. Lapidus and M. M. Savel'ev, *Izv. Akad. Nauk SSSR Ser. Khim.*, 1980, **2**, 335.
123. Osamu Fukugaki and Manabu Hanamoto, *Jpn. Kokai Tokkyo Koho* Patent No. 78 46,943; *Chem. Abstr.*, 1978, **89**, 108666y.
124. R. B. King, A. D. King, Jr., and M. Z. Iqbal, *J. Am. Chem. Soc.*, 1979, **101**(17), 4893.
125. Tamotsu Imai, U.S. Patent No. 4,179,469; *Chem. Abstr.*, 1980, **92**, 146249j.
126. Tamotsu Imai, U.S. Patent No. 4,219,684; *Chem. Abstr.*, 1980, **93**, 204056c.
127. Richard W. Goetz, Lawrence D. Meixsel, and David W. Smith, U.S. Patent No. 4,291,179; *Chem. Abstr.*, 1981, **95**, 203328x.
128. Kazuhiro Muruyama, Kazutoshi Terada, and Yoshinori Yamamota, *Chem. Lett.*, 1981, **7**, 839.
129. A. L. Lapidus, M. M. Savel'ev, L. T. Kondrat'ev, and E. V. Yastrebova, *Izv. Akad. Nauk SSSR Ser. Khim.*, 1981, **2**, 474.
130. Robert G. Salomon, Mary F. Salomon, and Joseph L. C. Kachinski, *J. Am. Chem. Soc.*, 1977, **99**(4), 1043.
131. Miriam Sohn, Jochanan Blum, and Jack Halpern, *J. Am. Chem. Soc.*, 1979, **101**(10), 2694.
132. Kiyotomi Kaneda, *et al.*, *J. Mol. Catal.*, 1980, **9**(2), 227.
133. J. A. Osborn, F. H. Jardine, J. F. Young, and G. Wilkinson, *J. Chem. Soc. A*, 1966, 1711.
134. J. A. Osborn and G. Wilkinson, *Inorg. Synth.*, 1967, **10**, 68.
135. M. Polievka and I. Uhlar, *Petrochemia*, 1979, **19**(1–2), 21.
136. Ronald Grigg, Thomas R. B. Mitchell, and Ashok Ramasubbu, *J. Chem. Soc. Chem. Commun.*, 1979, **15**, 669.
137. Shinji Murai and Noboru Sonoda, *Angew. Chem. Int. Ed. Engl.*, 1979, **18**(11), 837.
138. Emilios P. Antoniades, U.S. Patent No. 4,194,056; *Chem. Abstr.*, 1980, **93**, 7665z.
139. Charles M. Bartish, U.S. Patent No. 4,102,920; *Chem. Abstr.*, 1979, **90**, 54480h.
140. R. C. Larock, K. Oertle, and G. F. Potter, *J. Am. Chem. Soc.*, 1980, **102**(1), 190.
141. R. Grigg, T. R. B. Mitchell, S. Sutthivaiyakit, and N. Tongpenyai, *J. Chem. Soc. Chem. Commun.*, 1981, **12**, 611.
142. Hironori Arakawa and Yoshihiro Sugi, *Chem. Lett.*, 1981, **9**, 1323.
143. C. Graillat, H. Jacobelli, M. Bartholin, and A. Guyot, *Rev. Port. Quim.*, 1977, **19**(1–4), 279.
144. G. Innorta, A. Modelli, F. Scagnolari, and A. Foffani, *J. Organomet. Chem.*, 1980, **185**(3), 403.
145. Helmut Pscheidl, Enno Moeller, Hans Ulrich Juergens, and Detlef Haberland, Ger. (East) Patent No. 138,153; *Chem. Abstr.*, 1980, **92**, 65370q.
146. Harukichi Hashimoto and Yoshio Inoue, *Jpn. Kokai Tokkyo Koho* Patent No. 76 138,614; *Chem. Abstr.*, 1977, **87**, 67853v.
147. Shunkichi Hashimoto and Yoshio Inoue, *Jpn. Kokai Tokkyo Koho* Patent No. 78 07,612; *Chem. Abstr.*, 1978, **88**, 152051x.
148. Dieter Oppelt, Herbert Schuster, Joachim Thoermer, and Rudolf Braden, Ger. Offen. Patent No. 2,539,132; *Chem. Abstr.*, 1977, **87**, 168798m.
149. C. Fragale, M. Gargano, T. Gomes, and M. Rossi, *J. Am. Oil Chem. Soc.*, 1979, **56**(4), 498.
150. Bàlint, Szilàrd Tòrös, Jòzsef Bakos, and Làszlò Markò, *J. Organomet. Chem.*, 1979, **175**(2), 229.
151. V. Z. Sharf, B. M. Savchenko, V. N. Krutii, and L. Kh. Freidlin, *Izv. Akad. Nauk SSSR Ser. Khim.*, 1980, **2**, 456.

152. B. M. Savchenko, V. Z. Sharf, V. N. Krutii, and L. Kh. Freidlin, *Izv. Akad. Nauk SSSR Ser. Khim.,* 1979, **11**, 2632.
153. Hugh M. Dickers, Robert N. Haszeldine, Leslie S. Malkin, Adrian P. Mather, and R. V. Parish (1980). *J. Chem. Soc. Dalton Trans.,* 1980, **2**, 308.
154. Gabriela Kuncova and Vaclav Chvalovsky (1980), *Collect. Czech. Chem. Commun.* **45**(7), 2085.
155. M. G. Voronkov, N. N. Vlasova, S. A. Bol'shakova, and S. V. Kirpichenko, *J. Organomet. Chem.,* 1980, **190**(4), 335.
156. Hamao Watanabe, Muneo Asami, and Yoichiro Nagai, *J. Organomet. Chem.,* 1980, **195**(3), 363.
157. M. Tuner, J. Von Jouanne, H. D. Brauer, and H. Kelm, *J. Mol. Catal.,* 1979, **5**(6), 425.
158. M. Tuner, J. Von Jouanne, H. D. Brauer, and H. Kelm, *J. Mol. Catal.,* 1979, **5**(6), 433.
159. Amikam Zoran, Yoel Sasson, and Jochanan Blum, *J. Org. Chem.,* 1981, **46**(2), 255.
160. K. M. Nicholas, *J. Organomet. Chem.,* 1980, **188**(1), C10.
161. Ferdinand Hagedorn, Laszlo Imre, and Karlfried Wedemeyer, Ger. Offen. Patent No. 2,805,402; *Chem. Abstr.,* 1979, **91**, 175011k.
162. James M. Reuter, Amitabha Sinha, and Robert G. Salomon, *J. Org. Chem.,* 1978, **43**(12), 2438.
163. U. Bersellini, M. Catellani, G. P. Chiusoli, and G. Salerno, *Fundam. Res. Homogenous Catal.,* 1979, **3**, 893.
164. F. H. Jardine, *Prog. Inorg. Chem.,* 1981, **28**, 63.
165. Motoo Kawamata, Kazushi Ohshima, Akihide Kudoh, Makoto Kotani, Ger. Offen. Patent No. 2,852,245; *Chem. Abstr.,* 1979, **91**, 107795u.
166. Tadamitsu Kiyoura and Yasuo Kogure, *Jpn. Kokai Tokkyo Koho* Patent No. 79 27,529; *Chem. Abstr.,* 1979, **91**, 39119z.
167. G. V. Isagulyants, M. I. Rosengart, and V. G. Bryukhanov, *Izv. Akad. Nauk SSSR Ser. Khim.,* 1980, **4**, 870.
168. G. Sengupta, J. S. Bariar, S. N. Chaudhuri, S. C. Sinha, and S. P. Sen, *Indian J. Technol.,* 1980, **18**(7), 282.
169. George D. Lockyer, Jr., Dennis E. Burd, Richard F. Sweeney, Bernard Sukornick, and Harry E. Ulmer, U.S. Patent No. 4,187,386; *Chem. Abstr.,* 1980, **92**, 180726x.
170. Tamotsu Imai, U.S. Patent No. 4,180,690; *Chem. Abstr.,* 1980, **92**, 94033n.
171. Gregor H. Riesser, Ger. Offen. Patent No. 2,629,635; *Chem. Abstr.,* 1977, **86**, 156186b.
172. Franco Rivetti and Ugo Romano, *J. Organomet. Chem.,* 1979, **174**(2), 221.
173. Frederic H. Hoppstock and Kenneth J. Frech, U.S. Patent No. 4,108,919; *Chem. Abstr.,* 1979, **90**, 137237k.
174. Lambert Schaper and Swan Tiong Sie, *Eur. Patent Application No.* 18,683; *Chem. Abstr.,* 1981, **94**, 139414x.
175. Leon F. Koniz and John H. Estes, U.S. Patent No. 4,171,329; *Chem. Abstr.,* 1980, **92**, 6218y.
176. Edward Zienkiewicz, Pol. Patent No. 98,039; *Chem. Abstr.,* 1979, **91**, 107780k.
177. Y. S. Khodakov, P. A. Makarov, G. Delzer, and Kh. M. Minachev, *J. Catal.,* 1980, **61**(1), 184.
178. V. A. Taranushich, A. P. Savost'yanov, K. G. Il'in, L. V. Stainov, and V. B. Il'in, U.S.S.R. Patent No. 703,132; *Chem. Abstr.,* 1980, **92**, 83324g.
179. Philip Varghese and Eduardo E. Wolf, *J. Catal.,* 1979, **59**(1), 100.
180. Jean Michel Tatibouet and Jean Eugene Germain, *C. R. Hebd. Seances Acad. Sci. Ser. C,* 1979, **289**(11), 301.
181. V. I. Marshneva and L. M. Karnatovskaya, *Kinet. Katal.,* 1980, **21**(2), 444.
182. Farid Azizian and James S. Pizey, *J. Chem. Technol. Biotechnol.,* 1980, **30**(8), 429.

183. Youssef Louisi, J. Rajagopal Roa, and Milos Ralek, *Chem. Ing. Tech.*, 1976, **48**(6), 544.
184. A. Simecek, J. Vosolsoke, and J. Svergo, *Chem. Prum.*, 1980, **30**(9), 463.
185. Clarence E. Habermann, U.S. Patent No. 4,153,581; *Chem. Abstr.*, 1979, **91**, 38905r.
186. Yannick Le Goff, Michel Senes, and Christian Hamon, *Eur. Patent Application No.* 2,630; *Chem. Abstr.*, 1980, **92**, 6043n.
187. George J. Antos, John F. Flagg, U.S. Patent No. 4,207,171; *Chem. Abstr.*, 1980, **93**, 167857s.
188. British Petroleum Co. Ltd. Belgium Patent No. 869,280; *Chem. Abstr.*, 1979, **91**, 76744v.
189. P. I. Bel'kevich *et al.*, U.S.S.R. Patent No. 697, 177; *Chem. Abstr.*, 1980, **92**, 180722t.
190. Michael J. Antal, U.S. Patent No. 4,155,835; *Chem. Abstr.*, 1979, **91**, 76718q.
191. Farhad Behbahany, Z. Sheikhrezai, M. Djalali, and S. Salajegheh (1980), *J. Catal.* **63**(2), 285.
192. Michel Blanchard, Dominique Vanhoue, and Annie Derovault, *J. Chem. Res. Synop.*, 1979, **12**, 404.
193. Michel Blanchard, Dominique Vanhove, Francis Petit, and Andre Mortreux, *J. Chem. Soc. Chem. Commun.*, 1980, **19**, 908.
194. Dan Fraenkel and Bruce C. Gates (1980), *J. Am. Chem. Soc.* **102**(7), 2478.
195. Dominique Vanhove, Pierre Makambo, and Michel Blanchard, *J. Chem. Soc. Chem. Commun.*, 1979, **14**, 605.
196. E. K. Reiff, Jr., Report (FE–2028–14), 1977.
197. I-Der Huang and Catherine McCooey, U.S. Patent No. 4,225,458; *Chem. Abstr.*, 1980, **93**, 226479v.
198. Masatsugu Kajitani *et al.*, *Bull. Chem. Soc. Jpn*, 1979, **52**(8), 2343.
199. Walter A. Butte, Jr., U.S. Patent No. 4,186,146; *Chem. Abstr.*, 1980, **92**, 198082u.
200. Steven D. Daniels, *Res. Discl.*, 1981, **202**, 92.
201. C. Joseph Machiels and Robert B. Anderson, *J. Catal.*, 1979, **58**(2), 260.
202. Joseph C. Machiels and Robert B. Anderson, *Prepr. Can. Symp. Catal.*, 1977, **5**, 13.
203. Nobert Goetz, Walter Himmele, and Leopold Hupfer, Ger. Offen. Patent No. 2,830,998; *Chem. Abstr.*, 1980, **93**, 8184d.
204. Roger John Heath Cowles, Brit. Patent Application No. 2,002,359; *Chem. Abstr.*, 1979, **91**, 157448d.
205. M. S. Brodskii, Yu. A. Yalter, M. Ya. Gervits, and A. E. Kruglik, *Kinet. Katal.*, 1979, **20**(2), 341.
206. Sumio Umemura and Yasuo Nakamura, *Jpn. Kokai Tokkyo Koho* Patent No. 79,141,724; *Chem. Abstr.*, 1980, **92**, 94870h.
207. Mutsumi Matsumoto and Yoshimasa Seo, *Jpn. Kokai Tokkyo Koho* Patent No. 80 02,619; *Chem. Abstr.*, 1980, **92**, 147479w.
208. Nicholas C. Huie, George E. Kuhlman, Houssam M. Naim, David A. Palmer, and Hobe Schroeder, U.S. Patent No. 4,125,053; *Chem. Abstr.*, 1980, **93**, 204292b.
209. Hans Martin Weitz, Ger. Offen. Patent No. 2,854,154; *Chem. Abstr.*, 1980, **93**, 149822x.
210. Paul Gilmont and Arthur A. Blanchard, *Inorg. Synth.*, 1946, **2**, 238.
211. Arnulf P. Hagan, Tammy S. Miller, Donald L. Terrell, Bennett Hutchinson, and R. L. Hance, *Inorg. Chem.*, 1978, **17**(5), 1369.
212. G. K. Magomedov, O. V. Shkol'nik, A. V. Medvedeva, V. G. Syrkin, and G. U. Druzhkova, *Koord. Khim.*, 1981, **7**(7), 1081.
213. J. Drapier, M. T. Hoornaerts, A. J. Hubert, and P. Teyssie, *J. Mol. Catal.*, 1981, **11**(1), 53.
214. Romeu J. Daroda, J. Richard Blackborow, and Geoffrey Wilkinson, *J. Chem. Soc. Chem. Commun.*, 1980, **22**, 1098.
215. A. L. Lapidus, A. Yu. Krylova, L. T. Kondrot'ev, *Izv. Akad. Nauk SSSR Ser. Khim.*, 1980, **6**, 1432.

216. Aatonin Deluzarche, Gerard Jenner, Alain Kiennemann, and Fouad Abou Samra, *Erdoel Kohle Erdgas Petrochem.*, 1979, **32**(9), 436.
217. Miroslav Novotny, U.S. Patent No. 4,283,582; *Chem. Abstr.*, 1981, **95**, 168537r.
218. Wayne R. Pretzer and Thaddeus P. Kobylinski, *Ann. N.Y. Acad. Sci.*, 1980, **333**, 58.
219. Martin Barry Sherwin and Arthur Marvin Brownstein, British Patent Application No. 2,007,652; *Chem. Abstr.*, 1980, **92**, 58423f.
220. Donald E. Morris, U.S. Patent No. 4,158,668; *Chem. Abstr.*, 1979, **91**, 91180b.
221. V. A. Rybakov, A. M. Nalimov, and S. K. Ogorodnikov, *Jpn. Kokai Tokkyo Koho* Patent No. 79 106,411; *Chem. Abstr.*, 1980, **92**, 6073x.
222. Noboru Sonoda, Hiroshi Tanaka, Akira Hirano, and Hideo Haitani, *Jpn. Kokai Tokkyo Koho* Patent No. 79 55,515; *Chem. Abstr.*, 1979, **91**, 157228b.
223. Takeshi Onoda and Keisuke Wada, *Jpn. Kokai Tokkyo Koho* Patent No. 79 112,817; *Chem. Abstr.*, 1980, **92**, 22077g.
224. David Moy, U.S. Patent No. 4,266,070; *Chem. Abstr.*, 1981, **95**, 80526a.
225. Thomas A. Haase and Frederick A. Pesa, U.S. Patent No. 4,235,744; *Chem. Abstr.*, 1981, **94**, 139247v.
226. Robert Stern, Andre Hirschauer, Dominique Commereuc, and Yves Chauvin, British Patent Application No. 2,000,132; *Chem. Abstr.*, 1979, **91**, 192831v.
227. Robert Stern, Andre Hirschauer, Dominique Commereuc, and Yves Chauvin, U.S. Patent No. 4,264,515; *Chem. Abstr.*, 1981, **95**, 80181j.
228. Shinji Murai, Toshikazu Kato, Noboru Sonoda, Yoshio Seki, Kazuaki Kawamoto, *Angew. Chem. Int. Ed. Engl.*, 1979, **18**(5), 393.
229. Yoshio Seki, Shinji Murai, and Noboru Sonoda, *Angew. Chem. Int. Ed. Engl.*, 1978, **17**(2), 119.
230. Hidetaka Kojima, Shingo Oda, Takushi Yokoyama, and Yasukazu Murakami, Ger. Offen. Patent No. 2,901,347; *Chem. Abstr.*, 1979, **91**, 140341r.
231. William E. Smith, Can. Patent No. 1,051,037; *Chem. Abstr.*, 1979, **91**, 19894v.
232. Kazuhisa Murata and Akio Matsuda, *Chem. Lett.*, 1980, **1**, 11.
233. Jean Gauthier-Lafaye, Robert Perron, Eur. Patent Application No. 22,735; *Chem. Abstr.*, 1981, **94**, 191699h.
234. John L. Cox and Wayne A. Wilcox, U.S. Patent No. 4,155,832; *Chem. Abstr.*, 1979, **91**, 76788n.
235. Harold M. Feder and Jerome W. Rathke, U.S. Patent No. 4,152,248; *Chem. Abstr.*, 1979, **91**, 76792j.
236. Howard Alper and Madhuban Gopal, *J. Organomet. Chem.*, 1981, **219**(1), 125.
237. D. Slocum, *Catal. Org. Synth.*, 1978, **7**, 245.
238. Shinji Murai and Noboru Sonoda, *Angew. Chem. Int. Ed. Engl.*, 1979, **18**(11), 837.
239. William Edward Smith, Ger. Offen. Patent No. 2,716,035; *Chem. Abstr.*, 1978, **88**, 22384x.
240. Julius Strauss, Herbert Huebner, Heinz Mueller, and Engelbert Krempl, Ger. Offen. Patent No. 2,535,073; *Chem. Abstr.*, 1977, **86**, 170856b.
241. John Halstead Atkinson and Fred Dean, Brit. Patent No. 1,556,102; *Chem. Abstr.*, 1980, **93**, 46185n.
242. John A. Patterson, Wheeler C. Crawford, and James R. Wilson, U.S. Patent No. 4,163,761; *Chem. Abstr.*, 1979, **91**, 124206v.
243. Walter H. Seitzer, U.S. Patent No. 4,139,550; *Chem. Abstr.*, 1979, **91**, 109950q.
244. Leo J. Frainier and Herbert Fineberg, Ger. Offen. Patent No. 3,007,139; *Chem. Abstr.*, 1980, **93**, 210955m.
245. John M. Hasman, U.S. Patent No. 4,133,822; *Chem. Abstr.*, 1979, **90**, 106066z.
246. John M. Hasman, U.S. Patent No. 4,158,665; *Chem. Abstr.*, 1979, **91**, 106879n.

247. H. Sumathi Vedanayagam, V. V. S. Mani, M. M. Paulose, and M. R. Subbaram, *Indian J. Technol.,* 1975, **13**(5), 230.
248. Victor P. Kurkov, U.S. Patent No. 4,183,866; *Chem. Abstr.,* 1980, **92**, 215271z.
249. Norman S. Boodman and Jack W. Walter, U.S. Patent No. 4,075,254; *Chem. Abstr.,* 1978, **88**, 169750x.
250. Fumiji Miya and Morio Matsuda, *Jpn. Kokai Tokkyo Koho* Patent No. 79 13,484; *Chem. Abstr.,* 1979, **90**, 210796y.
251. William J. Ehmann, U.S. Patent No. 4,160,786; *Chem. Abstr.,* 1979, **91**, 175558u.
252. Akio Tamura, Yoshio Kinsho, and Takayuki Yoshida, Eur. Patent Application No. 23,699; *Chem. Abstr.,* 1981, **95**, 42486e.
253. Irwin J. Gardner, U.S. Patent No. 4,145,492; *Chem. Abstr.,* 1979, **91**, 58470y.
254. Kazuo Asakawa, Yasuo Yamamoto, Shuji Ebata, and Tadashi Nakamura, Ger. Offen. Patent No. 3,004,514; *Chem. Abstr.,* 1980, **93**, 226472n.
255. J. B. Wilkes, Belgium Patent No. 873,167; *Chem. Abstr.,* 1979, **91**, 140328s.
256. Jerzy Haber and Haline Piekarska-Sadowska, *Bull. Acad. Pol. Sci. Ser. Sci. Chim.,* 1978, **26**(12), 975.
257. Mitsuo Kojima, Harunori Soba, and Fuminori Kaneko, *Jpn. Kokai Tokkyo Koho* Patent No. 79 32,191; *Chem. Abstr.,* 1979, **91**, 28067e.
258. Libor Cerveny, Antonin Marhoul, Michal Zugarek, and Vlastimil Ruzicka, *Chem. Prum.,* 1980, **30**(1), 28.
259. Bernd Schroeder, Ruetger Neeff, and Rudolf Braden, Ger. Offen. Patent No. 2,830,456; *Chem. Abstr.,* 1980, **92**, 215151k.
260. Z. R. Ismagilov, N. M. Dobrynkin, and V. V. Popovskii (1979). *React. Kinet. Catal. Lett.* **11**(2), 125.
261. Sargis Khoobiar, U.S. Patent No. 4,169,070; *Chem. Abstr.,* 1980, **92**, 6080x.
262. Tetsuro Seiyama, Yusaku Takita, Shosuke Jwange, Takashi Maebara, and Noboru Yamazoe, *Jpn. Kokai Tokkyo Koho* Patent No. 79 132,515; *Chem. Abstr.,* 1980, **92**, 146271k.
263. H. Paetow and L. Riekert, *Rev. Port. Quim.,* 1977, **19**(1–4), 16.
264. F. R. Hartley, *J. Organomet. Chem. Rev. A,* 1970, **6**, 119.
265. Edward T. Sabourin and Charles M. Selwitz, U.S. Patent No. 4,204,078; *Chem. Abstr.,* 1981, **94**, 30335w.
266. Timothy F. Murray and Jack R. Norton, *J. Am. Chem. Soc.,* 1979, **101**(15), 4107.
267. A. Cowell and J. K. Stille, *Tetrahedron Lett.,* 1979, **2**, 133.
268. Richard F. Heck, U.S. Patent No. 3,988,358; *Chem. Abstr.,* 1977, **86**, 29509j.
269. Donald Valentine, Jr., Jefferson W. Tilley, and Ronald A. LeMahieu, *J. Org. Chem.,* 1981, **46**(22), 4614.
270. A. Cowell and J. K. Stille, *J. Am. Chem. Soc.,* 1980, **102**(12), 4193.
271. Toshio Sato, Kouichi Naruse, Masashi Enokiya, and Tamotsu Fujisawa, *Chem. Lett.,* 1981, **8**, 1135.
272. Stanislas Galaj, Yvette Guichon, and Yves Louis Pascal, *C. R. Hebd. Seances Acad. Sci. Ser. C* **288**(22), 541.
273. Ei-ichi Negishi, Anthony O. King, and Nobuhisa Okukado, *J. Org. Chem.,* 1977, **42**(10), 1821.
274. Hisashi Okamura and Hisashi Takei, *Tetrahedron Lett.,* 1979, **36**, 3425.
275. Charles U. Pittman, Jr. and Quock Y. Ng, U.S. Patent No. 4,258,206; *Chem. Abstr.,* 1981, **94**, 208332h.
276. Y. Becker, A. Eisenstadt, and J. K. Stille, *J. Org. Chem.,* 1980, **45**(11), 2145.
277. Yuki Fujii and John C. Bailar, Jr., *J. Catal.,* 1978, **55**(2), 146.
278. Jiro Tsuji and Tomio Yamakawa, *Tetrahedron Lett.,* 1979, **20**(7), 613.

279. Charles Tanielian, Alain Kiennemann, and Temel Osparpucu, *Can. J. Chem.,* 1979, **57**(15), 2022.
280. P. S. Hallman, T. A. Stephenson, and G. Wilkinson, *Inorg. Synth.,* 1970, **12,** 238.
281. T. A. Stephenson and G. Wilkinson, *J. Inorg. Nucl. Chem.,* 1966, **28,** 945.
282. Jean Claude Brosse, Christian Pinazzi, and Daniel Derouet, Fr. Demande Patent No. 2,425,451; *Chem. Abstr.,* 1980, **92,** 147537p.
283. Hideyuki Matsumoto, Taichi Nakano, Keizo Takasu, and Yoichiro Nagai, *J. Org. Chem.,* 1978, **43**(9), 1734.
284. Ilan Pri-Bar, Ouri Buchman, Herbert Schumann, Heinz J. Kroth, and Jochanan Blum, *J. Org. Chem.,* 1980, **45**(22), 4418.
285. Bui-The-Khai, Carlo Concilio, and Gianni Porzi, *J. Org. Chem.,* 1980, **46**(8), 1759.
286. R. A. Sanchez-Delgado and O. L. De Ochoa, *J. Mol. Catal.,* 1979, **6**(4), 303.
287. V. Z. Sharf, L. Kh. Freidlin, B. M. Savchenko, and V. N. Krutii, *Izv. Akad. Nauk SSSR Ser. Khim.,* 1979, **5,** 1134.
288. V. Z. Sharf, L. Kh. Freidlin, B. M. Savchenko, and V. N. Krutii, *Izv. Akad. Nauk SSSR Ser. Khim.,* 1979, **6,** 1393.
289. Barbara Graser and Hannelore Steigerwald, *J. Organomet. Chem.,* 1980, **193**(3), C67.
290. Walter Strohmeier, Manfred Michel, and Luise Weigelt, *Z. Naturforsch. B. Anorg. Chem. Org. Chem.,* 1980, **35B**(5), 648.
291. Walter Strohmeier and Luise Weigelt, *J. Organomet. Chem.,* 1979, **171**(1), 121.
292. John F. Knifton, U.S. Patent No. 4,169,853; *Chem. Abstr.,* 1980, **92,** 6221u.
293. M. M. Taqui Khan and Rafeeq Mohiuddin, *Proc. Natl. Symp. Catal. 4th,* p. 374, 1978.
294. James E. Lyons, *J. Chem. Soc. Chem. Commun.,* 1975, **11,** 412.
295. Bogdan Marciniec and Jacek Gulinski, *J. Mol. Catal.,* 1980, **10**(1), 123.
296. Masakatsu Matsumoto and Satoru Ito, *J. Chem. Soc. Chem. Commun.,* 1981, **17,** 907.
297. Charles Tanielian, Alain Kiennemann, and Temel Osparpucu, *Can. J. Chem.,* 1980, **58**(24), 2813.
298. James M. Reuter and Robert G. Salomon, *J. Org. Chem.,* 1977, **42**(21), 3360.
299. S. H. H. Chaston and F. G. A. Stone, *J. Chem. Soc. A,* 1969, 500.
300. P. Chini and S. Martinengo. *Inorg. Chim. Acta,* 1969, **3,** 315.
301. B. R. James and G. L. Rempel, *Chem. Ind.,* 1971, 1036.
302. B. R. James, G. L. Rempel, and W. K. Teo, *Inorg. Synth.,* 1976, **16,** 49.
303. Richard M. Laine, David W. Thomas, and Lewis W. Carey, *J. Org. Chem.,* 1979, **44**(26), 4964.
304. Takashi Ohara, Tadao Kondo, Norio Takaya, and Kunihiro Kubota, *Jpn. Kokai Tokkyo Koho* Patent No. 79 115,309; *Chem. Abstr.,* 1980, **92,** 58230r.
305. Richard M. Laine, *J. Org. Chem.,* 1980, **45**(16), 3370.
306. William E. Smith, U.S. Patent No. 4,123,444; *Chem. Abstr.,* 1979, **90,** 87245p.
307. William E. Smith, U.S. Patent No. 4,139,542; *Chem. Abstr.,* 1979, **90,** 151967e.
308. T. Cole, R. Ramage, K. Cann, R. Pettit, *J. Am. Chem. Soc.,* 1980, **102**(19), 6182.
309. Kiyotomi Kaneda, Masahiko Hiraki, Toshinobu Imanara, and Shiichiro Teranishi, *J. Mol. Catal.,* 1981, **12**(3), 385.
310. Takaya Mise, Pangbu Hong, and Hiroshi Yamazaki, *Chem. Lett.,* 1980, **4,** 439.
311. William J. Thomson and Richard M. Laine, *ACS Symp. Ser.,* 1981, **152;** Catal. Act. Carbon Monoxide p. 133.
312. Gary D. Mercer, J. Shing Shu, Thomas B. Rauchfuss, and D. Max Roundhill, *J. Am. Chem. Soc.,* 1975, **97**(7), 1967.
313. Gary D. Mercer, William B. Beaulien, and D. Max Roundhill, *J. Am. Chem. Soc.,* 1977, **99**(20), 6551.
314. D. Max Roundhill, Mark K. Dickson, Nagaraj S. Dixit, and B. P. Sudha-Dixit, *J. Am. Chem. Soc.,* 1980, **102**(17), 5538.

315. Charles Ungermann *et al.*, *J. Am. Chem. Soc.*, 1979, **101**(20), 5922.
316. W. Hieber, *Z. Anorg. Allegm. Chem.*, 1932, **204**, 165.
317. R. B. King, and F. G. A. Stone, *Inorg. Synth.*, 1963, **7**, 193.
318. D. Ballivet-Tkatchenko, N. D. Chau, H. Mozzanega, M. C. Roux, and I. Tkatchenko, *ACS Symp. Ser.*, 1981, **152;** *Catal. Act. Carbon Monoxide*, p. 187.
319. Dominique Commereuc, Y. Chauvin, F. Hugues, J. M. Basset, and D. Olivier, *J. Chem. Soc. Chem. Commun.*, 1980, **4**, 154.
320. Francois Hugues, Bernard Besson, and Jean M. Basset; *J. Chem. Soc. Chem. Commun.*, 1980, **15**, 719.
321. W. Keim, M. Berger, and J. Schlupp, *J. Catal.*, 1982, **61**(2), 359.
322. Howard Alper and Khaled E. Hachem, *J. Am. Chem. Soc.*, 1981, **103**(21), 907.
323. Laszlo Marko, Mazin A. Radhi, and Irma Otvos, *J. Organomet. Chem.*, 1981, **218**(3), 369.
324. Yu. I. Kozorezov and Yu. V. Ryabtseva, *Zh. Prikl. Khim. Leningrad*, 1979, **52**(10), 2191.
325. Kazushi Arata and Makoto Hino, *Chem. Lett.*, 1980, **12**, 1479.
326. Masao Saito, Tetsuo Aoyama, Shigeru Horie, and Kazuo Takada, *Brit. Patent Application No.* 2,000,121; *Chem. Abstr.*, 1979, **91**, 174844x.
327. Kozo Sano and Tekeo Ikarashi, Ger. Offen. Patent No. 2,921,551; *Chem. Abstr.*, 1980, **92**, 75892a.
328. R. B. Valitov, G. G. Garifzyanov, B. E. Prusenko, U. B. Imashev, and D. L. Rakhmankulov, U.S.S.R. Patent No. 759,499; *Chem. Abstr.*, 1980, **93**, 240235u.
329. Makoto Hino and Kazushi Arata, *Chem. Lett.*, 1979, **5**, 477.
330. Jean Pierre Anquetil, Jacques Vanrenterghem, Jean Claude Clement, Ger. Offen. Patent No. 2,648,443; *Chem. Abstr.*, 1977, **87**, 40058c.
331. William M. Castor and Pete J. Menegos, U.S. Patent No. 4,139,497. *Chem. Abstr.*, 1979, **90**, 152831t.
332. Gregor H. Riesser, U.S. Patent No. 4,152,300; *Chem. Abstr.*, 1979, **91**, 91324b.
333. S. A. Mamedova, F. V. Aliev, and T. G. Alkhazov, *Neftekhimiya*, 1977, **17**(5), 683.
334. Warren G. Schlinger and William L. Slater, U.S. Patent No. 4,172,842; *Chem. Abstr.*, 1980, **92**, 61582n.
335. Hans Joachim Scholl and Armin Zenner, Ger. Offen. Patent No. 2,808,980; *Chem. Abstr.*, 1980, **92**, 6290r.
336. Takeshi Okutani, Shinichi Yokoyama, Ryoichi Yoshida, Yuji Yoshida, and Tadao Ishii, *Jpn. Kokai Tokkyo Koho* Patent No. 79 124,006; *Chem. Abstr.*, 1980, **92**, 79364c.
337. Yu. M. Zhorov, E. V. Morozova, and G. M. Panchenkov, U.S.S.R. Patent No. 551,315; *Chem. Abstr.*, 1977, **86**, 172141p.
338. Kunitoshi Aoki, Makoto Honda, Tetsuro Dozono, Tsutomu Katsumata, Ger. Offen. Patent No. 2,856,413; *Chem. Abstr.*, 1979, **91**, 108458y.
339. Shigeo Kamimura, Masakatsu Hatano, Masayoshi Murayama, and Kazunori Oshima, *Jpn. Kokai Tokkyo Koho* Patent No. 79 22,320; *Chem. Abstr.*, 1979, **90**, 187875n.
340. Kanji Sakata, Fumio Ueda, Makoto Misono, and Yukio Yoneda, *Bull. Chem. Soc. Jpn.*, 1980, **53**(2), 324.
341. Asis Das and A. K. Kar, *Indian J. Technol.*, 1980, **18**(7), 267.
342. Angelo Joseph Magistro, U.S. Patent No. 4,100,211; *Chem. Abstr.*, 1978, **89**, 198250k.
343. Makoto Hino and Kazushi Arata, *Chem. Lett.*, 1979, **9**, 1141.
344. Makoto Hino and Kazushi Arata, *Chem. Lett.*, 1980, **8**, 963.
345. Peter Ettmayer and Gerhard Jangg, *Montash*, 1961, **92**, 834.
346. Walter Von Reppe, *Ann.*, 1953, **582**, 116.
347. Albert E. Wallis and Stanley C. Townshend, U.S. Patent No. 2,378,053; *Chem. Abstr.*, 1945, **39**, 5056.
348. Johannes Elzinga and Hepke Hogeveen, *J. Chem. Soc. Chem. Commun.*, 1977, **20**, 705.

349. B. Duane Dombek and Robert J. Angelici, *J. Catal.,* 1977, **48**(1–3), 433.
350. Masao Saito, Tetsuo Aoyama, Shigeru Horie, and Kazuo Takada, Ger. Offen. Patent No. 2,827,633; *Chem. Abstr.,* 1979, **90**, 137288c.
351. D. Ballivet-Tkatchenko, G. Coudurier, H. Mozzanega, and I. Tkatchenko, *Fundam. Res. Homogenous Catal.,* 1979, **3**, 257.
352. Thomas E. Kiovsky and Milton M. Wald, Ger. Offen. Patent No. 2,603,892; *Chem. Abstr.,* 1977, **86**, 57874t.
353. Romeu J. Daroda, J. Richard Blackborow, and Geoffrey Wilkinson, *J. Chem. Soc. Chem. Commun.,* 1980, **22**, 1101.
354. Franz Scheidl, Ger. Offen. Patent No. 2,815,580; *Chem. Abstr.,* 1980, **92**, 93911k.
355. Walter Rebafka, Ger. Offen. Patent No. 2,803,987; *Chem. Abstr.,* 1979, **91**, 175179w.
356. James L. Graff, Robert D. Sanner, and Mark S. Wrighton, "Report" (TR–12), p. 14, order No. AD–A061871.
357. Charles U. Pittman, Jr., William D. Honnick, Mark S. Wrighton, and Robert D. Sanner, *Fundam. Res. Homogenous Catal.,* 1979, **3**, 603.
358. W. H. Breckenridge, Gerald T. Bida, and Werner S. Kolln, *J. Phys. Chem.,* 1979, **83**(9), 1150.
359. Allen D. King, Jr., R. B. King, and D. B. Yang, *J. Am. Chem. Soc.,* 1980, **102**(3), 1028.
360. Howard Alper, *Org. Synth. Met. Carbon.,* 1977, **2**, 545.
361. R. B. King, A. D. King, Jr., and D. B. Yang, *Amer. Chem. Soc. Symp. Ser. Catal. Act. Carbon Monoxide,* 1981, **152**, 123.
362. Ryoji Noyori, *Ann. N.Y. Acad. Sci.,* 1977, **295**, 225.
363. Nikolaus Von Kutepow, *Ull. Ency. Tech. Chem. 4 Aufl.,* 1975, **10**, 411.
364. Robert A. Waghorne and Howard L. Mitchell III, U.S. Patent No. 4,172,810; *Chem. Abstr.,* 1980, **92**, 146398g.
365. Lee Hilfman, U.S. Patent No. 4,194,058; *Chem. Abstr.,* 1980, **93**, 71244p.
366. Sumio Unemura, Nagaaki Takamits, Hiroshi Yoshida, and Haruzo Ikezawa, *Jpn. Kokai Tokkyo Koho* Patent No. 79 81,233; *Chem. Abstr.,* 1980, **92**, 22280t.
367. Alan Brenner and Dennis A. Hucul, *Chem. Uses Molybdenum. Proc. Int. Conf. 3rd,* p. 194.
368. James T. Cobb, Jr. and Robert C. Streeter, *Ind. Eng. Chem. Process Des. Dev.,* 1979, **18**(4), 672.
369. Gisela Olive and Salvador Olive, U.S. Patent No. 4,179,462; *Chem. Abstr.,* 1980, **92**, 93920n.
370. P. Villeger, Joel Barrault, Jacques Barbier, Ginette Leclercq, and Raymond Maurel, *Bull. Soc. Chim. Fr.,* 1979, **9–10**(Pt. 1), 413.
371. Alan Brenner, *J. Mol. Catal.,* 1979, **5**(2), 157.
372. Claus Schliebener and Hans Juergen Wernicke, Ger. Offen. Patent No. 2,806,854; *Chem. Abstr.,* 1980, **92**, 8740m.
373. V. V. Abalyaeva, G. I. Kozub, M. L. Khidekel, U.S.S.R. Patent No. 695,948; *Chem. Abstr.,* 1980, **92**, 180717v.
374. V. L. Chernobrivets, *Katal. Katal.,* 1979, **17**, 46.
375. Toshihiko Kumazawa, Takeshi Yamamoto, and Hiroshi Otanaka, *Jpn. Kokai Tokkyo Koho* Patent No. 79 125,606; *Chem. Abstr.,* 1980, **92**, 110522d.
376. Tsutomu Katsumata, Tetsuro Dozono, and Makoto Honda, *Jpn. Kokai Tokkyo Koho* Patent No. 79,126,698; *Chem. Abstr.,* 1980, **92**, 128435n.
377. Yasuo Tokitoh, Noriaki Yoshimura, and Masuhiko Tamura, Ger. Offen. Patent No. 2,935,535. *Chem. Abstr.,* 1980, **93**, 71023r.
378. K. N. Anisimov and A. N. Nesmeyanov, *Dokl. Akad. Nauk SSSR,,* 1940, **26**, 58.
379. Malcolm H. Chisholm, F. Albert Cotton, Michael W. Extine, and Raymond L. Kelly, *J. Am. Chem. Soc.,* 1979, **101**(25), 7645.
380. D. T. Hurd, U.S. Patent No. 2,554,194; *Chem. Abstr.,* 1951, **45**, 7314, 8728.

381. D. T. Hurd, U.S. Patent No. 2,557,744; 1951.
382. H. E. Podall, J. H. Dunn, and H. Shapiro, *J. Am. Chem. Soc.,* 1960, **82,** 1325.
383. Michael J. Doyle, James G. Davidson, *J. Org. Chem.,* 1980, **45**(8), 1539.
384. Michele Pralus, Jean P. Schirmann, and Serge Y. Delavarenne, Ger. Offen. 2,616,934. *Chem. Abstr.* (1977). **86,** 55272w.
385. S. Ivanov, R. Boeva, and S. Tanielyan, *React. Kinet. Catal. Lett.,* 1976, **5**(3), 297.
386. M. R. Musaev, E. E. Gaidarova, and R. E. Akhmedova, *Dokl. Akad. Nauk Az. SSR,* 1980, **36**(7), 51.
387. L. N. Khabibullina, V. S. Gumerova, V. P. Yur'ev, and S. R. Rafikov, *Dokl. Akad. Nauk SSSR,* 1981, **256**(5), 1138.
388. Robert G. Bowman and Robert L. Burwell, Jr., *J. Catal.,* 1980, **63**(2), 463.
389. G. Lauterbach, G. Posselt, R. Schaefer, and D. Schnurpfeil, *J. Prakt. Chem.,* 1981, **323**(1), 101.
390. Abdullah Hassan and Roger J. Card, *Sains Malays.,* 1977, **6**(2), 125.
391. André Mortreux and Michel Blanchard, *J. Chem. Soc. Chem. Commun.,* 1974, **19,** 786.
392. Sukanya Devarajan, David R. M. Walton, and G. Jeffrey Leigh, *J. Organomet. Chem.,* 1979, **181**(1), 99.
393. R. B. King, C. C. Frazier, R. M. Hanes, and A. D. King, Jr., *J. Am. Chem. Soc.,* 1978, **100**(9), 2925.
394. A. D. King, Jr., R. B. King, and D. B. Yang, *J. Am. Chem. Soc.,* 1981, **103**(10), 2699.
395. G. P. Balabanov *et al.,* U.S. Patent No. 4,207,212; *Chem. Abstr.,* 1980, **93,** 150873w.
396. Horst Andraczek *et al.,* Ger. (East) Patent No. 132,131; *Chem. Abstr.,* 1979, **91,** 76705h.
397. D. Ralston *et al., Fuel Process. Technol.,* 1978, **1**(2), 143.
398. Giovanni Bagnasco, Paolo Ciambelli, Silvestro Crescitelli, and Gennard Russo, *Chim. Ind. Milan,* 1979, **61**(11), 795.
399. R. Maggiore, N. Giordano, C. Crisafulli, F. Castelli, L. Solarino, and J. C. J. Bart, *J. Catal.,* 1979, **60**(2), 193.
400. Howard D. Simpson, U.S. Patent No. 4,179,410; *Chem. Abstr.,* 1980, **92,** 131885q.
401. James G. Williamson, Christopher S. John, and Charles Kemball, *J. Chem. Soc. Faraday Trans. 1,* 1980, **76**(6), 1366.
402. Richard D. Cramer, U.S. Patent No. 4,161,609; *Chem. Abstr.,* 1980, **92,** 129595b.
403. Clarence D. Chang and William H. Lang, U.S. Patent No. 4,177,202; *Chem. Abstr.,* 1980, **92,** 131845b.
404. Craig B. Murchison and Dewey A. Murdick, U.S. Patent No. 4,151,190; *Chem. Abstr.,* 1979, **91,** 41855e.
405. M. Saito and R. B. Anderson, *J. Catal.,* 1980, **63**(2), 438.
406. E. Furimsky, R. Ranganathan, and B. I. Parsons, *Prepr. Can. Symp. Catal. 5th,* 444, 1977.
407. Orville Deacy Frampton, Ger. Offen. Patent No. 2,945,591; *Chem. Abstr.,* 1980, **93,** 149769k.
408. I. Halasz and G. Gati, *React. Kinet. Catal. Lett.,* 1979, **11**(1), 5.
409. Robert K. Grasselli, Arthur F. Miller, and Harley F. Hardman, U.S. Patent No. 4,192,776; *Chem. Abstr.,* 1980, **92,** 221638r.
410. Sumio Umemura, Kyoji Ohdan, Tukuo Matsuzaki, and Taizo Uda, Ger. Offen. Patent No. 2,720,946; *Chem. Abstr.,* 1978, **88,** 51341m.
411. Serge R. Dolhyj and Ernest C. Milberger, U.S. Patent No. 4,165,300; *Chem. Abstr.,* 1979, **91,** 158326z.
412. Tsutomo Katsumata, Makoto Honda, and Tetsuo Dozono, *Jpn. Kokai Tokkyo Koho* Patent No. 79 71,799; *Chem. Abstr.,* 1979, **91,** 142724s.
413. James Edward Lyons, George Suld, and Robert Wesley Shinn, Belgium Patent No. 875,110; *Chem. Abstr.,* 1980, **92,** 6227a.
414. Thomas H. Vanderspurt, U.S. Patent No. 4,184,981; *Chem. Abstr.,* 1980, **92,** 146281p.

415. Serge R. Dolhyj, Robert K. Grasselli, and Dev D. Suresh, U.S. Patent No. 4,213,917; *Chem. Abstr.,* 1980, **93,** 185989r.
416. P. Geneste, M. Bonnet, C. Frouin, and D. Levache, *J. Catal.,* 1980, **61**(1), 277.
417. G. N. Maslyanskii, G. L. Rabinovich, L. M. Treiger, B. Kh. Gokhman, and V. D. Seleznev, Ger. Offen. Patent No. 2,738,024; *Chem. Abstr.,* 1979, **91,** 60012u.
418. V. H. J. DeBeer and G. C. A. Schuit, *Prep. Catal. Proc. Int. Symp.,* 1976, 343.
419. Gerd Leston, U.S. Patent No. 4,158,101; *Chem. Abstr.,* 1979, **91,** 91353k.
420. Quintin W. Decker and Erich Marcus, U.S. Patent No. 4,190,601; *Chem. Abstr.,* 1980, **93,** 25895d.
421. Matti Harkonen and Pekka Nuojua, *Kem. Kemi,* 1980, **7**(3), 98.
422. Antal Sarkany and Pal Tetenyi, *J. Chem. Soc. Chem. Commun.,* 1980, **11,** 525.
423. George J. Antos, U.S. Patent No. 4,207,172; *Chem. Abstr.,* 1981, **94,** 46929d.
424. William L. Kehl and Angelo A. Montagna, U.S. Patent No. 4,081,353; *Chem. Abstr.,* 1978, **89,** 92187e.
425. T. Inui, M. Funabiki, and Y. Takegami, *React. Kinet. Catal. Lett.,* 1979, **12**(3), 287.
426. C. B. Lee and R. B. Anderson, *Prepr. Can. Symp. Catal.,* 1979, **6,** 160.
427. Umar M. U. Ahmad and John K. Pargeter, U.S. Patent No. 4,196,100; *Chem. Abstr.,* 1980, **93,** 75612e.
428. J. A. Dalmon and G. A. Martin, *Stud. Surf. Sci. Catal.,* 1981, **7**(Pt. A); *New Horiz. Catal.,* p. 402.
429. Philippe Villeger, Ginette Leclercq, and Raymond Maurel, *Bull. Soc. Chim. Fr.,* 1979, **9-10**(Pt. 1), 406.
430. P. N. Galich, V. S. Gutyrya, and A. A. Galinski, *Proc. Int. Conf. Zeolites 5th,* p. 661, 1980.
431. Phillipe Gallois, Jean-Jacques Brunet, and Paul Caubere, *J. Org. Chem.,* 1980, **45**(10), 1946.
432. Waldo R. De Thomas and Eugene V. Hort, U.S. Patent No. 4,182,721; *Chem. Abstr.,* 1980, **92,** 146589v.
433. Kurt Halcour, Paul Losacker, Helmut Waldmann, and Wulf Schwerdtel, Ger. Offen. Patent No. 2,823,165; *Chem. Abstr.,* 1980, **92,** 93965f.
434. Hal D. Burge, Dermont J. Collins, and Burton H. Davis, *Ind. Eng. Chem. Prod. Res. Dev.,* 1980, **19**(3), 389.
435. Susumu Tomidokoro, Michito Sato, and Daini Saika, British Patent Application No. 2,025,408; *Chem. Abstr.,* 1980, **93,** 7652t.
436. Dale D. Dixon and Randall J. Daughenbaugh, U.S. Patent No. 4,260,748; *Chem. Abstr.,* 1981, **95,** 43139f.
437. C. Joseph Machiels and Robert B. Anderson, *J. Catal.,* 1979, **60**(2), 339.
438. Juan G. Andrade, Wilhelm F. Maier, Lothar Zapf, and Paul Von Rague Schleyer, *Synthesis,* 1980, **10,** 802.
439. Mahavir Prashad, M. Seth, and A. P. Bhaduri, *Indian J. Chem. Sect. B,* 1980, **19B**(5), 393.
440. A. Elattar, W. E. Wallace, and R. S. Craig, *Adv. Chem. Ser.,* 1979, **178,** 7.
441. G. H. Watson, *Rep. ICTIS/TR IEA Coal Res. ICTIS-TR,* 1980, **09,** 56.
442. R. A. Schunn, *Inorg. Synth.,* 1974, **15,** 5.
443. John H. Nelson, Peter N. Howells, George C. DeLullo, and George L. Landen, *J. Org. Chem.,* 1980, **45**(7), 1246.
444. Emilian Angelescu, Anca Angelescu, and Ioan V. Nicolescu, *Rev. Chim. Bucharest,* 1979, **30**(6), 523.
445. Yasumasa Sakakibara, Koichi Aiba, Mutsuji Saki, and Norito Uchino, *Bull. Inst. Chem. Res. Kyoto Univ.,* 1979, **57**(3), 240.
446. Hiroyuki Morikawa, Shoji Kitazume, and Satoshi Ohtaka, *Jpn. Kokai Tokkyo Koho* Patent No. 79 151,948; *Chem. Abstr.,* 1980, **92,** 198010u.

447. Karel Sporka, Jiri Hanika, Vlastimil Ruzicka, and Vladimir Tomek, Czech. Patent No. 178,047; *Chem. Abstr.,* 1980, **92,** 6151w.
448. Eiichi Ibuki, Shigeru Ozasa, Yasuhiro Fujioka, Motofumi Okada, and Katsutoshi Terada, *Bull. Chem. Soc. Jpn.,* 1980, **53**(3), 821.
449. Tamio Hayashi, Yoshio Katsuro, and Makoto Kumada, *Tetrahedron Lett.,* 1980, **21**(40), 3915.
450. Kurt Madeja *et al.,* Ger. (East) Patent No. 134,054; *Chem. Abstr.,* 1979, **91,** 28062z.
451. Petr Svoboda, Jiri Hetflejs, and Jiri Soucek, *Chem. Prum.,* 1980, **30**(2), 89.
452. Volker Ladenberger, Klaus Bronstert, Gerhard Fahrbach, and Wolfgang Groh, Ger. Offen. Patent No. 2,748,884; *Chem. Abstr.,* 1979, **91,** 5721z.
453. Takashi Onishi, Yoshiji Fujita, and Takashi Nishida, *Chem. Lett.,* 1979, **7,** 765.
454. Donald Floyd Birkelbach and George Willie Knight, Eur. Patent Application No. 9,160; *Chem. Abstr.,* 1980, **93,** 95907h.
455. Jin Sun Yoo and Henry Erickson, U.S. Patent No. 3,992,323; *Chem. Abstr.,* 1977, **86,** 73408z.
456. Donald V. Hillegass, U.S. Patent No. 4,151,345; *Chem. Abstr.,* 1979, **91,** 5715a.
457. W. L. Gilliland and A. A. Blanchard, *Inorg. Synth.,* 1946, **2,** 234.
458. L. Mond, C. Langer, and F. Quincke, *J. Chem. Soc.,* 1890, 749.
459. Jan A. K. Du Plessis and Corrie J. Du Toit, *Afr. J. Chem.,* 1979, **32**(4), 147.
460. Marco Foa and Luigi Cassar, *Gazz. Chim. Ital.,* 1979, **109**(12), 619.
461. Wilbur E. Martin and Michael F. Farona, *J. Organomet. Chem.,* 1981, **206**(3), 393.
462. Jean Gauthier-Lafaye and Robert Perron, Eur. Patent Application No. 18,927; *Chem. Abstr.,* 1981, **94,** 156327p.
463. Brian F. G. Johnson, Jack Lewis, *Inorg. Synth.,* 1972, **15,** 93.
464. John Frederick Knifton, Brit. Patent Application No. 2,024,811; *Chem. Abstr.,* 1980, **93,** 167662z.
465. Henry W. Choi and E. L. Muetterties, *Inorg. Chem.,* 1981, **20**(8), 2664.
466. R. A. Schunn, G. C. Demitras, H. W. Choi, and E. L. Muetterties, *Inorg. Chem.,* 1981, **20**(11), 4023.
467. Mark S. Wrighton, James L. Graff, Carol L. Reichel, and Robert D. Sanner, Report (TR – 16), 1979.
468. Youval Shvo and Richard M. Laine, *J. Chem. Soc. Chem. Commun.,* 1980, **16,** 753.
469. Yoshio Ono and Atsuo Murata, *Jpn. Kokai Tokkyo Koho* Patient No. 80 51,042; *Chem. Abstr.,* 1980, **93,** 167870r.
470. Haven Sylvester Kesling and Lee Randall Zehner, Brit. Patent Application No. 2,000,495; *Chem. Abstr.,* 1980, **92,** 6082z.
471. G. Gardos and A. Redey, *Hung. J. Ind. Chem.,* 1979, **7**(4), 411.
472. Gian Paolo Chiusoli, William Giroldini, and Giuseppe Salerno, *Gazz. Chim. Ital.,* 1980, **110**(7–8), 371.
473. M. Albert Vannice and Robert L. Garten, U.S. Patent No. 4,093,643; *Chem. Abstr.,* 1978, **89,** 165951m.
474. Shun-Ichi Murahashi and Tomonari Watanabe, *J. Am. Chem. Soc.,* 1979, **101**(24), 7429.
475. Francis P. Daly, *J. Catal.,* 1980, **61**(2), 528.
476. Udo Birkenstock and Herbert Schmidt, Ger. Offen. Patent No. 2,848,978; *Chem. Abstr.,* 1980, **93,** 102064m.
477. Karl F. Cossaboon, U.S. Patent No. 4,185,036; *Chem. Abstr.,* 1980, **92,** 146418p.
478. S. Asad Al-Ammar and Geoffrey Webb, *J. Chem. Soc. Faraday Trans. 1,* 1900, **75**(8) (1979).
479. Charles A. Drake, U.S. Patent No. 4,216,169; *Chem. Abstr.,* 1980, **93,** 167611g.
480. Klaus Mainusch, Reinhard Schorsch, and Bernard Scheppinghoff, Ger. Offen. Patent No. 2,839,134; *Chem. Abstr.,* 1980, **93,** 70990s.

481. George C. Accrombessi, Patrick Geneste, and Jean-Louise Olive, *J. Org. Chem.,* 1980, **45**(21), 4139.
482. Takeshi Onoda, Keisuke Wada, Keiichi Sato and Yukio Kasori, *Jpn. Kokai Tokkyo Koho* Patent No. 79 151,903; *Chem. Abstr.,* 1980, **92**, 180665b.
483. Howard A. J. Carless and David J. Haywood, *J. Chem. Soc. Chem. Commun.,* 1980, **20**, 980.
484. James A. Hinnenkamp and John A. Scheben, U.S. Patent No. 4,188,490; *Chem. Abstr.,* 1980, **92**, 164504t.
485. Yohei Fukuoka, Akihiro Tamura, Ryoichi Mitsui, and Yoshio Suzuki, *Jpn. Kokai Tokkyo Koho* Patent No. 79 73,725; *Chem. Abstr.,* 1979, **91**, 141426j.
486. M. K. Starchevskii, M. N. Vargaftik, and I. I. Moiseev, *Kinet. Katal.,* 1979, **20**(5), 1163.
487. Kuniaki Takahashi, Bunji Shinoda, and Shoichi Takahashi, *Jpn. Kokai Tokkyo Koho* Patent No. 79 128,529; *Chem. Abstr.,* 1980, **92**, 128548b.
488. Helmut Fiege and Karlfried Wedemeyer, Ger. Offen. Patent No. 2,851,788; *Chem. Abstr.,* 1980, **93**, 167944t.
489. Hitoshi Yuasa and Hirosuke Imai, Ger. Offen. Patent No. 2,842,264; *Chem. Abstr.,* 1979, **91**, 21441p.
490. J. A. Peters and H. Van Bekkum, *Recl. Trav. Chim. Pays-Bas.,* 1981, **160**(1), 21.
491. Shun-Ichi Murahashi, Tsumoru Hirano, and Tsuneo Yano, *J. Am. Chem. Soc.,* 1978, **100**(1), 348.
492. Hisashi Saito, *Shokubai,* 1981, **23**(2), 139.
493. Barry Trost, *Acc. Chem. Res.,* 1980, **13**(11), 385.
494. Jiro Tsuji, *Am. N.Y. Acad. Sci.,* 1980, **333**, 250.
495. T. A. Stephenson, S. M. Morehouse, A. R. Powell, J. P. Heffer, and G. Wilkinson, *J. Chem. Soc.,* 1965, 3632.
496. Yuzo Fujiwara, Osamu Maruyama, Michiaki Yoshidomi, and Hiroshi Taniguchi, *J. Org. Chem.,* 1981, **46**(5), 851.
497. Joseph E. Plevyak, James E. Dickerson, and Richard F. Heck, *J. Org. Chem.,* 1979, **44**(23), 4078.
498. Miwako Mori, Katsumi Chiba, Ohta Nahoko, and Yoshio Ban, *Heterocycles,* 1979, **13**, 329 (Spec. Issue).
499. Yuzo Fujiwara, Tomio Kawauchi, and Hiroshi Taniguchi, *J. Chem. Soc. Chem. Commun.,* 1980, **5**, 220.
500. Kiyoshi Kikukawa, Kiyoshi Kono, Kazuhiko Nagira, Fumio Wada, and Tsutomu Matsuda, *Tetrahedron Lett.,* 1980, **21**(30), 2877.
501. Kazuhiko Nagira, Kiyoshi Kikukawa, Fumio Wada, and Tsutomu Matsuda, *J. Org. Chem.,* 1980, **45**(12), 2365.
502. André J. Anciaux, André J. Hubert, Alfred F. Noels, N. Petiniot, and Philippe Teyssié, *J. Org. Chem.,* 1980, **45**(4), 695.
503. Marc O. Terpko and Richard F. Heck, *J. Am. Chem. Soc.,* 1979, **101**(18), 5281.
504. Ronald Grigg, Thomas R. B. Mitchell, and Ashok Ramasubbu, *J. Chem. Soc. Chem. Commun.,* 1980, **1**, 27.
505. Nicholas A. Cortese and Richard F. Heck, *J. Org. Chem.,* 1977, **42**(22), 3491.
506. Tadashi Okamoto and Shinzaburo Oka, *Bull. Chem. Soc. Jpn.,* 1981, **54**(4), 1265.
507. Hong-Son Ryang and Christopher S. Foote, *J. Am. Chem. Soc.,* 1980, **102**(6), 2129.
508. Friedrich Wunder, Therese Quadflieg, Guenter Roscher, and Guenther Heck, Eur. Patent Application No. 4,079; *Chem. Abstr.,* 1980, **92**, 41363p.
509. Hideo Nagashima and Jiro Tsuji, *Chem. Lett.,* 1981, **8**, 1171.
510. Alan J. Chalk and S. A. Magennis, *Ann. N.Y. Acad. Sci.,* 1980, **333**, 286.
511. Fumiaki Kawamoto, Shozo Tanaka, and Yoshihiro Honma, *Jpn. Kokai Tokkya Koho* Patent No. 79 122,204; *Chem. Abstr.,* 1980, **92**, 93914p.

512. Robert G. Brown and John M. Davidson, *Adv. Chem. Ser.*, 1974, **132**, 49.
513. Richard F. Heck, *Acc. Chem. Res.*, 1979, **12**(4), 146.
514. Georg Brauer, "Handbook of preparative inorganic chemistry," p. 1502. Academic Press, New York, 1965.
515. Tomiya Isshiki, Yasuhiko Kijima, and Yuh Miyauchi, U.S. Patent No. 4,212,989, *Chem. Abstr.*, 1980, **93**, 238850x.
516. Hajime Yoshida, Nobuyuki Sugita, Kiyoshi Kudo, and Yoshimasa Takezaki, *Bull. Chem. Soc. Jpn.*, 1976, **49**(8), 2245.
517. Atlantic Richfield Co. Belgium Patent No. 870,745; *Chem. Abstr.*, 1979, **91**, 74213x.
518. Jiro Tsuji, Mitsuo Takahashi, and Takashi Takahashi (1980). *Tetrahedron Lett.*, 1980, **21**(9), 849.
519. Robert Becker, Johann Grolig, and Christian Rasp, Ger. Offen. Patent No. 2,908,251; *Chem. Abstr.*, 1981, **94**, 65349f.
520. Robert J. Harvey, U.S. Patent No. 4,260,781; *Chem. Abstr.*, 1981, **95**, 61821b.
521. Byron R. Cotter, U.S. Patent No. 4,211,721; *Chem. Abstr.*, 1981, **94**, 15412a.
522. Katsuharu Miyata, Makoto Aiga, and Seiji Hasegawa, Eur. Patent Application No. 0,815; *Chem. Abstr.*, 1980, **92**, 22292y.
523. D. E. James, L. F. Hines, and J. K. Stille, *J. Am. Chem. Soc.*, 1976, **98**(7), 1806.
524. J. K. Stille and R. Divakaruni, *J. Org. Chem.*, 1979, **44**(20), 3474.
525. Idemitsu Kosan Co., Ltd. Fr. Demande Patent No. 2,376,116; *Chem. Abstr.*, 1979, **91**, 4962s.
526. Paul R. Stapp, U.S. Patent No. 4,237,071; *Chem. Abstr.*, 1981, **94**, 120872y.
527. Hubert Mimoun, Ger. Offen. Patent No. 2,920,678; *Chem. Abstr.*, 1980, **92**, 128393x.
528. Masanobu Hidai, Haruyoshi Mizuta, Kiyomiki Hirai, and Yasuzo Uchida, *Bull. Chem. Soc. Jpn.*, 1980, **53**(7), 2091.
529. V. A. Golodov, Yu. L. Sheludyakou, and D. V. Sokolsky, *Fund. Res. Homogen. Catal.*, 1979, **3**, 239.
530. Timothy Paul Murtha and Ernst Adolph Zuech, Eur. Patent Application No. 6,401; *Chem. Abstr.*, 1980, **93**, 94956m.
531. Siegfried Engels, *Z. Chem.*, 1980, **20**(6), 212.
532. Alan Ivor Foster, John James McCarroll, and Stephen Robert Tennison, Brit. Patent Application No. 2,012,607; *Chem. Abstr.*, 1980, **92**, 44536h.
533. Lee E. Trimble and Rudolf Johannes H. Voorhoeve, Belgium Patent No. 876,483; *Chem. Abstr.*, 1980, **92**, 131532x.
534. Kenzie Nozaki, Eur. Patent Application No. 4,410; *Chem. Abstr.*, 1980, **92**, 75814b.
535. Eliahu Licht, Yehoshua Schächter, and Herman Pines, *J. Catal.*, 1980, **61**(1), 109.
536. Roy John Sampson, British Patent No. 1,505,826; *Chem. Abstr.*, 1979, **90**, 41171h.
537. George J. Antos, U.S. Patent No. 4,172,853; *Chem. Abstr.*, 1980, **92**, 146241a.
538. George J. Antos, U.S. Patent No. 4,216,346; *Chem. Abstr.*, 1980, **93**, 167599j.
539. Ernst Ingo Leupold *et al.*, Ger. Offen. Patent No. 2,846,148; *Chem. Abstr.*, 1980, **93**, 113949e.
540. F. Solymosi and A. Erdohelyi, *J. Mol. Catal.*, 1980, **8**(4), 471.
541. Emil Albin Broger, Eur. Patent Application No. 5,747; *Chem. Abstr.*, 1980, **93**, 114702f.
542. Marc Bonnet, Patrick Geneste, and Marcel Rodriguez, *J. Org. Chem.*, 1980, **45**(1), 40.
543. T. H. Vanderspurt, *Ann. N.Y. Acad. Sci.*, 1980, **333**, 155.
544. Axel Kleeman, Juergen Martens, and Horst Weigel, Ger. Offen. Patent No. 2,947,825; *Chem. Abstr.*, 1981, **95**, 98325u.
545. G. Leclercq, L. Leclercq, and R. Maurel, *Bull. Soc. Chim. Belg.*, 1979, **88**(7–8), 599.
546. Allen I. Feinstein, Ralph J. Bertolacini, Pae K. Kim, U.S. Patent No. 4,177,219; *Chem. Abstr.*, 1980, **92**, 94031k.
547. Mihaly Bartok and Jozsef Czombos, *J. Chem. Soc. Chem. Commun.*, **3**, 106 (1981).

548. Laurence Bagnell and Edward A. Jeffrey, *Aust. J. Chem.,* 1981, **34**(3), 697.
549. Babu Y. Rao, John T. Nolan, Jr., and John H. Estes, U.S. Patent No. 4,113,789; *Chem. Abstr.,* 1979, **90,** 41173k.
550. Roy T. Mitsche and George N. Pope, U.S. Patent No. 4,179,581; *Chem. Abstr.,* 1980, **92,** 198073s.
551. Helmut Fiege, Karlfried Wedemeyer, Kurt Bauer, Reiner Moelleken, Ger. Offen. Patent No. 2,824,407; *Chem. Abstr.,* 1980, **92,** 180833e.
552. L. G. Morozov and V. A. Druz, *Kinet. Katal.,* 1980, **21**(4), 1071.
553. V. V. Mahajani, *Indian Chem. Eng.,* 1979, **21**(3), 45.
554. Paul Whitfield Tamm, Ger. Offen. Patent No. 2,736,996; *Chem. Abstr.,* 1978, **89,** 149292s.
555. Serge R. Dolhyj and Louis J. Velenyi, U.S. Patent No. 4,219,687; *Chem. Abstr.,* 1980, **93,** 239007q.
556. Nobuo Takahashi, Yuichi Orikasa, and Tatsuaki Yashima, *J. Catal.,* 1979, **59**(1), 61.
557. Philippe Courty, Germain Martino, and Jean Francois Le Page, Ger. Offen. Patent No. 2,856,863; *Chem. Abstr.,* 1979, **91,** 140512x.
558. D. C. Grenoble, *J. Catal.,* 1978, **51**(2), 203.
559. Yun-Yang Haung, U.S. Patent No. 4,210,597; *Chem. Abstr.,* 1980, **93,** 167621k.
560. Friedrich A. Wunder, Hans Juergen Arpe, Ernst Ingo Leupold, and Hans Joachim Schmidt, Ger. Offen. Patent No. 2,814,427; *Chem. Abstr.,* 1980, **92,** 79362a.
561. Masaru Ichikawa and Yasuo Kido, *Jpn. Kokai Tokkyo Koho* Patent No. 79 41,293; *Chem. Abstr.,* 1979, **91,** 56355r.
562. Murray S. Cohen, Jaan G. Noltes, and Gerard Van Koten, Ger. Offen. Patent No. 2,836,898; *Chem. Abstr.,* 1979, **91,** 140427y.
563. J. Baskeyfield Lewis, John D. Bell, John Townend, and Simon D. A. Pollock, U.S. Patent No. 4,182,916; *Chem. Abstr.,* 1980, **92,** 180869w.
564. Jean Pierre Brunelle, Ger. Offen. Patent No. 2,745,456; *Chem. Abstr.,* 1978, **89,** 132225z.
565. Susan N. Anderson and Fred Basolo, *Inorg. Synth.,* 1963, **7,** 214.
566. S. M. Jörgensen, *Z. Anorg. Allgem. Chem.,* 1903, **34,** 85.
567. Steven E. Diamond, Andrew Szalkiewicz, and Frank Mares, *J. Am. Chem. Soc.,* 1979, **101**(2), 490.
568. Yasuo Sado and Kiyohiko Tajima, *Jpn. Kokai Tokkyo Koho* Patent No. 79 92,913; *Chem. Abstr.,* 1979, **91,** 210906e.
569. Heinz Erpenbach, Klaus Gehrmann, Hans Klaus Kuebbeler, and Klaus Schmitz, Ger. Offen. Patent No. 2,836,084; *Chem. Abstr.,* 1980, **93,** 45992m.
570. Richard V. Procelli, Vijay S. Bhise, and Arnold J. Shapiro, Ger. Offen. Patent No. 2,940,752; *Chem. Abstr.,* 1980, **93,** 71046a.
571. Tomiya Isshiki, Yasuhiko Kijima, and Yuh Miyauchi, Ger. Offen. Patent No. 2,847,241; *Chem. Abstr.,* 1979, **91,** 19920a.
572. Tatsuaki Yashima, Yuichi Orikasa, Nobuo Takahashi, and Nobuyoshi Hara, *J. Catal.,* 1979, **59**(1), 53.
573. Clarence D. Chang, U.S. Patent No. 4,161,489; *Chem. Abstr.,* 1979, **91,** 160238r.
574. Masaru Ichikawa and Koichi Shikakura, *Stud. Surf. Sci. Catal.,* 1981, **7**(Pt. B); *New Horiz. Catal.,* p. 925.
575. Takaya Mise, Pangbu Hong, and Hiroshi Yamazaki, *Chem. Lett.,* 1981, **7,** 993.
576. Tamotsu Imai, U.S. Patent No. 4,220,764; *Chem. Abstr.,* 1980, **93,** 239429d.
577. John M. Holovka and Edward Hurley, Jr., U.S. Patent No. 4,169,110; *Chem. Abstr.,* 1980, **92,** 75858u.
578. Miroslav Novotny and Lowell R. Anderson, U.S. Patent No. 4,223,001; *Chem. Abstr.,* 1980, **93,** 222609q.

579. Shigeto Suzuki, U.S. Patent No. 4,156,686; *Chem. Abstr.,* 1979, **91,** 107681d.
580. John G. Ekerdt and Alexis T. Bell, *J. Catal.,* 1979, **58**(2), 170.
581. N. M. Gupta, V. S. Kamble, K. Annaji Rao, and R. M. Iyer, *J. Catal.,* 1979, **60**(1), 57.
582. Werner O. Haag and Tracy J. Huang, U.S. Patent No. 4,157,338; *Chem. Abstr.,* 1979, **91,** 195752f.
583. R. C. Streeter and E. R. Tucci, *Hydrocarbon Process. Int. Ed.,* 1980, **59**(4), 107.
584. M. A. Vannice and R. L. Garten, *J. Catal.,* 1980, **63**(1), 255.
585. M. Albert Vannice and Samuel J. Tauster, U.S. Patent No. 4,171,320; *Chem. Abstr.,* 1980, **92,** 44581u.
586. Wayne R. Pretzer, Thaddeus P. Kobylinski, and John E. Bozik, U.S. Patent No. 4,133,966. *Chem. Abstr.,* 1979, **90,** 120998m.
587. P. G. J. Koopman, A. P. G. Kieboom, H. Van Bekkum, and J. W. E. Coenen, *Carbon,* 1979, **17**(5), 399.
588. Tetsumi Suzuki, Seiichi Hino, Satoru Igarashi, Mitsuru Tanaka, and Hiromu Kobayashi, *Jpn. Kokai Tokkyo Koho* Patent No. 79 40,897; *Chem. Abstr.,* 1979, **91,** 58116n.
589. Charles A. Drake and Timothy P. Murtha, U.S. Patent No. 4,215,019; *Chem. Abstr.,* 1980, **93,** 238817s.
590. Anatoli Onopchenko, Edward T. Sabourin, and Charles M. Selwitz, *J. Org. Chem.,* 1979, **44**(8), 1233.
591. S. Galvagno, J. Schwank, and G. Parravano, *J. Catal.,* 1980, **61**(1), 223.
592. Tamechika Yamamoto, *Chem. Econ. Eng. Rev.,* 1979, **11**(5), 14.
593. A. Ozaki, K. Urabe, K. Shimazaki, and S. Sumiya, *Stud. Surf. Sci. Catal. 3 Prep. Catal.,* 1979, **2,** 381.
594. Shigeru Ozaki, *Jpn. Kokai Tokkyo Koho* 79 119,386; *Chem. Abstr.,* 1980, **92,** 61203q.
595. R. C. Everson and D. T. Thompson, *Platinum Met. Rev.,* 1981, **25**(2), 50.
596. Tomoyuki Inui, Maski Funabiki, and Yoshinobu Takegami, *Ind. Eng. Chem. Prod. Res. Dev.,* 1980, **19**(3), 385.
597. Anders Nielsen, *Catal. Rev. Sci. Eng.,* 1981, **23**(1), 17.
598. Atsumu Ozaki, *Acc. Chem. Res.,* 1981, **14**(1), 16.
599. J. L. Dawes, J. D. Holmes, *Inorg. Nucl. Chem. Lett.,* 1971, **7,** 847.
600. Brian F. G. Johnson and Jack Lewis, *Inorg. Synth.,* 1972, **13,** 92.
601. G. Braca, G. Sbrana, and P. Pino, *Chem. Ind. Milan,* 1964, **46,** 206.
602. John S. Bradley, *J. Am. Chem. Soc.,* 1979, **101**(24), 7419.
603. B. Duane Dombek, *J. Am. Chem. Soc.,* 1980, **102**(22), 6855.
604. Michael J. Doyle, Arjan P. Kouwenhoven, Cornelis A. Schapp, and Bart Van Oort, *J. Organomet. Chem.,* 1979, **174**(3), C55.
605. R. B. King, A. D. King, Jr., and K. Tanaka, *J. Mol. Catal.,* 1980, **10**(1), 75.
606. John F. Knifton (1981). *J. Mol. Catal.,* 1981, **11**(1), 91.
607. John F. Knifton, *ACS Symp. Ser.,* 1981, **152;** *Catal. Act. Carbon Monoxide,* p. 225.
608. Yoshio Seki, Kenji Takeshita, Kazuaki Kawamoto, Shinji Murai, and Noboru Sonoda, *Angew. Chem. Int. Ed. Engl.,* 1980, **19**(11), 928.
609. R. A. Sanchez-Delgado, I. Duran, J. Monfort, and E. Rodriguez, *J. Mol. Catal.,* 1981, **11**(2–3), 193.
610. Howard Alper and Shiyamalie Amaratunga, *Tetrahedron Lett.,* 1980, **21**(27), 2603.
611. Richard M. Laine, *J. Am. Chem. Soc.,* 1978, **100**(20), 6451.
612. Duane B. Dombek, *Am. Chem. Soc. Symp. Ser. Catal. Act. Carbon. Monoxide,* 1981, **152,** 213.
613. Tomoji Kawai and Tadayoshi Sakata, *Nature London,* 1979, **282**(5736), 283.
614. John Kiwi, Enrico Borgarello, Ezio Pelizzetti, Mario Visca, and Michael Graetzel, *Angew. Chem.,* 1980, **92**(8), 663.

615. John Frederick Knifton, Ger. Offen. Patent No. 3,020,383; *Chem. Abstr.,* 1981, **95,** 80174j.
616. John F. Knifton, *J. Chem. Soc. Chem. Commun.,* 1981, **2,** 41.
617. Dennis Earl Jackman, Eur. Patent Application No. 11,207; *Chem. Abstr.,* 1980, **93,** 204077k.
618. L. E. Aneke, J. J. J. Den Ridder, P. J. Van den Berg, *Recl. Trav. Chim. Pays-Bas,* 1981, **100**(6), 236.
619. James C. Cannon, U.S. Patent No. 4,235,757; *Chem. Abstr.,* 1981, **94,** 103146x.
620. Hisao Kanoh, Takao Nishimura, and Akimi Ayame, *J. Catal.,* 1979, **57**(3), 372.
621. Francesco Pignataro, *Chim. Ind. Milan,* 1980, **62**(5), 429.
622. William W. Prichard, U.S. Patent No. 4,209,651; *Chem. Abstr.,* 1980, **93,** 185747k.
623. Hans Diem, Christian Dudeck, Gunter Lehmann, Guenther Matthias, and Nobert Petri, Ger. Offen. Patent No. 2,803,318; *Chem. Abstr.,* 1979, **91,** 140344u.
624. Wolfgang Sauer, Werner Fliege, Christian Dudeck, and Nobert Petri, Eur. Patent Application No. 4,881; *Chem. Abstr.,* 1980, **92,** 128572e.
625. Toshihiko Hashizume, Shigeo Kimura, and Koichi Kurata, *Jpn. Kokai Tokkyo Koho* Patent No.; *Chem. Abstr.,* 1977, **87,** 67853v.
626. R. W. Clayton and S. V. Norval, *Catalysis London,* 1980, **3,** 70.
627. Shuichi Kagawa, Masakazu Iwamoto, and Tetsuro Sieyama, *Chem. Tech,* 1981, **11**(7), 426.
628. D. R. Coulson, *Inorg. Synth.,* 1972, **13,** 121.
629. P. Roffia, G. Gregorio, F. Conti, G. F. Pregaglia, and R. Ugo, *J. Mol. Catal.,* 1977, **2**(3), 191.
630. Hideyuki Matsumoto *et al., J. Organomet. Chem.,* 1980, **199**(1), 43.
631. Yoichiro Nagai, Hamao Watanabe, and Michio Yoshibayashi, *Jpn. Kokai Tokkyo Koho* Patent No. 79 88,224; *Chem. Abstr.,* 1979, **91,** 175513a.
632. Hamao Watanabe, Mitsunobu Kobayashi, Kazuaki Higuchi, and Yoichiro Nagai, *J. Organomet. Chem.,* 1980, **186**(1), 51.
633. Toshikazu Hirao, Toshio Masunaga, Yoshiki Ohshiro, and Toshio Agawa, *Tetrahedron Lett.,* 1980, **21**(37), 3595.
634. Kenzie Nozaki, Eur. Patent Application No. 8,139; *Chem. Abstr.,* 1980, **93,** 70980p.
635. Noriaki Yoshimura and Masuhiko Tamura, Ger. Offen. Patent No. 3,034,098; *Chem. Abstr.,* 1981, **95,** 61442d.
636. Paul Binger, John McMeeking, and Ulf Schuchardt, *Chem. Ber.,* 1980, **113**(7), 2372.
637. James R. Dalton and Steve L. Regen, *J. Org. Chem.,* 1979, **44**(24), 4443.
638. Shun-Ichi Murahashi, Masaaki Yamamura, Kin-ichi Yanagisawa, Nobuaki Mita, and Kaoru Kondo, *J. Org. Chem.,* 1979, **44**(14), 2408.
639. Akio Minato, Kohei Tamao, Tamio Hayashi, Keizo Suzuki, and Makoto Kumada, *Tetrahedron Lett.,* 1980, **21**(9), 845.
640. N. Miyaura, T. Yanagi, and A. Suzuki, *Synth. Commun.,* 1981, **11**(7), 513.
641. Barry M. Trost, Thomas R. Verhoevan, and Joseph M. Fortunak, *Tetrahedron Lett.,* 1979, **20**(25), 2301.
642. Yoshinao Tamaru, Youichi Yamamoto, Yoshimi Yamada, and Zen-ichi Yoshida, *Tetrahedron Lett.,* 1979, **20**(16), 1401.
643. Paul Helquist, *Tetrahedron Lett.,* 1978, **22,** 1913.
644. Francois Guibe, Pierre Four, and Henriette Riviere, *J. Chem. Soc. Chem. Commun.* **10,** 432.
645. Roberts O. Hutchins, Keith Learn, and Robert P. Fulton, *Tetrahedron Lett.,* 1980, **21**(1), 27.
646. Tadakatsu Mandai, Susumu Hashio, Jiso Goto, and Mikio Kawada, *Tetrahedron Lett.,* 1981, **22**(23), 2187.

647. Y. Tamaru, R. Suzuki, M. Kagotani, and Z. Yoshida, *Tetrahedron Lett.*, 1980, **21**(39), 3787.
648. P. E. Cattermole and A. G. Osborne, *Inorg. Synth.*, 1977, **17**, 115.
649. S. Martinengo, P. Chini, and G. Giordano, *J. Organomet. Chem.*, 1971, **27**, 389.
649a. S. Martinengo, G. Giordano, and P. Chini, *Inorg. Synth.*, 1980, **20**, 209.
650. G. K. Magomedov and O. V. Shkol'nik, *Zh. Obshch. Khim.*, 1980, **50**(5), 1103.
651. Pangbu Hong, Bo-Re Cho, and Hiroshi Yamazaki, *Chem. Lett.*, 1979, **4**, 339.
652. Pangbu Hong and Hiroshi Yamazaki, *Chem. Lett.*, 1979, **11**, 1335.
653. A. Lapidus, M. M. Savel'ev, L. T. Kondrat'ev, and E. V. Yastrebova, *Izv. Akad. Nauk SSSR Ser. Khim.*, 1981, **7**, 1564.
654. Masaru Ichikawa, *J. Catal.*, 1979, **59**(1), 67.
655. R. B. King, A. D. King, Jr., M. Z. Iqbal, and K. Tanaka, *Ann. N.Y. Acad. Sci.*, 1980, **333**, 74.
656. Mark K. Dickson, B. P. Sudha, and D. Max Roundhill, *J. Organomet. Chem.*, 1980, **190**(2), C43.
657. Tadashi Sato, Hirokazu Kaneko, and Shinichi Yamaguchi, *J. Org. Chem.*, 1980, **45**(19), 3778.
658. Masahito Segi, Tadashi Nakajima, and Sohei Suga, *Bull. Chem. Soc. Jpn.*, 1980, **58**(5), 1465.
659. Lowell D. Markley, U.S. Patent No. 4,188,346; *Chem. Abstr.*, 1980, **92**, 198079y.
660. Fumie Sato, Yoshikuni Mori, and Masao Sato, *Tetrahedron Lett.*, 1979, **20**(16), 1405.
661. K. Kijenski and W. Skupinski, *React. Kinet. Catal. Lett.*, 1980, **14**(2), 247.
662. Akihiro Sato, Masami Tachibana, and Kazutsune Kikuta, Ger. Offen. Patent No. 2,918,089; *Chem. Abstr.*, 1980, **92**, 181899t.
663. Gerhard Staiger, Ger. Offen. Patent No. 2,831,829; *Chem. Abstr.*, 1980, **92**, 164524z.
664. Akio Itoh, Nobuyuki Kuroda, Kazuo Matsuura, Mitsuji Miyoshi, and Takeichi Shiraishi, U.S. Patent No. 4,209,601; *Chem. Abstr.*, 1980, **93**, 205349n.
665. Nicholas Marios Karayannis and John Stephen Skryantz, Eur. Patent Application No. 15,645; *Chem. Abstr.*, 1980, **93**, 221327j.
666. Nobuyuki Kuroda, Kazuo Matsuura, Mitsuji Miyoshi, and Takeichi Shiraishi, U.S. Patent No. 4,202,953; *Chem. Abstr.*, 1980, **93**, 150914k.
667. K. M. Abd El-Salaam and E. A. Hassan, *Surf. Tech.*, 1979, **9**(3), 195.
668. Kh. M. Minachev, G. V. Antoshin, D. Klisurski, N. K. Guin, and N. Ts. Abadzhieva, *React. Kinet. Catal. Lett.*, 1979, **10**(2), 163.
669. Jean Michel Tatibouet and Jean Eugene Germain, *C. R. Hebd. Seances Acad. Sci. Ser. C,* 1979, **289**(12), 305.
670. C. N. Kenney, *Catal. London*, 1980, **3**, 123.
671. Reginald Gregory, Ger. Offen. Patent No. 2,626,424; *Chem. Abstr.*, 1977, **87**, 41649h.
672. D. A. Kondrat'ev, A. A. Dergachev, T. N. Bondarenko, and Kh. M. Minachev, *Izv. Akad. Nauk SSSR Ser. Khim.*, 1980, **6**, 1318.
673. Nobert F. Cywinski, U.S. Patent No. 4,181,809; *Chem. Abstr.*, 1980, **92**, 197927m.
674. D. M. Kavishwar, Y. C. Bhattacharyulu, and K. A. Venkatachalam, *Indian J. Technol.*, 1980, **18**(6), 232.
675. Clarence Dayton Chang and William Harry Lang, Ger. Offen. Patent No. 2,912,067; *Chem. Abstr.*, 1980, **92**, 149901b.
676. Martin Post, Maria Franciscus, and Lambert Schaper, Ger. Offen. Patent No. 2,947,931; *Chem. Abstr.*, 1980, **93**, 204243m.
677. Lambert Schaper and Swan Tiong Sie, Ger. Offen. Patent No. 2,846,254; *Chem. Abstr.*, 1979, **91**, 60023y.
678. Alvin B. Stiles, U.S. Patent No. 4,111,847; *Chem. Abstr.*, 1979, **90**, 93156b.
679. H. Boerma, *Prep. Catal. Proc. Int. Symp.*, 1976, **105**.

680. Dev. D. Suresh, Noel J. Bremer, and Robert K. Grasselli, Ger. Offen. Patent No. 2,634,606; *Chem. Abstr.,* 1977, **86,** 122009f.
681. Ube Industries, Ltd., *Fr. Demande* Patent No. 2,407,906; *Chem. Abstr.,* 1979, **91,** 174825s.
682. Helmut Scharf, Ger. Offen. Patent No. 2,802,672; *Chem. Abstr.,* 1979, **91,** 157227x.
683. T. Van Herwijnen and W. A. DeLong, *J. Catal.,* 1980, **63**(1), 83.
684. C. S. John, *Catal. London,* 1980, **3,** 169.
685. Hermann Irngartinger, Annette Goldmann, Raymond Schappert, Philip Garner, and Paul Dowd, *J. Chem. Soc. Chem. Commun.,* 1981, **10,** 455.
686. John E. Dabrowski, U.S. Patent No. 4,225,732; *Chem. Abstr.,* 1981, **94,** 15392u.
687. Timothy P. Murtha, U.S. Patent No. 4,152,362; *Chem. Abstr.,* 1979, **91,** 91323a.
688. Lynn H. Slaugh, U.S. Patent No. 4,229,320; *Chem. Abstr.,* 1981, **94,** 102986j.
689. Masato Tanaka, *Synthesis,* 1981, **1,** 47.
690. Erlind M. Thosteinson, Bernard Duane Dombek, and Rocco Anthony Fiato, Ger. Offen. Patent No. 3,043,112; *Chem. Abstr.,* 1981, **95,** 80202s.
691. Linda S. Benner, Patrick Perkins, and K. Peter C. Vollhardt, *ACS Symp. Ser.,* 1981, **152;** Catal. Act. Carbon Monoxide, p. 165.
692. Ernst R. F. Gesing, Jim A. Sinclair, and K. Peter C. Vollhardt, *J. Chem. Soc. Chem. Commun.,* 1980, **6,** 286.
693. Leon Hagelee, Robert West, Joseph Calabrese, and John Norman, *J. Am. Chem. Soc.,* 1979, **101**(17), 4888.
694. Kenzie Nozaki, U.S. Patent No. 4,180,694; *Chem. Abstr.,* 1980, **92,** 163552v.
695. Michele Pralus, Jean P. Schirmann, and Serge Y. Delavarenne, Ger. Offen. Patent No. 2,605,041; *Chem. Abstr.,* 1979, **86,** 16526n.
696. J. R. Blickensderfer, R. J. Hoxmeier, and H. D. Kaesz, *Inorg. Chem.,* 1979, **18**(12), 3606.
697. Leonard Kaplan and Roy L. Pruett, U.S. Patent No. 4,199,521; *Chem. Abstr.,* 1980, **93,** 132071d.
698. Jóse L. Vidal and W. E. Walker, *Inorg. Chem.,* 1980, **19**(4), 896.
699. Anna M. Lennertz, Jüergen Laege, Manfred J. Mirbach, and Alfons Saus, *J. Organomet. Chem.,* 1979, **171**(2), 203.
700. Isao Koga, Yohji Terui, Masuhito Ohgushi, Tohru Kitahara, and Kenichi Watanabe, U.S. Patent No. 4,292,433; *Chem. Abstr.,* 1981, **95,** 204149b.
701. Andrew Cornish, Michael F. Lappert, Grace L. Filatovs, and Terrence A. Nile, *J. Organomet. Chem.,* 1979, **172**(2), 153.
702. Yu. S. Levkovskii, N. M. Ryutina, and F. K. Shmidt, *Kinet. Katal.,* 1980, **21**(3), 797.
703. Derek H. R. Barton, Stephen G. Davies, and William B. Motherwell, *Synthesis,* 1979, **4,** 265.
704. E. N. Frankel, R. A. Awl, and J. P. Friedrich, *J. Am. Oil Chem. Soc.,* 1979, **56**(12), 965.
705. Kazushi Arata and Kozo Tanabe, *Bull. Chem. Soc. Jpn.,* 1980, **53**(2), 299.
706. G. A. El-Shobaky, M. M. Selim, and I. F. Hewaidy, *Surf. Technol.,* 1980, **10**(1), 55.
707. P. S. Radhakrishnamurti and P. C. Misra, *Indian J. Chem. Sec. A,* 1980, **19**A(5), 427.
708. George A. Moczygemba and Henry L. Hsieh, U.S. Patent No. 4,223,116; *Chem. Abstr.,* 1980, **93,** 205368t.
709. Gilbert Balavoine and Francois Guibe, *Tetrahedron Lett.,* 1979, **41,** 3949.
710. U. M. Dzhemilev, F. A. Selimov, and G. A. Tolstikov, *Izv. Akad. Nauk SSSR, Ser. Khim.,* 1980, **2,** 348.
711. J. Beger and F. Meier, *J. Prakt. Chem.,* 1980, **322**(1), 69.
712. Wilhelm Keim and Michael Röper, *J. Org. Chem.,* 1981, **46**(18), 3702.
713. Wolfgang Sucrow and Fritz Lübbe, *Angew. Chem. Int. Ed. Engl.,* 1979, **18**(2), 149.

CHAPTER 3

Hydrogenations—General and Selective

A. B. STILES

University of Delaware
Center for Catalytic Science and Technology
Colburn Laboratory
Newark, Delaware

I. Introduction

A. GAS PURITY

A primary requirement for a hydrogenation reaction is consideration of the gases and the environment involved. It may seem elementary, but it is essential that the hydrogen be pure or that it contain only impurities that can be tolerated by the reaction involved. Inasmuch as pure hydrogen is nonexistent, the consideration becomes one of minimizing the impurity level. What was considered a pure gas 25–30 yr ago would now be considered an unacceptably impure gas. Improvements in purification and analytical procedures have made it possible to reduce impurity levels by more than 90% and have established entirely different criteria concerning the purity of a gas. Methods of purification will be touched on subsequently in this chapter.

In considering pure hydrogen, we shall first discuss hydrogen used in either laboratory- or bench-scale applications, which is typically cylinder or bottled hydrogen. A grave error, which is frequently committed, is to assume that cylinder hydrogen is pure. This should not be considered axiomatic, nor should any cylinder gas (carbon monoxide, nitrogen, or any of the other less frequently used gases) be considered pure. Therefore, it is essential for reproducible data that all cylinder gases be purified by the person using the hydrogen as part of an experimental setup.

The most frequently encountered impurity in cylinder hydrogen is a small quantity of oxygen, but halides, such as chlorine, and sulfur compounds are almost as frequently encountered. Hydrogen derived from natural gas or light hydrocarbons can also contain halide, sulfur, alkali, carbon dioxide, nitrogen, and even acetylene and ethylene.

Hydrogen derived from coal, lignite, and heavy crudes or residual oil can contain sulfur, arsenic, antimony, selenium, mercury, vanadium, nickel, and essentially anything else present in coal ash. Hydrogen can also contain lubricating oil from compressors or pipe joints or from contaminated tubing used in assemblying equipment or apparatus.

Frequently hydrogen contains water vapor and, in fact, some purification procedures convert oxygen impurities in hydrogen to water vapor. In some hydrogenations water vapor is a poison, whereas in others it is an activator. Usually water vapor is a poison, but it is frequently temporary; its deactivating effect may be overcome by raising the temperature of the reaction involved.

During the generation of hydrogen, high temperatures ($700-1400°C$) frequently are encountered, and at these high temperatures the small amount of nitrogen generally present in natural gas used in deriving the hydrogen is converted in small quantities to ammonia. These trace quantities of ammonia can be extremely harmful to certain hydrogenation catalysts.

Hydrogen can also contain acidic constituents such as carbon dioxide, formic acid, COS, SO_2, and H_2S, and these can be harmful to the reaction. It can be said parenthetically at this time that frequently the material being hydrogenated (coconut oil, for example) can become rancid, producing organic acids; a source of even more harmful mineral acids is the occasional need for such a product to be acid-washed in its processing, with the result that a small amount of acid is retained in the material being hydrogenated. This can be fatal to a hydrogenation reaction and may be difficult to detect because chemical analysis may indicate that there is no acidity in the material being processed.

B. TEMPERATURE CONTROL

A further consideration of major importance in hydrogenations is control of temperature. Essentially, all hydrogenations are extremely exothermic and, as a consequence, heat removal is necessary to avoid decomposition or overhydrogenation of the material being hydrogenated. For example, in the hydrogenation of nitro compounds the reaction is very exothermic and the temperature must be controlled in a comparatively narrow range, so extreme precautions must be taken to employ a heat sink or to provide unusually good heat-exchange conditions. This factor will be dealt with in more detail subsequently in this chapter.

C. VAPOR PHASE HYDROGENATIONS

In the preceding paragraph, the problem of temperature control was emphasized, and this leads naturally to the next vital factor relating to the conditions under which hydrogenations are conducted — whether in the vapor phase or in the liquid phase. The vast majority of industrial hydrogenations are conducted in the vapor phase. This intensifies the heat removal problem. Reactions ordinarily conducted in the vapor phase are hydrogenation of carbon monoxide to various oxygenated products, nitrogen to ammonia, carbon monoxide to methane, acetylene selectively to ethylene, and butadiene selectively to butene.

Temperature control in the first-mentioned reaction, hydrogenation of carbon monoxide to various oxygenated products, is effected by various means. The primary method is to limit the quantity of carbon monoxide in the stream so that the reaction will be limited by the quantity of reactants available, the hydrogen in this case being the heat sink for the heat of reaction. In the case of an ammonia reaction, the equilibrium is less favored at higher temperatures and, as a consequence, the reaction becomes self-extinguishing as the temperature rises. Although the methanol reaction is also less favored at high temperatures, a competing reaction, such as methane formation, then begins to produce a major reaction product. As a consequence of the exothermic methanation reaction, the temperature rise becomes even more rapid than when methanol is the product. The competing methanation reaction must be absolutely avoided.

In the purification steps mentioned, the amount of acetylene or butadiene is usually comparatively small, so that the heat of reaction is easily absorbed by the total quantity of gases being processed.

D. LIQUID PHASE HYDROGENATIONS

Temperature control in liquid phase reactions is ordinarily effected either by internal cooling coils within the reactor or by means of recycled product also acting as a heat sink. Liquid phase reactions, such as the hydrogenation of benzene to cyclohexane, phenol to cyclohexanol, and nitrobenzene to aniline, are all extremely exothermic; and the temperature is ordinarily controlled by feeding a diluent nonreactant along with a reactant. In most cases the diluent is a recycled product itself, such as cyclohexane, aniline, or cyclohexanol. A completely inert material can also be added to the incoming liquid streams, with cyclohexane, hexane, or water being occasionally utilized. More frequently a liquid component that absorbs heat by vaporization is used. Compounds that fall in this category are anhydrous ammonia, dimethyl amine, dimethyl ether, and butane or hexane.

An additional way of controlling temperature rise and heat dissipation is to have a large quantity of an inert gas in the hydrogen stream. This gas can be methane or nitrogen, but one of the factors that should be kept clearly in mind is that neither of these gases can be considered absolutely inert under all circumstances. Nitrogen, as a matter of fact, is very reactive with nickel and certain other hydrogenation catalysts. It can be absorbed to deactivate the catalyst or it can be reacted to form ammonia, which in some cases may react even further to form amines in the hydrogenated product.

II.   Catalyst Types

A.   THE IRON GROUP

When one speaks of types of catalysts being used for a given hydrogenation, one usually describes a catalyst by simply saying that it is a nickel catalyst or a precious metal catalyst; it can also sometimes be said that it is an iron group catalyst. These terms are very nondescriptive and usually are as misleading as they are informative. An iron group catalyst, for example, can be nickel, iron, or cobalt, and it can be in any of several different forms. Usually such catalysts are supported, which means that they are precipitated or impregnated onto a support, such as diatomaceous earth (kieselguhr), powdered silica, powdered activated carbon, magnesium oxide, a rare earth element, alumina, or a molecular sieve. (There are many types of alumina, each having its own good or adverse effect on the resultant catalyst.) It is not the intent of this chapter to describe catalyst preparation, which is very complex, but only to indicate that, in describing a catalyst as a nickel catalyst, one is not really defining it at all; even identifying the support does not indicate what the performance of the catalyst will be. A catalyst is greatly affected by the method of preparation as well as by the type of support and promoters either intentionally or unintentionally present under the precipitation conditions. The method of reducing and/or stabilizing a catalyst can also have a profound effect on catalyst performance, both from an activity and a selectivity viewpoint.

B.   PRECIOUS METALS

Precious metals are in a class by themselves, and the two most frequently used are platinum and palladium, with rhodium following closely behind.

Rhodium finds a major application in the hydrogenation of carbon monoxide to selected monobasic, dibasic, and tribasic alcohols. Precious metals are frequently supported on activated carbon, utilizing specific preparative conditions and also carefully defining the type of activated carbon and physical characteristics. Activated carbon is one of the most difficult supports to define, and such a definition usually involves identifying the carbon source, such as lignite or coconut char, and the preparative or activating conditions. Carbon can be activated by a number of procedures, all of which introduce different characteristics into the carbon and different characteristics into the catalyst when the carbon is used as the support.

C. COPPER CHROMITE

Copper chromite is a frequently used catalyst in both vapor and liquid phase hydrogenations. Copper chromite can be made from either hexavalent or trivalent chromium, each producing individual characteristics in the finished catalyst. These chromites are ordinarily used for ester hydrogenation, and they are very selective in avoiding overhydrogenation of the esters to hydrocarbons; they are also effective in hydrogenating unsaturated esters (oleic acid esters) to unsaturated alcohols, thus preserving the double bond(s).

D. METAL SULFIDES

Metal sulfides are sometimes used for hydrogenations, molybdenum sulfide being one frequently applied and mentioned in the literature. Molybdenum sulfide, coupled with nickel or cobalt (also as a sulfide) on alumina, is used for hydrogenations that involve hydrodenitrogenation and hydrodesulfurization. This work has been much studied and widely reported. Existing catalysts of this type have been quite successfully utilized with the types of nitrogen and sulfur compounds ordinarily found in petroleum-derived hydrocarbons. However, with the impending advent of the use of shale oil, oil from tar sands, heavy crudes, and coal-derived liquids that contain substantially more aromatic and refractory nitrogen and sulfur compounds, the problem is becoming more severe and these catalysts will probably undergo major developmental changes to obtain catalysts satisfactory for these fuel sources.

E. ORGANOMETALLIC COMPLEXES

Organometallic complexes (clusters) such as those identified as substituted triphenylphosphines also are hydrogenation catalysts and are useful for a small number of specialized reactions such as those involving pharmaceuticals and specialty processing. It is obvious that because of their organic moiety they must be used at comparatively low temperatures, but they do have amazing selectivity resembling that of enzymes.

F. ACTIVATED ALLOY CATALYSTS

Catalysts called alloy or Raney catalysts are derived by first making an aluminum or magnesium alloy of the metal in question, such as nickel, cobalt, or copper, and then dissolving away a part of the alloy, such as the aluminum or magnesium component, with the result that highly active, almost atomically dispersed metals result. So-called Raney nickel is an extremely reactive but nondirective catalyst. It also is very susceptible to thermal or chemical poisoning and deactivation. This, however, does not prevent its extensive application in oil-hardening processes in which its settling characteristics are particularly desirable. Another attractive characteristic of the alloy-type catalyst is that it can be activated at low temperatures in simple facilities and without the need for a separate reducer and reduction step. A separate reduction and stabilization procedure thus is avoided, with substantial savings in some operations.

G. PREPARATIVE CONDITIONS

The preparative conditions for essentially all catalysts, particularly those used for hydrogenation, are usually well-guarded trade secrets. As previously stated, the chemical composition of a catalyst is used very frequently as a method for identifying a given catalyst. Thus, one can say that a given reaction employs a nickel catalyst, but unless much more detail, usually involving preparative conditions, is disclosed, the type of catalyst actually employed has not been described. This applies also to Raney and alloy-type catalysts.

There are many ways of activating an alloy catalyst, some being much more effective than others for a given reaction and some in particular leaving a substantial residue of undissolved aluminum or aluminum oxide and alkali subsurface on the exposed metal. This aluminum and alkali can have a very marked effect on the performance of the catalyst from the viewpoints of both activity and selectivity. Storage of the activated catalyst also presents a problem in that the nickel is so extremely reactive that many liquids (water and methanol) used as vehicles for immersing the catalyst can react with it with an adverse effect.

H. METAL—SUPPORT INTERACTION

It was mentioned previously that a catalyst can be used with various types of supports and that the support itself can have a major effect on the performance of the catalyst. The literature is becoming increasingly devoted to this substantial effect. This effect can frequently be attributed to a solid state reaction, such as nickel oxide reacting with aluminum oxide to form nickel aluminate which is a spinel having characteristics entirely different from those of elemental nickel supported on alumina. There was also a tendency in the earlier literature to identify alumina only as such, whereas there are many species of alumina, such as alpha, gamma, and eta, to name only a few. The crystalline species and chemical reactivity of these various aluminas strongly affect pseudomorph, epitaxial, and solid state reactions with the catalytic entity.

This species variation introduces a different concept of the influence of the support on the catalytic entity. This influence has in recent years been referred to as an epitaxial effect in which the spacing and crystal species of the support are mimicked by the catalyst entity, with the result that the spacing and crystal species of the catalytic entity are different from those normally observed on crystallization and have spacing characteristics of the catalytic entity itself. It is obvious that this has a substantial effect on the catalyst composite, with the result that its performance is entirely different when the catalyst is made in such a way as to promote epitaxial deposition of the catalytic component.

III. Reactor Types

A. TANK TYPE

The simplest type of reactor is the tanklike reactor that resembles a large, totally enclosed cylindrical vessel in which the catalyst is present as a single catalyst bed resting on a bottom support grid (Fig. 1). The problems with this reactor usually relate to distribution of the gases and the occurrence of extremes in temperature in the catalyst bed, with decomposition or no reaction taking place at these overheated or underheated spots. An improvement in this simple type of reactor can be achieved by the installation of baffle plates to effect better distribution of the gases, but the pressure drop is thus increased and incomplete reactions at the cold walls still are not avoided.

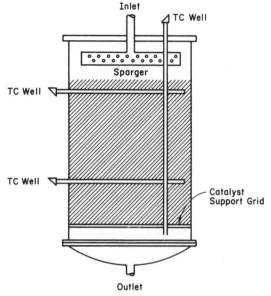

Fig. 1. A simple catalytic reactor.

B. MULTITRAY TYPE

The next and much more sophisticated type of reactor is the multitray type currently finding substantial use in many forms and modifications (Fig. 2). The multitray reactor permits better distribution, better cooling between trays, the introduction of cold or hot gas between trays, the use of different types of catalyst on different trays, selected catalyst changes on certain trays where catalysts have become deactivated, and the operation of different trays at different temperatures. The latter can be practiced in cases in which the equilibrium condition is favored by low temperatures. In this type of reactor, the bottom tray is maintained at a low temperature where the reaction is most favored by equilibrium or selectivity considerations.

C. MULTITUBULAR

Another type of reactor frequently used, but an expensive one both in which to charge and discharge the catalyst and to maintain, is the multitubular reactor (Fig. 3). These reactors are used in essentially any reaction that is extremely exothermic or endothermic, and inasmuch as hydrogenations are usually exothermic we will consider the benefit of this reactor in an

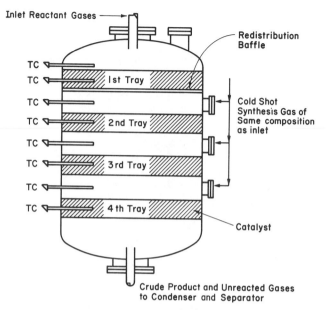

Fig. 2. A tray-type synthesis converter.

exothermic reaction. A tubular reactor can be designed in such a way that incoming gases pass through and are heated in tubes in the reaction zone. As they are heated, the incoming cold gases remove heat generated in the reaction zone, resulting in a near isothermal reaction. Figure 3 shows this diagramatically. A tubular reactor can also be designed in such a way that a fluid such as a boiling liquid can be on the cooling side of the tubes, thus effecting excellent temperature control and heat dissipation. The catalyst is on the "fire side" of the boiler tubes.

D. SLURRY TYPE

A frequently used type of reactor is the slurry system (Fig. 4). It can be operated either on a continuous basis or batchwise, with a slurry feed being added to the reactor either at the top or bottom and the product being drawn off at the opposite end. Gas flow is generally continuous and is designed in such a way as to usually move cocurrently with the liquid.

As mentioned previously for gas phase reactors, it is essential that the gas be distributed equally throughout the reactor. This is equally true for a slurry reactor and is necessary to prevent erratic temperature conditions conducive to product decomposition, on the one hand, or cold spots where the reaction is incomplete, on the other.

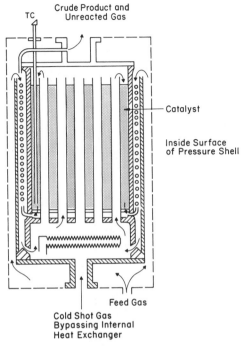

Fig. 3.   A tubular reactor.

E.   H$_2$ DEPLETION OR STARVATION

Pockets of gas from which the H$_2$ has been depleted can cause hydrogen starvation with localized product decomposition or dehydrogenation. In the slurry type of operation another transfer must take place: from the gas to the liquid and from the liquid to the catalyst. It is essential that the H$_2$ be readily available to the liquid at all points in the converter.

As a matter of fact, the reaction environment may be desirably modified by the incorporation or substitution of a liquid that more readily dissolves large volumes of hydrogen than does the liquid normally present in the reaction medium. The literature provides information on the solubility of hydrogen in various liquids, and these data indicate what liquids are the most suitable for the reaction under study.

F.   SAFETY

There are several characteristics of hydrogenations that make these reactions particularly in need of careful safety evaluation. The first consideration

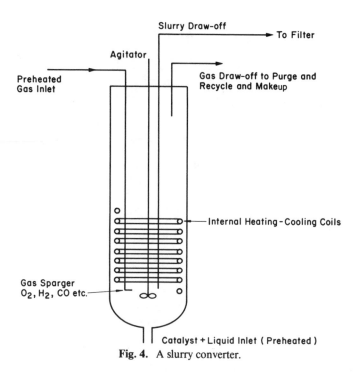

Fig. 4. A slurry converter.

is, of course, the hydrogen itself, which has an explosive range of 4% hydrogen in air to 4% oxygen in hydrogen. This is the widest range of all combustible gases, and the problem is made more severe by the fact that hydrogen has a low kindling point and is also easily ignited by surfaces that would otherwise not be considered catalytic. Iron rust, for example, is a good catalyst for the oxidation of hydrogen and the catalytic initiation of this reaction.

Hydrogen under pressure also diffuses readily through metals and even more rapidly through plastic tubing. Because hydrogen diffuses so rapidly through metals, it is utilized in a commercial process for the purification of hydrogen by passing impure hydrogen in contact with a silver–palladium alloy membrane that permits the hydrogen to pass through but does not permit the impurities to pass.

A compensating characteristic of hydrogen, however, is that it diffuses rapidly, is low-density, and as a consequence rises rapidly to the point of escape. It can, however, be entrapped in an enclosed, confined space, and this factor should be very carefully considered in the design of equipment or housing for equipment. A factor in the use of hydrogen that is extremely important is that it tends to cause embrittlement of carbon-bearing steel at temperatures above about 300°C. The hydrogen reacts with the carbon or carbide in the steel, causing removal of the carbon by the formation of methane. The weakened steel piping or reactor walls are then rendered brittle and fracture under relatively mild pressure conditions.

There are many metallurgical schemes for averting this problem and also for detecting it. These methods are described in the literature and should be referred to for future consideration.

G. CATALYST HANDLING

The last, but very significant, safety factor to be considered in catalytic hydrogenations is the handling of the catalyst, which itself causes hydrogen to react with atmospheric oxygen at room temperature. Even so-called stabilized catalysts can ignite a hydrogen pocket in a reactor, with severe consequences. The most acceptable method of handling catalysts is usually as a catalyst slurry which is always kept wet and is never exposed as a dry powder to the atmosphere.

H. HANDLING USED CATALYST—DISPOSAL

Used hydrogenation catalysts frequently are extremely pyrophoric (ignite spontaneously) and, when combined with an organic liquid, spontaneously oxidize and ignite not only themselves but also the organic liquid. The problem can be severe where even small quantities of used catalyst are discarded. Because most metals now have a high scrap value, this facet of the problem is avoided. Frequently, however, when catalysts are removed from process streams by filtration, they must be handled by personnel who are then exposed to hot organic wet catalysts, which can present a severe handling hazard. Proper protective equipment and clothing are mandatory.

An added problem with copper chromite catalysts is that not only do they ignite spontaneously but they also oxidize and liberate hexavalent chromium which is water-soluble and is extremely toxic in all sewage disposal systems. Hexavalent chromium causes immediate deactivation of the bacterial agents even when chromates are present only to the extent of a fractional part per million.

IV. Hydrogenations—General

The simplest hydrogenation, at least by reputation, is hydrogenation of the double bond, and in this group the conversion of ethylene to ethane is the simplest.

$$C_2H_4 + H_2 \rightleftharpoons C_2H_6$$

However, one should be immediately dismiss the idea that the hydrogenation of ethylene to ethane is as a matter of fact a simple and straightforward reaction. It can easily be conducted at temperatures as low as $-100°C$, with many types of platinum or palladium catalysts and, of course, at extremely high temperatures where the the reaction is very fast; but if the temperature exceeds approximately 700°C and the pressure is atmospheric, then dehydrogenation also becomes a factor. The thermal dehydrogenation of ethane is a commercial route to ethylene and is treated in Chapter 4.

It is well to point out emphatically that the reaction is not a simple one. That this so-called simple hydrogenation is not really simple should be thought of as indicative of how extremely complex all hydrogenations are. A classic experiment that has been repeated and recorded on many occasions, one as recently as 1980 in the Russian literature, is to deuterate a nickel catalyst and then pass ethylene over the catalyst and determine the products formed from this hydrogenation. The ethylene is nondeuterated when it reaches the catalyst. When it is adsorbed on the catalyst, reacts, and is then desorbed from the catalyst, all forms of ethane are observed. There is ethane with six normal hydrogen atoms and ethane with six deuterium atoms, and all combinations possible in this range. This shows that, when ethylene is adsorbed at a deuterated site, hydrogen passes from the ethylene onto the surface of the catalyst and deuterium passes into the ethylene, indicating that the adsorption process is much more complex than had ever been supposed before these classic experiments were made. The only purpose of reviewing these studies is to show the complexity of what we consider a very simple hydrogenation and how complex the more intricate hydrogenations must be by comparison.

A. HYDROGENATION OF BENZENE TO CYCLOHEXANE

The hydrogenation of benzene to cyclohexane is practiced commercially to derive cyclohexane for various chemical uses. Commercial operations employ a nickel catalyst, fixed bed Raney nickel being particularly suitable. The pressure can be from 100 to 1000 psi, an average of 700 to 800 psi being typical. The temperature is $200 \pm 30°C$, and the liquid space velocity is approximately 3. The feed composition is approximately 20% benzene and 80% cyclohexane, the heat being dissipated both by the boil-off of cyclohexane and by the specific heat of the liquids. The cyclohexane that boils off is then recycled in part as a component of the feed stream, and the remainder is withdrawn as the product. The reaction can be conducted both in a fixed bed tray-type system (Fig. 2) or as a slurry-type operation (Fig. 4), both of which have previously been described.

B. HARDENING EDIBLE AND NONEDIBLE OILS

One of the major applications of hydrogenation is in the hardening of both industrial and edible oils. This operation is conducted very widely in both large and small plants and in unsophisticated plants. Because most of these operations are in unsophisticated plants, it is imperative that the catalyst be in a form that will not creat industrial hazards. As a consequence, the nickel catalyst most frequently used in this operation is provided either as an oil suspension or as a flake or a block in the hardened oil. In any of these forms, the possibility of spontaneous combustion is essentially eliminated. Oil is hardened by saturating a certain number of double bonds in cottonseed oil, soybean oil, or another suitable vegetable oil. The reactor can be of the type in Fig. 2 or Fig. 4.

Problems sometimes encountered in this reaction are rancidity of the oil, the acid of which attacks the catalyst and deactivates it; overhydrogenation which makes a wax instead of a spreadable product from the oil; the typical handling operations involved in disposing of spent catalyst; and the already mentioned combustibility problems. A problem that cropped up about 10 yr ago was the suspicion or accusation that the nickel remaining in hardened edible oil caused cancer. Conditions for hydrogenation vary widely, with the following covering the range.

Catalyst	Activated Raney nickel; 30–50% reduced Ni on kieselguhr (diatomaceous earth)
Quantity	5% with 0.1% makeup
Pressure	10–500 psi
Agitation	gas or mechanical
Time	2-hr turnaround
Temperature	80–150°C

D. HYDROGENATION OF NITRO GROUPS TO AMINES

This section considers the hydrogenation of nitro groups to amines, for example, the hydrogenation of nitrobenzene to aniline. This reaction is extremely exothermic, with the consequences that the heat must be efficiently removed and the temperature must remain in a specified, carefully controlled range. Nitrohydrogenations can be conducted at temperatures as low as 75°C to as high as 200°C, but at higher temperatures there may be a problem involving deamination and/or polymerization caused by deamination. The catalyst most frequently used is nickel on kieselguhr, but platinum or palladium on carbon moderated with certain other components can also be used.

The type of converter used can be the slurry type or the multitray type already referred to, and the reaction is ordinarily conducted in the liquid phase with a heavy recycle of aniline to act as a heat sink. Although this reaction is not in our classification of selective hydrogenations, it must be conducted in such a way that the nitro group is hydrogenated without hydrogenating the ring. Hydrogenation of the ring, of course, is relatively easy and extremely exothermic. To prevent ring hydrogenation elevated temperatures must be avoided, particularly in the presence of an active nickel catalyst. It is imperative that the temperature be maintained at a relatively low, preselected level and that the agitation or catalyst and gas distribution in the converter be adequate to prevent the formation of pockets or areas in which excessive temperatures can be reached. Reactors of the types illustrated in Figs 2 and 4 can be used.

As previously stated, nitrohydrogenation is facile, and the nitrobenzene is usually poison-free; consequently the catalyst can be used over and over again. As a result, catalyst consumption in this reaction is extremely small.

However, when one is using a precious metal catalyst, it is obvious that it is mandatory that consumption be small. Small amounts of sulfur, halide, and other well-known poisons must not be present in the nitrobenzene. This, of course, means that the benzene to be nitrated must be very carefully specified to avoid these contaminants.

It should be pointed out that nitro groups on aromatics having also halide or sulfur-containing substituent groups can be hydrogenated with proper control of the hydrogenation conditions. Certain phosphates, secondary amines, and heterocyclic compounds can be added to avoid both poisoning of the catalyst by the offending groups and also hydrodesulfurization or hydrodechlorination of the desired substituent groups. Another scheme consists of increasing the pressure and lowering the temperature so that the net hydrogenation rate remains the same, and frequently by lowering the temperature the tendency toward halide or sulfur poisoning can be avoided. The following conditions are employed.

Catalyst	Carbon-supported precious metals; Second choice Ni
Quantity of catalyst	0.1–0.5%
Catalyst consumption	3–10 ppm
Pressure	100–2000 psi
Temperature	75–200°C

Nitrobenzene can also be hydrogenated in the vapor phase in a fluid bed reactor using a reduced copper catalyst supported on silica or alumina. The reactor is a vessel much like the entrained catalyst reactor of Fig. 3, except that the reaction space is larger in diameter and shorter in length. The catalyst is first separated from the reactor effluent by a cyclone separator and finally by bag filters. The heat of reaction is removed both by the large excess of H_2 and by the catalyst solids. The conditions of reaction are as follows.

Temperature	180–230°C
Pressure	20–40 psi
Catalyst	10–15% CuO on SiO_2 or Al_2O_3 reduced to Cu
Reactor type	Fig. 3 modified substantially
Catalyst consumption	0.05–0.15 lb/100 lb aniline

D. HYDROGENATION OF NITRILES TO AMINES

Nitriles also can be hydrogenated to amines and simultaneously to an additional methylene group. This operation is frequently employed in certain polymer and agrichemical intermediate processing. The reaction is also extremely exothermic and must be handled with careful temperature control. Deamination can be a serious problem but can be avoided or minimized by incorporating anhydrous ammonia into the reaction environment. Deamination can also cause the formation of polymers which are absorbed on the catalyst with concomitant catalyst deactivation. Of the catalysts most frequently used for this reaction, cobalt as a finely divided material supported on kieselguhr or alumina is probably the first choice. Ruthenium supported on alumina or activated carbon is probably the second choice. The reaction conditions are usually comparatively mild: 500–525 psi partial pressure hydrogen and relatively low temperatures of 100–200°C, with the lower range being favored. The following conditions are employed.

Catalyst	Co on kieselguhr or Al_2O_3; Ru on Al_2O_3 or powdered carbon
Quantity of catalyst	0.5–5.0%
Catalyst consumption	0.01–1.0%
Pressure	300–2500 psi
Temperature	80–200°C
Anhydrous NH_3	5–100% of weight of nitriles

V. Hydrogenations—Selective

A. HYDROGENATION OF CO TO CH_4

The next reaction to be considered is the hydrogenation of carbon monoxide to methane. This is probably one of the most straightforward and by-product-free reactions we shall consider. It can be conducted over a very wide range of conditions of pressure and temperature using a nickel-on-alumina or a nickel-on-chromium catalyst. Ruthenium is also a very effective catalyst; however, at elevated temperatures, Ru may produce Fischer–Tropsch products.

The reaction conditions are such that the temperature can range from 125°C to as high as 800 to 900°C, but at higher temperatures the conversion of methane to CO and hydrogen is favored. This reaction has been extensively studied with the objective of converting coal to CO and hydrogen and these products to substitute natural gas (SNG, methane). When the objective is to obtain as complete a conversion to methane and the minimum amount of carbon monoxide in the product as possible, it is essential that the temperature be maintained at a level below 350°C. Either of the reactors depicted in Figs. 1 and 2 is most frequently used, and the following conditions are employed.

Catalyst	Ni on Al_2O_3 or Cr_2O_3, Ru on Al_2O_3
Pressure	10–1000 psi
Quantity of catalyst	5000–20,000 space velocity per hour
Temperature	100–800°C
Catalyst consumption	Very low, 2–7-yr life

B.   CARBON MONOXIDE TO METHANOL

The synthesis of methanol from CO and hydrogen is one of the large-volume chemical operations. It is possible and even likely that methanol and higher alcohols derived simultaneously by the hydrogenation of carbon monoxide will be some of the most important liquid fuels in the future. The synthesis of alcohols from CO and hydrogen can be much more selective than the Fischer–Tropsch process which is not considered in our discussions of catalyzed hydrogenation because it is treated separately in Chapter 5, this volume.

Methanol synthesis from CO and hydrogen is a process that has been engineered and designed to a high degree of development. There are presently plants that produce as much as 2500 tons/day of methanol and plants on the drawing board that will produce 5000 tons/day (and even 10,000 tons/day) in a single-line system. A single-line system has one gas-generating system, one compressor, one reactor, and one gas purification and distillation system.

There is much argumentation as to what the pressure should be for the most efficient operation. This is generally dictated by the pressure at which the gas is generated. Inasmuch as most technology now permits generation

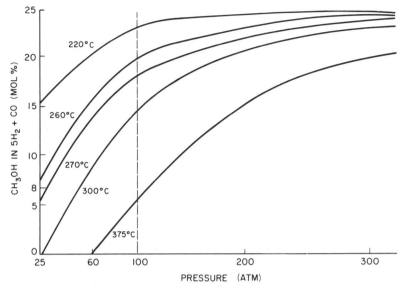

Fig. 5. Equilibrium conditions for methanol synthesis.

at only 750 psi, the trend has been toward this level or toward 1000 or 1500 psi. However, it is probable that, as larger plants are designed, the need will be to use higher pressures in order to minimize the size of the converters and the interconnecting piping.

The temperature of operation is as low as possible to favor methanol synthesis. Figure 5 indicates the equilibrium conditions for methanol synthesis from a typical CO/H_2 ratio.

There are two well-known synthesis conditions for methanol—one is a high-temperature, high-pressure system using a zinc chromite catalyst, whereas the other is a low-pressure, low-temperature system employing a copper–zinc oxide system stabilized with alumina or chromium. The productivity of an alcohol unit is measured on the laboratory scale in milliliters per milliliter of catalyst per hour, but for a plant-scale unit generally in tons per day per cubic foot of catalyst volume. A catalyst that produces 1 mm of methanol per milliliter of catalyst per hour on a laboratory scale is essentially identical to a catalyst that produces 1 ton/ft³ day in a plant converter. Typical commercial plants currently produce between 0.5 and 1 ton/ft³ day. There are, however, catalysts that have been developed that will produce 10 to 25 times this volume of methanol.

The major difficulty, however, in utilizing these highly productive catalysts is that at the higher production rate the heat generated per unit volume of converter is so great that extraction or dissipation of the heat becomes an

TABLE I

Characteristics of Alcohol Synthesis Systems

Characteristic	High pressure and high temperature	Low temperature and high or low pressure	Methanol and higher alcohols
Catalyst	$ZnO-Cr_2O_3$	$CuO-ZnO-Al_2O_3$	$ZnO-CuO-Mn_2O_3$, $K_2O\ Al_2O_3$
Product	MeOh (97%)	MeOH (99%)	MeOH (75%)
Ethanol and higher alcohols (%)	1–2	<1	10–25
Dimethyl ether (%)	2–4	<1	3–6
Temperature (°C)	325–400	200–280	275–400
Pressure (psi)	2500–7500	750–5000	1500–5000
Production rate (tons/ft³/day)	0.7–1.2	1.0–4.0	0.7–2.5
Catalyst life (yr)	2+	1–2	0.4–1

extremely difficult problem. In other words, present-day catalysts are more advanced than the engineering of alcohol units. The converter system described in U.S. Patent No. 4,235,899 is designed to accommodate these more active catalysts.

Catalysts that produce higher alcohols are slightly different from those used for methanol synthesis and usually contain a small amount of manganese and potash as well as other ingredients. The pressure and temperature of operation also are usually higher for higher alcohols than for methanol. Reaction conditions for the two cases are compared in Table I.

Catalyst life for a methanol unit of this type is usually 1–2 yr but has been known to extend to as long as 6 or 7 yr. Higher alcohols synthesis, however, involves more deterioration, and the catalyst generally does not last more than 6 months, or even less if higher alcohols are produced in major quantities. Methanol synthesis is described in detail in Chapter 6 of this volume.

C. NITROGEN HYDROGENATION TO AMMONIA

This process was first developed about 1906, and early ammonia plants operated with essentially the same catalyst currently being used, which is a fused iron oxide containing alumina, calcium, potash, and possibly some silica as promoters. Patents have been issued for the use of cerium oxide as a catalyst promoter. If this is an improvement, it is the only one that has been made in the last 75 yr. This is not a record of which catalysis scientists can be proud.

The pressure of operation is as low as 2000 psi at present; earlier plants (during World War II) operated at 5000 to 12,000 psi. Ammonia is another of the major commodity products of the chemical industry; currently 70 million tons/yr are produced throughout the world. Most ammonia goes into fertilizers, but a large amount is oxidized to nitric acid and nitrate derivatives. The reactor used for ammonia synthesis is very similar to that used for alcohol synthesis (Figs. 1–3). See Chapter 4, Volume 3 for more information.

Poisons that deactivate ammonia synthesis catalysts are CO, water vapor, sulfur and halide compounds, and lubricating oil.

D. HYDROGENATION OF MALEIC AND FUMARIC ACIDS

Maleic and fumaric acids can both be hydrogenated to succinic acids that in turn can be hydrogenated to butanediol. Hydrogenation of the double bond is very simple, but hydrogenation of the carboxylic entities is much more difficult. This reaction usually requires a high hydrogen pressure on the order of 5000 psi and a noble metal catalyst, preferably supported ruthenium. This process has limited, if any, commercialization, but many patents have appeared in the literature, particularly in Japan, claiming that it is a viable and economic process.

E. · HYDROGENATION OF FURAN TO TETRAHYDROFURAN

Furan, which can be derived by the decarbonylation of furfuraldehyde, is easily hydrogenated to tetrahydrofuran either in slurry-type or fixed bed operations as indicated in Table II.

F. HYDROGENATION OF NAPHTHALENE TO TETRALIN

This reaction also is a very easy hydrogenation to accomplish and is usually performed at temperatures of 150 to 180°C and pressures of 300 to 2000 psi. Tetralin is used as a solvent; in the donor solvent coal liquefaction system it is a source of hydrogen for coal hydrogenation.

A serious problem in the hydrogenation of naphthalene is the tendency for the naphthalene to sublime and collect as a solid in draw-off lines, pressure sensor lines, and rupture disk lines. These lines can be kept clear by

TABLE II

Hydrogenation Conditions for Furan to Tetrahydrofuran

Conditions	Slurry-type operation	Fixed bed operation
Reactor type	Fig. 4	Figs. 1 and 2
Catalyst type	Supported Ni	Lump activated Ni–Al$_3$ alloy
Temperature (°C)	80–160	80–140
Pressure (psi)	400–2000	750–5000
Liquid hourly space velocity (LHSV)	1–3	1–3
Catalyst life or consumption	0.001–0.01%	1–2 yr per charge
Poisons	Typical + H$_2$O	Typical + H$_2$O

heating them to above the condensation temperature or by back-flushing with tetralin. The conditions of reaction are as follows.

Catalyst	Usually Ni on kieselguhr (slurry); lump NiAl$_3$ activated alloy (fixed bed)
Quantity of catalyst	0.3–2.0%
Catalyst consumption	0.01–0.0001%
Pressure	300–2000 psi
Temperature	80–175°C
Liquid hourly space velocity	1–3
Types of converters	Fig. 2 or 4

G. HYDROGENATION OF GLUCOSE TO SORBITOL

This operation is conducted at comparatively high pressures—2000 psi to as high as 5000 psi, the high pressure being employed to minimize temperature and exposure time, which must be low to avoid charring of the sugar. The following conditions are employed.

Catalyst	Copper chromite or promoted copper chromite
Quantity of catalyst	2.5–10% (slurry)
Catalyst consumption	0.1–0.01%
Pressure	2000–5000 psi
Temperature	90–150°C
Range of dilution	Sugar/water ratio 1 : 5
Type of reactor	Fig. 4

H. HYDROGENATION OF ETHYL ANTHRAQUINONE TO ETHYL HYDROANTHRAQUINONE

This is a step in a preferred route to hydrogen peroxide. The hydrogenation is performed at rather low pressures, 25–200 psi, and low temperatures to avoid overreduction to tetrahydroanthrahydroquinones. The conditions of reaction are as follows.

Catalyst	Pd on a support such as alumina, carbon, or a rare earth element
Catalyst quantity	1–5%
Catalyst consumption	0.1–1.0% (excluding regeneration)
Pressure	25–200 psi
Temperature	60–100°C
Type of reactor	Fig. 4

I. SELECTIVE HYDROGENATION OF ACETYLENE TO ETHYLENE IN THE PRESENCE OF EXCESS HYDROGEN

Acetylene is a contaminant in petroleum products such as propane, ethane, and butane, which are cracked to produce ethylene for polyethylene or ethylene glycol manufacture. The acetylene interferes with subsequent operations and can be removed catalytically or by scrubbing. The catalytic removal of acetylene, its hydrogenation to and not beyond ethylene, represents one of the best examples of selective catalysis. This reaction is performed on a commercial scale at approximately 200–250°C over nickel catalysts sulfided to a very exact degree. Because during the operation of the process the sulfur tends to be hydrogenated and mobilized from the catalyst, there must be continuous feeding of sulfur to the reaction environment to replace the sulfur that is lost and thereby maintain selectivity. The pressure of hydrogenation can be 200 psi to as high as 1000 psi, and the catalyst is generally a nickel–alumina material sometimes containing a small amount of cobalt and occasionally chromium. The value of chromium is questionable because such an ingredient tends to make catalysts resistant to sulfurization and facilitates mobilization of the sulfur, thus causing rapid loss of selectivity.

Palladium sulfide is also used for this process, usually not with an excess but only a stoichiometric quantity of hydrogen. Otherwise, the acetylene and some of the ethylene itself may be hydrogenated to ethane. Butadiene can also be selectively hydrogenated to butene via a catalyst under conditions very similar to those previously described. Phenylacetylene can also be hydrogenated selectively to styrene similarly. The conditions of reaction are as follows.

Catalyst	Ni on Al_2O_3 sulfided to 20–25 mol% based on Ni content; Pd on Al_2O_3, sulfided
Space velocity	2000–5000 v/v per hour
Catalyst consumption	Very low
Pressure	10–500 psi
Temperature	100–250°C
Special conditions	Continuous addition of H_2S or other sulfiding agent at about 1 ppm level in inlet gas
Type of reactor	Figs. 1–3

J.   HYDROGENATION OF ESTERS

Organic acid esters, such as coconut oil, peanut oil, cottonseed oil, palm oil, palm kernel oil, and other vegetable oils, can be hydrogenated in the liquid phase (slurry) to produce straight-chain alcohols in the C_8–C_{24} range. Most linear alcohols are derived instead by Ziegler chemistry from ethylene via aluminum alcoholate which is hydrolyzed to aluminum hydroxide and the corresponding alcohol.

As economic conditions are changing and coconut oil has not increased in price at the rate at which the price of ethylene has increased, it is entirely possibly that the hydrogenation of coconut oil will again become a competitive or even a better route to alcohols than the current Ziegler process. This is an interesting example of how old chemistry based on natural products may, because of the increase in the cost of petroleum products, become the preferred route to some chemicals.

In the synthesis of alcohols from coconut oil, the hydrogenation is conducted at about 5000 psi and approximately 300°C with a copper chromite-type catalyst. The catalyst is susceptible to deactivation by the acidity of the oil, and rancid oils require the use of a base in the catalyst, so a catalyst of the copper chromite–barium chromate type is used when the oils are acidic. This same type of barium chromate–copper chromite catalyst is also useful for hydrogenating the acids themselves, e.g., palmitic or steric acid.

A lower-boiling ester such as ethylacetate can be hydrogenated in the vapor phase in hydrogen or hydrogen–nitrogen mixtures over a fixed bed copper chromite either promoted or unpromoted. The following conditions are employed.

Catalyst	Copper chromite or copper chromite–barium chromate powder (slurry); pelleted copper chromite or barium chromate–copper chromite (fixed bed)
Space velocity	1–3 LHSV (slurry); 2000–8000 (vapor phase)
Catalyst consumption	0.1–1% (slurry); 0.1–0.5% (vapor phase)
Pressure	3000–5000 psi (slurry); 400–2000 psi (vapor phase)
Temperature	250–325°C (slurry); 150–250°C (vapor phase)
Special conditions	When acid is present in the reactor environment, the catalyst must contain a base
Type of reactor	Figs. 2 and 3 (vapor phase) and 4 (slurry)

K. HYDROGENATION OF PHENOL TO CYCLOHEXANOL

Phenol can be hydrogenated very easily to cyclohexanol using a nickel catalyst at approximately 500 to 2000 psi and at temperatures in the range 125–200°C. Cyclohexanol is used both as a solvent and as a highly efficient raw material for oxidizing to cyclohexanone and eventually to adipic acid. The conditions of reaction are as follows.

Catalyst	Reduced nickel on kieselguhr
Liquid hourly space velocity	2–5 (slurry)
Catalyst consumption	0.1–0.5%
Pressure	600–2000 psi
Temperature	100–200°C
Reactor type	Figs. 2 and 4

L. SELECTIVE HYDROGENATION OF A NITRO GROUP TO AN OXIME

In Section IV.C a procedure was described for the hydrogenation of nitro groups to amines. It is also possible to hydrogenate selectively only to the oxime stage. In the case of nitrocyclohexane, selective hydrogenation provides a route to caprolactam via Beckmann rearrangement of the oxime. In the hydrogenation of a nitro group to an oxime, a moderated platinum catalyst is used. A process was described in Section V.I for moderating a nickel catalyst by converting part of the nickel to sulfide. In the case in question the platinum catalyst is moderated by the coprecipitation of a lead compound in fractional percentages. This operation is performed at 70 to 150°C, a rather wide range, and 1000 psi. The following conditions are employed.

Catalyst	0.5% Pd or Pt with a Pb moderator, on carbon
Liquid hourly space velocity	0.5–1.0
Catalyst consumption	0.2–1.0%
Pressure	400–1500 psi
Temperature	70–140°C
Reactor type	Fig. 1, 2, or 4

M. HYDRODESULFURIZATION AND HYDRODENITROGENATION

In the purification of petroleum products, that is, the removal of sulfur and nitrogen from crude petroleum, the procedure is basically a hydrogenation designed to remove selectively sulfur and nitrogen from various types of crude oil by converting these impurities to H_2S and NH_3. These processes are considered selective hydrogenations because the objective is to remove the sulfur and the nitrogen from heterocyclic compounds, for example, without hydrogenating the aromatic portion of the hydrocarbon. The aromatic should not be hydrogenated because the unsaturated material has a higher octane rating and therefore should be preserved; furthermore, by hydrogenating one used hydrogen which, of course, imposes an unwanted cost.

The catalyst employed in this operation is usually cobalt, nickel or molybdenum oxide supported on γ-alumina, or a nickel or cobalt oxide with tungsten oxide also supported on γ-alumina and sometimes considered best for hydrodenitrogenation. These products are converted to the respective sulfides with H_2S, CS_2, COS, or even thiols and are used in the sulfide state.

The conditions employed are temperatures of approximately 275–450°C and pressures of 500 to 2500 psi. The catalyst frequently becomes very badly coked; regeneration is effected by reoxidation of the catalyst and the coke, the latter being removed as carbon dioxide and water vapor. The conditions of reaction are as follows.

Catalyst   CoMo oxides on Al$_2$O$_3$, sulfided;
                            NiMo oxides on Al$_2$O$_3$; NiW oxides
                            on Al$_2$O$_3$; CoW oxides on Al$_2$O$_3$;
                            others being investigated, all sulfided
Pressure   500–2500 psi
Temperature   275–450°C
Space velocity   Variable
Catalyst consumption   Variable
Reactor type   Figs. 1, 2, and 4.

This reaction is currently being intensively researched to adapt both the hydrodesulfurization and hydrodenitrogenation processes to the removal of sulfur and nitrogen from more aromatic (heterocyclic) and higher boiling crudes which also have a higher content of these impurities as well as ash. It is likely that significant changes in catalysts and processing conditions will emerge from these studies. Also see Chapter 4, Volume 1.

N. HYDROGENATION OF HALIDE AND SULFUR AND SULFATE-SUBSTITUTED NITRO COMPOUNDS TO THE CORRESPONDING AMINES

An obvious problem in processing organic materials of this type is the poisoning effect of the halide and the sulfur or sulfate component. Hydrogenation, however, can be effected at low temperatures, say 80–100°C, but this may require relatively high pressures of 1500–3000 psi. The operation can be improved by the incorporation of an amine into the hydrogenating mixture, which seems to form a weak bond with the acidic substituent group and thus prevents it from reacting unfavorably with the catalyst. A precious metal, such as supported platinum or palladium, or a mixture thereof, is frequently used for this operation, and the support generally is carbon or titanium because of its applicability in acidic media.

O. HYDROGENATION OF CARBON MONOXIDE TO VARIOUS PRODUCTS

This hydrogenation is such an extensive subject that it would be impossible to include it in this chapter except as a notation. One of the operations that has received the most attention is the conversion of CO and hydrogen to ethylene glycol and other di- and trihydroxy compounds employing as catalysts organic rhodium complexes and extremely high pressures. The high pressures seem to be a major deterrent to commercialization of the process and, consequently, extensive work has been done to reduce the pressures to a more economic level, such as 1000 to 5000 psi. Patents have appeared for this range, but it is not known at this time what the prognosis is for commercial applications of the process.

P. HYDROGENATION OF UNSATURATED ESTERS TO UNSATURATED ALCOHOLS

This process is the third in a family of hydrogenations of natural oils such as coconut oil, tallow, palm oil, cottonseed oil, soybean oil, and others. We have already discussed oil hardening with a nickel catalyst to produce edible and industrial fats; this reaction does not cleave the ester linkage. The second process is the hydrogenolysis whereby saturated fatty acid esters are converted to saturated alcohols.

The next process in this family is the hydrogenolysis of an unsaturated oil to unsaturated alcohols. That is, the unsaturation of the alcohol is largely preserved despite the hydrogenolysis reaction. The selectivity of copper chromite for the ester hydrogenolysis without double bond saturation is well demonstrated. The reaction conditions are as follows.

Catalyst	Copper chromite, supported and/or promoted
Liquid hourly space velocity	0.5–2
Catalyst consumption	0.1–1.0% of product
Pressure	2500–5000 psi
Temperature	225–325°C
Problem	Active Fe or Ni must be avoided in the catalyst
Reactor types	Figs. 3 and 4

Q. PURIFICATION OF CRUDE TEREPHTHALIC ACID STREAMS

Crude terephthalic acid can be purified by two commercial routes, one being esterification to the methyl ester, which is then transesterified to the ethylene terephthalate ester polymer. A second method is to hydrogenate selectively the impurities in the crude with a palladium-on-carbon catalyst. The hydrogenated impurities are more readily removed from the acid by extraction or washing. This is mentioned only as another example of catalytic purification by selective hydrogenation of the impurity(ies).

Many further hydrogenations could be described, but the purpose of this chapter is not to give details of hydrogenation of many similar types but to give representatives of each type. With these representative types one skilled in the art can readily translate the data given to the data needed.

CHAPTER 4

Dehydrogenation and Oxidative Dehydrogenation

A. B. STILES
University of Delaware
Center for Catalytic Science and Technology
Colburn Laboratory
Newark, Delaware

I. Introduction

Dehydrogenation is generally one of the least specific of chemical reactions; oxidative dehydrogenation also is relatively nonselective. Many thermal dehydrogenations are conducted in the absence of intentionally added catalysts but, in contrast, essentially all oxidative dehydrogenations are conducted in a catalytic environment. This may convey the impression that, inasmuch as some dehydrogenations are effected with no catalyst at all that, when catalysts are employed, catalyst composition and manufacturing procedures are not critical. Nothing could be further from the truth, because in certain oxidative dehydrogenations the state of oxidation and the partial pressure of oxygen in the system must be very carefully controlled, as well as catalyst composition and catalyst preparation procedures, to ensure the conversion and yields that must be obtained. Although most oxidative dehydrogenations and dehydrogenations are nonspecific, there are a few that are extremely specific, and it is for these that catalysts must be carefully controlled as to composition and treatment in the reactor.

A. PURPOSE OF DEHYDROGENATIONS

The purposes of dehydrogenation are really twofold. The first is to produce hydrogen for various applications, and the second is to produce chemicals. This does not mean that the two processes cannot be combined;

for example, in the dehydrogenation of ethane to produce ethylene, sufficient hydrogen is produced that it can be separated for use in hydrogenation or combined with synthesis gas ($CO + H_2$). It may be, however, that the separation to produce hydrogen is too costly and as a result the hydrogen is simply removed from the ethylene by distillation. If the hydrogen stream is sufficiently pure, it can be used for hydrogenation purposes. However, often sufficient methane and other diluents are present, and as a result the hydrogen is not usable.

In oxidative dehydrogenations not only is the desired product, such as acetylene, obtained, but there may also be produced simultaneously sufficient carbon monoxide and hydrogen for a side stream which can be converted to alcohols. These are commercial companion processes practical in several locations. Dehydrogenation and oxidative dehydrogenation frequently not only produce desired products directly but also side streams which are of economic importance and consequently are a plus factor when considering the process for adaptation.

B.   CHARACTERISTICS OF REACTIONS

Dehydrogenations, whether catalytic or thermal, are extremely endothermic. The bond between hydrogen and carbon and between carbon and carbon is extremely strong, and as a result the energy input must be high to effect the desired dehydrogenation. Furthermore, dehydrogenation is frequently an equilibrium reaction so that, if one performs the dehydrogenation at high temperatures, as is ordinarily the case, the tendency for rehydrogenation to occur as the product cools and makes contact with activated walls (such as stainless steel) leads to recombination of the products to form the components of the original feed stream or something similar thereto. As a consequence, it is generally a requirement that the effluent following a dehydrogenation reaction be quickly quenched by direct contact with water or a heat exchanger (with noncatalytic walls) so as to minimize this recombination of products.

Because such a large amount of heat must be applied to the endothermic reaction, reactions are ordinarily performed in tubular-type reactors that will subsequently be described. Carbon deposition can be a severe problem as a result of cracking. Carbon can be deposited to the extent that it plugs the tubes or forms a hard cake that severely restricts heat exchange and causes tube overheating and deterioration.

Another important aspect of dehydrogenation is the oxidative dehydrogenation process. This operation, in contrast to simple dehydrogenation, is extremely exothermic and must also be performed in a tubular reactor to dissipate the heat load. Some of the problems previously mentioned in relation to the thermal dehydrogenation step are avoided in the oxidative scheme. The most obvious is the avoidance of external heat and the problem of heat transfer from the exterior to the interior reaction zone. Furthermore, carbon deposition is much less of a problem in oxidative dehydrogenation. An additional and very important factor is that, inasmuch as water and carbon monoxide are formed in oxidative dehydrogenation, the possibility of recombining the products to form the material being fed is minimized. It is obvious that, if hydrogen is produced along with an olefinic material, even though water and carbon monoxide are present, there is still a possibility of recombining these products and thus loss of the desired product.

C. TEMPERATURE CONTROL

The importance of temperature control cannot be overemphasized. Too low temperatures mean an incomplete reaction, particularly in thermal dehydrogenation, but many other difficulties are encountered if the temperature is allowed to exceed the optimum. If the temperature exceeds the optimum, decomposition and carbon deposition can occur, with attendant destruction of the catalyst, and in some cases carbon deposition is so severe that it can even distort the reactor. The effect on the catalyst can be very complex. The obvious effect is to cause sintering or melting and loss of surface area. Other conditions that can develop are interaction of a support with the catalytic material and interaction of the catalytic material with itself to form a species that is noncatalytic.

Excessive temperatures can also cause a weakening of construction materials with the effect that the equipment is distorted and in some cases may even rupture. In an endothermic reaction, the problem is heat flux through the walls of the converter so that heat can be supplied to a reaction occurring internally. This usually means that the firing must be conducted extremely cautiously, and in some cases the flame must impinge as gently as possible against the metal to prevent scaling or abrading. Scaling very often occurs when the flame is oxidizing in nature, and under these conditions the metal can be slightly oxidized with the result that scaling and spalling of the metal occur.

When the reaction is an oxidative dehydrogenation and the internal temperature is excessive, much of the difficulty previously described also occurs internally, with the same adverse effects on the catalyst or construction materials. If the reaction is an exceptionally exothermic one, then comparatively small tubes must be used to facilitate the heat exchange. These tubes usually are in bundles of 500 to 5000 or even more; heat exchange usually is severely affected adversely if the tubes become warped or bent, as can occur with small tubes. Heat exchange then becomes much less effective, with tubes touching in some cases or spaced apart too distantly in others, thus compounding the difficulties of heat removal, carbon formation, and carbon removal.

As described in the paragraph above, carbon deposition is a severe problem whether the reaction is oxidative or thermal dehydrogenation. After experience has shown the optimum conditions for minimizing carbon deposition, usually the next step is to determine the conditions under which the carbon can be harmlessly removed.

The simplest and probably the most frequently used process for carbon removal is the addition of small quantities of oxygen to the system while the reaction is suspended. Control of the oxygen content usually is to a level of about 0.5% in an inert gas. Because of the large volumes involved, the inert gas may be quite costly, so regeneration is effected by recycling the off-gas from the regeneration and adding small quantities of oxygen to the system. Then the content of oxygen in the gas stream tends to rise, and the amount of carbon dioxide in the exit stream tends to decrease, with the effect that eventually one obtains the O_2 level of pure air and the equipment becomes suitable for entering, if all other safety requirements have been met. One should bear in mind, however, that in some cases this high O_2 level may adversely affect the metals of contruction and cause scaling and weakening to an unacceptable extent.

Instead of using dilute oxygen for oxidation there are cases in which steam or water vapor is employed. This is a much safer procedure in that the water vapor produces an endothermic reaction forming carbon dioxide, carbon monoxide, and hydrogen. From a safety viewpoint, it is essential that the carbon monoxide be recognized as an extremely poisonous effluent. Steam with carbon dioxide or carbon dioxide alone can also act as an oxidant for the carbon.

In some cases, it is possible to remove carbonaceous material by passing steam plus a small amount of hydrogen through the catalyst bed. The high temperature plus the steam may cause decomposition and cracking of some of the hydrocarbon. Hydrogen then hydrogenates any olefinic material and helps it to volatilize from the catalyst environment. In many cases in which the catalyst is regenerated and the carbon removed by oxidation, the catalyst must be re-reduced. However, treatment with steam and hydrogen tends to avoid this requirement and as a consequence may be the procedure of choice.

D. REACTION SPECIFICITY

As previously stated, both thermal and catalytic dehydrogenations are relatively nonspecific; that is, when the reaction occurs, there is a tendency for substantial by-product formation. This point will become even more apparent when we describe specific reactions subsequently in this chapter. It is usually unwise to generalize too much on any subject, and this is particularly true at present because there are some oxidative dehydrogenations that can be quite specific. These, however, are the exceptions, and in general the reactions are very nonspecific. An example is the dehydrogenation of naphtha to produce ethylene, which also results in a large variety of other products. Catalysts have been employed in this area, but the effect is still a shotgun-type product distribution.

E. PRODUCT SEPARATION

Here again it is unwise to generalize, but in many processes separation of the mixture of products is extremely difficult. Some of the tallest distillation columns in the world are those used for the separation of ethylene from the products formed after the cracking of propane. A chemical equation shows that propane plus heat produces ethylene plus methane ($C_3H_8 \rightarrow CH_4 + C_2H_4$). It appears from the stability and other characteristics of the products involved that a nice, clean reaction occurs. However, because of the free radical nature of the cracked products a large number of polymers of the ethylene or ethyne radical result. These range up to a liquid product, referred to as dripolene, that can be added to gasoline used as an automotive fuel.

Acetylene is also produced in the cracking reaction and is a nuisance in the product because small quantities of acetylene in the ethylene cause difficulties in its downstream utilization. The acetylene is usually removed by selective catalytic hydrogenation to ethylene; this process is covered in Chapter 3.

F. WATER VAPOR OR STEAM ADDITION TO DEHYDROGENATION PROCESSES

In the previous paragraphs the formation of carbon and residue deposits in the reactor was frequently mentioned. It was also mentioned that these deposits sometimes are removed by feeding through the reactor, after the reaction process is suspended, a quantity of steam that will react with these carbonaceous materials to produce volatile carbon dioxide, carbon monoxide, and hydrogen. In many dehydrogenation reactions steam is added along with the feed to be dehydrogenated to minimize carbon deposition. Steam is probably the most frequently employed decarbonization agent, but other factors are also often employed. These are simply enumerated at this time but will be described more completely subsequently in the chapter and constitute passivation of the construction metals of the reactor. Passivation of reactor walls is difficult to accomplish, but methods employed with various amounts of success include nitriding or carburizing, electroplating with a relatively nonreactive metal, and galvanizing with zinc. The latter obviously is effective only at comparatively low temperatures. Efforts have also been made to electroplate with gold or silver, but to the author's knowledge this has never been practiced on a commercial basis.

Catalysts themselves are sometimes moderated with a component such as sulfide, phosphate, or chloride, but here again these components may become quite volatile under extreme temperature conditions in some environments. Steam, for example, at high temperatures tends to mobilize all these components. Also rather frequently catalysts may be treated by exposure to relatively high temperatures to deactivate some of the more reactive sites which in turn then tend to prevent carbon deposition in the reactor environment. Sulfur is frequently an impurity in varying quantities in the gas stream or feed stream to be dehydrogenated. Although most workers associated with catalysts tend to fear and avoid sulfur in the stream, one should be cognizant of the fact that sulfur may react beneficially with the catalyst to produce a less reactive site having substantially less of a tendency toward carbon formation. This may also occur on the walls of the reactor where an incipient sulfide is formed, also minimizing carbon deposition at this location.

In addition to sulfur in producing this effect, other components in the gas stream can also be continuously added to minimize carbon deposition. These can be halides, organic or inorganic compounds, ammonium phosphate, phosphorus oxide, sulfides, sulfates, or organic sulfides such as thiols.

II. Reactor Characteristics

Inasmuch as heat exchange is a primary requirement in either endothermic or exothermic dehydrogenation, a typical reactor is of tubular design. This, in its simplest form, is shown in Fig. 1 and basically resembles very closely the boiler tubes in a steam–water boiler. This is the type used, for instance, in the cracking or dehydrogenation of propane and ethane. There is a second type of reactor, shown in Fig. 2, which is basically a bundle of tubes enclosed

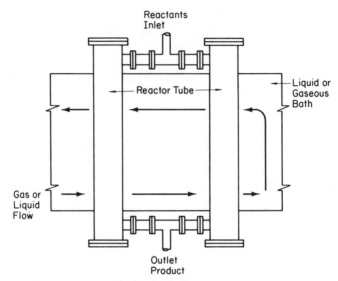

Fig. 1. A tubular reactor.

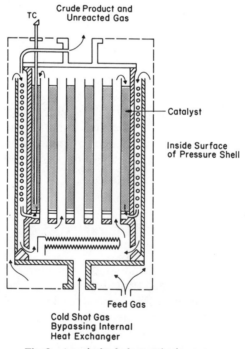

Fig. 2. A typical tubular synthesis reactor.

in a cylindrical heat exchange medium. This is the type of reactor used for methanol dehydrogenation and the dehydrogenation of propane plus ammonia to acrylonitrile.

A. MATERIALS OF CONSTRUCTION

The importance of the materials of construction has been touched upon in the preceding paragraphs and will be discussed as part of individual process reviews subsequently in this chapter.

The alloys frequently used in the construction of dehydrogenation reactors have a high nickel and chrome content (25–20%), and in some cases high Ni–Cr alloys of this general type also contain cobalt.

The general statement can be made that dehydrogenation reactions are essentially reversible. The reaction proceeds to the dehydrogenation products at high temperatures but reverses direction at lower temperatures. This reversal is enhanced with certain construction materials containing nickel, cobalt, and even iron.

An example of a reversible dehydrogenation reaction is the synthesis of acetylene at high temperatures. At lower temperatures and in the presence of hydrogen, acetylene can rehydrogenate to an olefin or a hydrocarbon. Avoidance of this unwanted reaction is effected either by a very rapid quench (water spray, Fig. 4) or a very rapid quench plus deactivated or passivated reactor walls. Means of passivating the reactor walls have previously been discussed. A rapid quench is usually obtained with a water spray or steam addition to lower the temperature to a value where rehydrogenation is avoided or is very slow.

B. FLUIDIZED BED REACTOR

The fluidized bed reactor has many characteristics that make it highly desirable for dehydrogenation reactions. First, at high temperatures there is very little contact of the reactants or the product with the reactor walls. During the reaction the catalyst is in contact with the reactants and products at all times until the catalyst and reaction products are separated. The fluidized bed has characteristics that permit the introduction of heat via the hot catalyst, which avoids the hot metallic surfaces and the adverse effect thereof. The catalyst can be specially fabricated for fluid bed applications with sufficient thermal stability and selectivity to minimize the reverse reactions.

Another factor favoring the fluidized reactor is that the catalyst can be regenerated easily, as frequently as necessary, e.g., every few minutes.

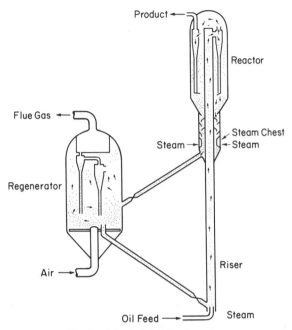

Fig. 3. A typical fluidized reactor.

Briefly, the cycle can be such that the reactant and hot catalyst meet as incoming streams, the reaction occurs in seconds, and the catalyst and reactant separate, the catalyst being cycled to a reoxidation chamber from which it emerges both decarbonized and reoxidized (regenerated) as well as very hot. This hot catalyst then goes into the reaction chamber where it again contacts the incoming reaction stream and repeats the cycle. As indicated, this cycle can be placed on essentially any preferred chronological scale.

The fluidized reactor shown in Fig. 3 is typical of the most modern petroleum processing units. There are several disadvantages that should be pointed out which, to a certain extent, counterbalance the advantages. A disadvantage is the relative nonflexibility of the unit in that the rate of gas flow must be very carefully monitored and must not exceed the optimum or the catalyst will be lifted from the reactor and entrained in the exit gas stream overloading the separator system. Furthermore, unless the through-put gas moves at a sufficiently rapid rate, the catalyst will not be fluidized or, if designed to be entrained, will not be conveyed to the proper location in the reactor and will settle into almost a static bed with poor reactant–catalyst contact. A further disadvantage is that the catalyst may be severely abraded during the course of its cycling.

For maximum efficiency, a rather narrow range of spheroidal particle sizes should be used. If the catalyst is smaller than this, it will be carried out with the product, resulting in severe economic loss. A final point must be made that, if the catalyst is designed for abrasion resistance but is made excessively hard, it will erode the equipment itself, necessitating expensive repairs and outages. A careful balance must be maintained among the abrasion characteristics of the catalyst, its hardness, and its effect on the durability of the converter walls. This type of reactor is used in petroleum processing and in the Standard Oil of Ohio process for acrylonitrile. The latter is an interesting application and is a unique departure from the typical dehydrogenation and oxidative dehydrogenation tubular reactors.

C. HEAT RECOVERY

Whether the dehydrogenation reaction is oxidative or thermal, it is almost universally a high-temperature reaction. This means that substantial amounts of energy are required either with internal combustion of the reactants or external combustion of fuel to supply the essential energy for the dehydrogenation reaction. It is axiomatic that, because of the large amount of energy required, the recovery of as much of this energy as possible is also required for economic reasons. Recovery of the heat is effected by heat exchangers which may be a major part of the overall equipment costs. The heat that is recovered in the heat exchanger can be employed to preheat the gases going into the reactor, or it can be recovered in an exchanger designed also as a boiler. There are many other heat recovery considerations that are individual to each operation, and they will be discussed more completely as they relate to the individual processes subsequently described in this chapter.

D. OPERATIONS THAT USUALLY OCCUR AT LOW PRESSURES

When one considers that in dehydrogenation reactions there is almost always an increase in the volume of gas during the reaction, it becomes axiomatic that the reaction be conducted at low pressures. There may be economic reasons for using pressures above atmospheric or even substantially above atmospheric, but ordinarily reactions are conducted at pressures as close to atmospheric as economically feasible. Because reactions are favored by low pressures, the question is naturally asked as to the feasibility of operating at subatmospheric pressures. The literature reports the synthesis of acetylene at subatmospheric pressures, but it is believed at the present time that no plants of this type are operating commercially.

Despite the fact that most of the operations are conducted at low pressures, this does not ensure that there are no problems with reactors failing as a result of pressure causing bulging or thinning of the walls. Although the pressures are generally low, the temperatures are extremely high and sufficiently high so that metal creep and weakness are experienced. Nickel and nickel cobalt alloys are most resistant to this creep problem, and as a consequence they are widely used together with molybdenum and chromium. In some tubular reactors, the tube must be supported by both the top and bottom to avoid elongation due to its own weight.

III. Types of Catalysts

For most high-temperature reactions, a metallic catalyst is used. This can be in the form of metal supported on a refractory support such as fused alumina or alumina–magnesia, alumina–silica such as mullite, magnesium aluminate (spinel), and a chromium–alumina-type refractory. Chromium may have a tendency toward catalytic effectiveness itself and as a consequence must be very carefully examined before it is used as a support. In contrast, if it is possible to develop a dual-functioning catalyst in which the metal and the support have dual and companion roles, in such cases chromium, for example, can be of substantial benefit. Metals used for dehydrogenation are copper, silver, and sometimes gold. The precious metals platinum, palladium, rhodium, ruthenium, and silver can be used at very high temperatures, but ordinarily silver is not sufficiently stable at temperatures above 700°C.

The most selective catalysts are probably mixed oxide-type catalysts, although this is a statement that perhaps should not be generalized on because silver can be a very directive catalyst. These quite often are grouped into a class identified as heteropoly acids. Heteropoly acids include vanadium oxide, tungstic oxide, and molybdenum oxide, and these compounds can be modified by phosphoric, sulfuric, arsenic, antimonic, boric, silicic, titanic, and ceric acids (oxides). These can be further modified by the inclusion of iron group metals such as cobalt, nickel, and iron, and still further by the incorporation of alkali or alkaline earth elements as oxides. One glance at this grouping could lead one to the conclusion that heteropoly acids include a major fraction of all elements in the periodic table. This is basically true, but only a few of them have interesting catalytic properties in the processes we shall discuss in this chapter.

The problem with metals generally is that they have a relatively low melting point and a strong tendency toward crystallization. This can have a severe effect on catalyst properties and as a consequence on the economics of the process in question. Metallic catalysts used in some processes generate whiskers that can cause a pressure drop, catalyst embrittlement, and general catalyst failure.

IV. Safety

There are a few relatively unique safety features of dehydrogenation and oxidative dehydrogenation that should be pointed out. The first to be considered is hydrogen, which has its own characteristics, making it a hazard to be carefully reckoned with. First, hydrogen has a very wide explosive range as previously pointed out. Second, it diffuses readily through metals or through joints that would be gas-tight to other gases. Third, it has a very low kindling temperature and can be ignited with a hot steam line, for instance, and its ignition is easily catalyzed so that iron rust or any roughened or activated surface tends to provide a point of ignition.

Carbon monoxide is similarly demanding of respect in regard to safety. It is very toxic and is a cumulative poison in the human body. It is odorless and about the same density as air so that it circulates quickly and readily with air. It also is combustible, but not to the extent of hydrogen, and an explosion caused by carbon monoxide is relatively low-level. The primary hazard relating to carbon monoxide are its lack of odor and that it frequently is associated with odorous gases such as hydrogen sulfide and ammonia. Because these gases mask the presence of carbon monoxide, it is possible to obtain a fatal exposure before the hazard is noted.

Another hazard of carbon monoxide is its tendency to react chemically with metals to form carbonyls. Carbon monoxide reacts with iron and nickel at temperatures of 25 to 175°C and at pressures from atmospheric to 1000 lb or more. Carbonyls are very dense gases and collect in vessels or in pits or low spots in buildings. They usually have a musty odor that may be mistaken for the normal odor in these low spots. Metallic carbonyls are almost impossible to remove from the human circulatory system. It is absolutely essential to avoid exposure to them through either inhalation or skin contact.

In an oxidative dehydrogenation operation it is axiomatic that the feed gases not enter the explosive range. This, of course, can occur during plant start-up or shutdown if the sequence or the rates are carelessly managed. Furthermore, if there is an interruption in the hydrocarbon feed during the course of the operation, the feed can then pass into the explosive range where equipment damage or injury to personnel, in an explosion or melt-down, can occur.

Other safety features are those typical of high-temperature operations and consist of equipment weakening due to scale formation or to the reactor passing into a temperature range exceeding the yield point of the metals. Many metals, especially superalloys, have a rather sharp temperature at which the tensile strength of the metal decreases very rapidly.

Another factor that is relatively unique to this operation is the decarbori-zation of steel or the embrittlement of steel by exposure to high tempera-tures and relatively high-pressure hydrogen. This action is explained on the basis of hydrogen penetrating the inner metallic crystal and converting any elemental carbon or combined carbon to methane, which then evolves as a gas bubble below the surface of the metal. This bubble enlarges, further weakening the already decarborized and weakened metal.

Equipment weakening can also result from the metals removal accompa-nying the formation of metal carbonyls in the process previously described. The removal of iron or nickel from alloys, even stainless steel alloys, can be so severe that weakening due to the thinning of tube walls, for example, can lead to rupture. In such situations, minimum thickness holes are frequently utilized so that, when there is a problem of potential rupture, a leak first occurs and is easily detected.

Nondestructive wall thickness measurements can be made with a number of available instruments based on wave transmission.

V.   Thermal Dehydrogenation

There is always a question as to whether there is really any reaction that is not catalytic. One could argue very effectively that in thermal dehydrogena-tion free radicals are generated at the tube walls or at some similar site where the dehydrogenation really occurs. One could also argue that there is particulate matter, ash, or elemental carbon that also acts as a catalyst. However, without belaboring the point, we shall summarily identify the next four reactions as thermal without identifying them as noncatalytic.

A.   PROPANE TO ETHYLENE

The conversion of propane to ethylene, shown chemically as follows,

$$C_3H_8 \rightarrow C_2H_4 + CH_4 \quad \text{endothermic}$$

is a commercially highly important process for the manufacture of ethylene for polyethylene and ethylene oxide synthesis. Propane alone or propane and butane, both of which are very good sources of ethylene, are fed into an externally heated tubular reactor such as that shown in Fig. 1. These tubes are situated in a furnace, permitting them to be heated efficiently in a multipass coil to temperatures of approximately 800 to 900 °C. The material of construction of the tubes is generally a stainless steel of high nickel and chromium content such as "25-20."

It is obvious that the reaction is favored by low pressure because there is an increase in volume during the reaction, so the reaction is effected at as low a pressure as is practical where the gases can still flow through at a suitably rapid rate. As previously stated, subatmospheric pressure is desirable, but in most cases this is impractical and therefore the reaction is conducted at a pressure of 5 to 25 psi. As is the case in all processes in which there is a strong tendency for reversal of the reaction, the gases are rapidly quenched and are maintained in an environment in which hydrogenation catalysts are avoided as much as possible. The reaction is not a clean one because, at the same time that methane and ethylene are produced, propylene, acetylene, hydrogen, butadiene, butane, and a liquid product called dripolene are also produced.

Some of the problems of this process have already been briefly noted, but to summarize them, each will be again mentioned along with others that have not yet been touched on. Probably the most serious problem is carbon deposition. This can produce many difficulties such as plugging of the reactor tubes and formation of an insulating layer of carbon on the walls, with the result that heat exchange and heat removal from the walls is not effective. This can both intensify the carbon deposition problem and cause warping and sagging of the reactor tubes because of overheating. One frequently used method of minimizing this severe problem is to feed steam and/or hydrogen simultaneously with the propane feed. The quantity of hydrogen or steam must not be sufficiently high to cause a reversal of the reaction to propane or ethane.

Tube failure can be a major cost and safety factor; it is generally caused by overheating or by scaling and is attributable to the choice of improper metals. Hydrogen embrittlement can be a factor, but because of the low pressure it is not as severe a factor as in other type of operations at elevated pressures.

It is possible under these high-temperature conditions to produce a carbon that is entrained in the gases and is very abrasive, and this abrasive carbon causes severe erosion at tube bends. This problem is usually handled by both minimizing carbon and linear gas velocity and designing the tube bends to maximize the radius of curvature and hence reduce centrifugal force against the tube walls.

Last to be considered is the problem of coproducts, which may be an economic benefit rather than a problem. This is the simultaneous synthesis of acetylene, propylene, diolefins such as butadiene, and other olefinic materials. As previously mentioned, a liquid product is also obtained, which is identified as dripolene and usually sold as a gasoline supplement.

B.   ETHANE TO ETHYLENE

The dehydrogenation of ethane to ethylene is performed in much the same manner as the aforementioned propane-to-ethylene conversion. The conditions required are somewhat more severe, so that the temperature is in a range of approximately 50°C higher than that used for propane. Ordinarily carbon deposition is not as severe with ethane as with propane, but both types of feeds can be very severely and adversely affected if there are small amounts of higher boiling materials in the hydrocarbon feed. Careful monitoring of the feed stream to avoid significant quantities of these higher boiling contaminants must be practiced.

In both processes sulfur is a common contaminant. There is not complete agreement on the effect of sulfur or sulfur compounds, but it is suspected that it tends, to a certain degree, to deactivate the tube walls and thus minimize carbon deposition. But there also is the possibility of an adverse effect because of the formation of metallic sulfide, which causes corrosion of the tube walls. Other contaminants that can adversely affect the reaction are small amounts of ash or brine, which may be entrained with the ethane even after careful fractionation and separation.

C.   NAPHTHA TO ETHYLENE

Naphtha is sometimes considered economically more suitable for conversion to ethylene than the previously mentioned two feed streams. With the ever-expanding demand for ethylene, various sources are always being investigated. With the market becoming tight for ethane, propane, and butane, other sources such as naphtha are cost-competitive and are adopted when it is believed that the naphtha will be sufficiently stable pricewise to justify the investment.

Naphtha cracking is effected in both tubular reactors, previously described, and fluidized bed reactors (Fig. 3). For a fluidized bed reactor the question becomes, Is the fluidized solid a catalyst or is it simply a heat transport medium? This is a moot question and will not be discussed here. The carbon, which is deposited on the heat transport solid, is removed by a typical regeneration. The heat transport solid can be sand, alumina, mullite, crushed and screened refractory of varying composition, or a natural occurring grainy material such as zircon sand, rutile, or even slag. Each of these can have special desirable properties that make its use attractive or beneficial. Care must be exercised to avoid iron group metals that can cause carbon deposition greater than that normally experienced.

After carbon removal in the regenerating system the hot sand is recycled to the reactor riser tube, where the reaction occurs, and then passes over into the separation zone for removal from the product and the return of the carbon-coated solid to the regeneration and reheating zone.

There is some evidence that there is a preference for one type of thermal carrier over another type. This lends credence to the contention that there is a catalytic effect. One can effectively also argue that there is no surface that is not catalytically effective under the relatively high-temperature conditions of this reaction.

A major factor in the cracking of naphtha is that there are many more products than with propane and ethane. As a consequence, the separation becomes even more difficult than for the latter two compounds. This factor, plus the unpredictable cost of naphtha, tends to militate against its adaptation as a raw material for the production of ethylene and propylene.

D. METHANE TO ACETYLENE

This reaction is in a class that is almost unique. It is an oxidative dehydrogenation and is not catalytic. It was previously mentioned that the dehydrogenation of ethane to ethylene is a relatively high-temperature operation. In contrast, the dehydrogenation of methane to acetylene is an extremely high-temperature reaction and is effected at temperatures in the range 1300–1600°C where the equilibrium for acetylene is most favored. It is apparent that metallic reactors cannot be successfully used in a partial oxidative reaction with natural gas or methane because the reaction takes place well above the melting point of stainless steel or any of the common metals. The reactors, therefore, are firebrick-lined, with heat exchange and heat removal occurring before the hot gases contact the metallic surfaces. Contact with metallic surfaces at lower temperatures is tolerable, and the final heat exchange is performed in a metal heat exchanger; however, cooling the gases after the extremely high-temperature reaction is accomplished by spraying water directly into the gas stream (Fig. 4), thus converting the spray into steam which becomes part of the product gas from which the heat must be economically and effectively removed. One of the most

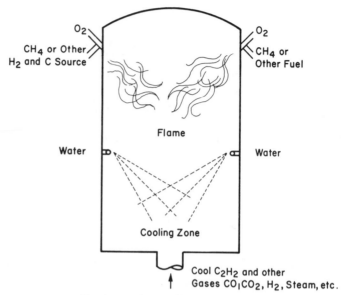

Fig. 4. A quick quench C_2H_2 generator.

critical operations in acetylene synthesis is the quick quench that prevents rehydrogenation of the acetylene to ethylene or ethane.

This operation is also one in which the volume increase during the course of the reaction and the superatmospheric pressure are harmful, and as a consequence the reaction is conducted at the lowest possible pressure which in most cases is 25 psi or less.

The synthesis of acetylene in an electric arc was practiced early in this century. It is mentioned not for its historical significance but from the standpoint that electrical energy may have a more stable price structure than that of fossil energy sources. If this is true, then electric arc production of acetylene may be a viable process in the future. Substantial improvements in the rotating arc process, in which an arc is rapidly rotated in a relatively small circle and gases are passed through the arc, allow greater contact with a larger quantity of gases.

Inasmuch as the feed streams in this partial combustion-type synthesis produce other materials such as large quantities of CO and hydrogen, the economic use of these by-products is essential. The usual coproduct from an acetylene operation of this kind consequently is methanol or methanol plus higher alcohols which have increasing use in chemical- and energy-related fields. The synthesis of methanol is considered separately in Chapter 6.

VI. Catalytic Dehydrogenation

A. METHANOL TO CARBON MONOXIDE AND HYDROGEN

The synthesis of methanol is based on the following reaction:

$$CO + 2H_2 \rightleftharpoons CH_3OH \quad \text{exothermic}$$

The reaction proceeds to the left at elevated temperatures and reduced pressures and to the right at high pressures and low temperatures, namely, $500+$ psi and $200-400°C$. The dehydrogenation occurs at ≤ 200 psi and $> 200°C$.

The purpose of the decomposition reaction is to produce CO and hydrogen synthesis gas for other types of hydrogenation or for carbonylation. Pure hydrogen can be derived from the conversion of carbon monoxide by the CO shift reaction $(CO + H_2O \rightarrow CO_2 + H_2)$ to carbon dioxide which is then scrubbed out; the resultant hydrogen then is high-purity except for a small amount of residual carbon monoxide which can be methanated $(CO + 3H_2 \rightarrow CH_4 + H_2O)$ to a harmless product.

Pure hydrogen is available for hydrogenations and thus provides a convenient source for small plants. As a matter of historical interest, hydrogen was produced this way during World War II for the filling of barrage balloons at the time of the Blitz. After barrage balloons were no longer used in World War II, the same hydrogen-generating system was employed to produce carbon dioxide which, as shown previously in the CO shift reaction, is a coproduct with the H_2. The carbon dioxide was compressed into fire extinguishers, and these in turn were used by the U.S. Air Force.

The methanol dissociation reaction is endothermic, and a tubular reactor, resembling that of Fig. 2, is used. The reaction temperature is $275-350°C$, and the feed consists of vaporized methanol and sufficient water vapor so that the carbon monoxide is converted by the CO shift reaction to carbon dioxide. The carbon dioxide is scrubbed, as previously indicated, in a separate alkylamine scrubber, and essentially pure hydrogen is derived. When the equipment was adapted to peacetime use, a methanation step was added that converted any residual carbon monoxide to methane which is, as previously noted, harmless in most hydrogenation reactions.

For those interested in the so-called hydrogen economy fuel system, the dissociation of methanol into hydrogen and potential hydrogen (the CO shift reaction) produces a hydrogen stream derived from an easily stored and transported liquid (methanol) under very simple conditions manageable possibly even in automobiles or homes.

B. METHANOL DEHYDROGENATION TO FORMALDEHYDE AND HYDROGEN

The dehydrogenation of methanol to formaldehyde and hydrogen is an endothermic reaction and is performed in a reactor, as shown in Fig. 2. It can be effected at 400 to 500°C over a metallic catalyst such as copper, silver, or a copper or silver alloy generally also containing silicon. Oxidative dehydrogenation processes for methanol will be described later in this chapter, and it will then be shown that direct dehydrogenation is not a feasible process and therefore is not presently commercially practiced. There is, however, a desirable feature of the dehydrogenation of methanol to formaldehyde, which is that completely anhydrous formaldehyde should theoretically be obtainable as well as a usable H_2 coproduct. Unfortunately, the facts are that the dehydrogenation is not sufficiently clean (note the dissociation process described earlier) to avoid water formation completely and the leakage of methanol which in many cases is as objectionable a product as water itself. The operation, of course, is conducted at low pressures.

C. CYCLOHEXANE DEHYDROGENATION TO BENZENE

Cyclohexane is an easily transportable liquid and also is nonpoisonous; as a consequence it is looked upon by those interested in the so-called hydrogen fuel economy system as an ideal carrier for hydrogen. The process whereby cyclohexane is dehydrogenated to benzene and hydrogen is effected at approximately 450 to 500°C over a silver or copper metallic catalyst as gauze or as metal distributed on a low-surface-area support. The converter is depicted in Fig. 2. The dehydrogenation is not complete, and some cyclohexane leaks through. This is not a problem ordinarily, but when benzene is the desired product, a separate distillation is required for the purification. In addition to the aforementioned silver and copper, platinum and palladium can also be used for the dehydrogenation.

The hydrogenation of benzene is a very simple, easily accomplished reaction, and as a consequence the hydrogen – benzene – cyclohexane cycle is a relatively attractive one.

D. ETHYLBENZENE DEHYDROGENATION TO STYRENE

This is a large-volume business of substantial commercial importance. Ethylbenzene can be readily made by reacting ethylene with benzene, and

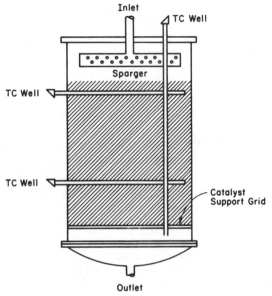

Fig. 5. A simple catalytic reactor.

dehydrogenation of the ethylbenzene is comparably simple, thus easily producing styrene monomer which is widely used in plastics and elastomers. Dehydrogenation is effected at about 425 – 500 °C over an iron oxide – potassium carbonate catalyst. The reactor is of the simple tank type illustrated in Fig. 5. The heat required for the dehydrogenation is supplied as sensible heat in the very large volume of incoming steam, which may be in as high as a ratio of 10 : 1 versus benzene. It should be recognized that the function of the steam is not only to supply heat but also to maintain the catalyst in a relatively high state of iron oxidation. Carbon deposition is also beneficially affected by the high steam/benzene ratio.

Styrene, although now manufactured primarily by the foregoing method, is also manufactured by an oxidative dehydrogenation process recently described in the literature and subsequently described in this chapter.

E.   ISOPROPYL ALCOHOL DEHYDROGENATION TO ACETONE

Although most acetone is derived from the oxidative splitting of isopropylbenzene, whereby not only acetone but also phenol is produced, it is interesting to note that, at one time, a large amount of acetone was made by the dehydrogenation of isopropyl alcohol. This is effected in a tubular reactor as shown in Fig. 2 at 270 ± 20°C and over a copper – zinc or a silver catalyst. The catalysts generally are these metals or metal alloys on a refractory support.

F. AMMONIA DEHYDROGENATION TO NITROGEN AND HYDROGEN

This is a process whereby hydrogen has been produced for hydrogenation and the derivation of N_2 for nitriding of metals in a reducing atmosphere. The anhydrous ammonia is passed through a tubular reactor like that shown in Fig. 2 at a temperature of approximately 600°C. The pressure is essentially atmospheric or as high as 25 psi, and catalysts that are effective are reduced iron oxide, nickel oxide, and elemental ruthenium on activated or α-alumina.

In addition to the derivation of hydrogen, this process also has been used for the derivation of nitrogen by burning the hydrogen produced to generate heat and for other applications; then the nitrogen of the ammonia, coupled with the nitrogen of the air used as an oxygen source, produces a very large volume of inert gas useful for the blanketing of tank cars, ships, tanks in tank farms, and many other uses.

G. TETRALIN DEHYDROGENATION TO NAPHTHALENE AND DECALIN DEHYDROGENATION TO TETRALIN

Tetralin is a hydrogen donor (carrier) in many currently considered coal liquefaction schemes. It is fed as a vehicle for pulverized coal, and two effects are achieved. The first is solubilization of the coal in the tetralin, and the second is hydrogenation of the coal by the transfer of hydrogen from the tetralin to the hydrogen-deficient coal. Inasmuch as the naphthalene that results is easily hydrogenated to tetralin and with somewhat more difficulty to decalin, this route to coal processing eliminates the need for passing hydrogen into the reactor where the coal is liquefied. Use of the hydrogen donor principle simplifies many of the gas addition and distribution problems that occur where it is necessary to meter and handle H_2.

Tetralin can be dehydrogenated at a comparatively low temperature of about 135°C, which is within the lower segment of the temperature range of the hydrogenation of coal.

Catalysts that are useful for this dehydrogenation are nickel, platinum, rhodium, copper, and copper oxide – zinc oxide or copper – zinc alloy in the form of a type of brass. In coal hydrogenation, coal ash appears to be an adequate catalyst.

VII. Oxidative Dehydrogenation

A. METHANOL CONVERSION TO FORMALDEHYDE

Combination of Dehydrogenation and an Oxidative Dehydrogenation Fuel-Rich System

There are two general processes for the conversion of methanol to formaldehyde by oxidative dehydrogenation. One is conducted with a deficiency of oxygen and an excess of methanol, whereas the other is conducted with just the reverse, an excess of oxygen and a low percentage of methanol. These two conditions are selected not only because of the catalyst characteristics

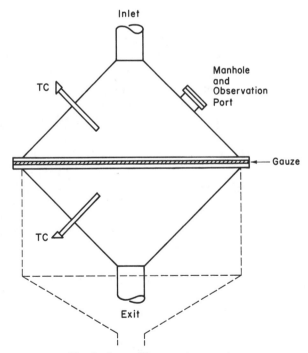

Fig. 6. A metallic gauze-type reactor.

but also because they represent positions just beyond the extremes of the explosive limits.

Fuel-rich dehydrogenation is conducted over a silver catalyst, usually silver gauze, and is accomplished in a reactor of the type depicted in Fig. 6. Usually the reactor is situated above a heat exchanger which is also a boiler generating steam. The products of the reaction at elevated temperatures (650–700°C) are cooled to about 150°C, and simultaneously large quantities of steam are generated. The silver-catalyzed methanol dehydrogenation is unique in that two reactions occur simultaneously, as indicated by

$$CH_3OH + O_2 \rightarrow HCHO + H_2O \quad \text{exothermic}$$
$$CH_3OH \rightarrow HCHO + H_2 \quad \text{endothermic}$$

The foregoing two reactions occur under essentially the same temperature and pressure conditions. The endothermicity and exothermicity balance, and the reaction is essentially 70% complete (unreacted methanol accounts for the remaining 30%). Approximately half of this 70% total reaction is effected by the exothermic oxidative dehydrogenation which in turn supports the endothermicity of the remaining reaction. The reaction is quite efficient, producing approximately a 95% yield from the 70% that is converted.

B. METHANOL TO HCHO VIA OXIDATIVE DEHYDROGENATION

Oxidative dehydrogenation is effected in a reactor of the type shown in Fig. 2. The tubes are approximately 1 in. in diameter and are housed in a heat transport liquid such as Dowtherm or a salt mixture consisting of sodium and potassium nitrates and nitrites (Hytec). This reaction is conducted with a methanol/air ratio of approximately 1 : 13 and a temperature in the range 290–425°C. The catalyst employed in this case is an iron molybdate having a ratio of approximately one iron molecule to three to five molybdenum molecules. It may also consist of bismuth molybdate, which will be described more fully subsequently in this chapter. The conversion in the case of the iron molybdate catalyst is approximately 100%, there being less than 1% unreacted methanol. The yield depends upon conditions and varies from approximately 90 to 96%.

Little or no formic acid is produced in either of the two previously described reactions. The advantages of each are, first, in the case of iron molybdate, the high conversion and the high yield. The disadvantage is that the catalyst does not function for as long a period as the silver gauze used in the fuel-rich system. The maintenance, however, for the tubular reactor used in an oxide-type system is two to three times that required for the silver gauze system. The catalyst change is extremely difficult and time-consuming, and pressure drop is a serious problem (usually the cause of shutdown).

The advantages of the silver gauze system are the low cost and low maintenance of the unit, but there is a severe penalty in that a large quantity of the methanol is not reacted and this must be dealt with by one of several methods. One technique is to distill and separate the methanol and recycle it. The second is to have a dual converter system in which air is added between the first converter and the second and the reaction is carried to completion in the second reactor. The problem with this is that the yield, instead of being in the 95% range as previously described, is in the 80% range. The penalty for the second stage consequently is very severe.

C. OXIDATIVE DEHYDROGENATION OF METHANE AND AMMONIA TO HCN

The next reaction to be considered is that of a hydrocarbon, methane in particular, plus ammonia and oxygen, producing HCN and H_2.

$$NH_3 + CH_4 + O_2 \rightarrow HCN + CO + CO_2 + H_2$$

This reaction is conducted in a reactor of the type depicted in Fig. 5. The reactor in this case is similar to the previously described silver gauze formaldehyde reactor situated in such a way that the effluent gases pass directly into a heat exchanger which is also a boiler. The temperature of the reaction is approximately 1250°C. The catalyst is typically platinum, rhodium gauze, or platinum–rhodium supported on a granular ceramic, making a particulate catalyst. The yield is approximately 90%, and the conversion is approximately 80%. A noteworthy parallel exists with the fuel-rich HCHO process in that in the oxidative synthesis of HCN the reaction is incomplete and efforts to make the conversion more nearly complete cause a sharp deterioration in the yield. It is possible that in both cases the phenomenon known as hydrogen poisoning, which a self-poisoning or product poisoning effect, is encountered. This is a major subject and will not be described more fully here.

D. DEHYDROGENATION OF BUTENE TO BUTADIENE

This reaction is one in which a tubular reactor (Fig. 2) is used almost exclusively. There are indications that a fluidized system such as that employed in the acrylonitrile route described below could be used. The catalyst of choice is of the bismuth molybdate type promoted in many different ways. This is evident from patents derived at various industrial laboratories such as those at Phillips Petroleum and Standard Oil of Ohio. The conditions of reaction are approximately 425–490°C with a diluted air–butene feed. Yields are approximately 80%, and utilization of the oxygen is approximately 100%.

E. PROPYLENE AND AMMONIA PLUS OXYGEN TO ACRYLONITRILE

This is one of the reactions that has probably been researched as much as any of the other well-known and economically important reactions. There are many patents in this field for different types of catalysts for the reaction, all of them having characteristics slightly different from the others but basically all producing yields based on ammonia utilization of approximately 70%. Crude acrylonitrile always contains a small amount of acetonitrile. Unreacted ammonia and unreacted propene may present both disposal and economic problems, and as a consequence there has been a considerable effort to increase the yields from both of these ingredients.

The reaction can be conducted either in a fluidized bed of the type shown in Fig. 3, with major adaptations, or in a fixed bed reactor such as the tubular variety shown in Fig. 2. Both processes are conducted in the vapor phase over a bismuth phosphomolybdate-type catalyst but with many modifications via the inclusion of colloidal silica, alkali, iron, cobalt, and rare earth metals.

During the course of use, the catalyst tends to be reduced to a catalytically inferior form, and it is consequently necessary to oxidize and regenerate it repeatedly.

F. CYCLOHEXANE DEHYDROGENATION TO CYCLOHEXANOL AND CYCLOHEXANONE

A purist might argue that this is not a dehydrogenation reaction, but inasmuch as hydrogen is removed from the cyclohexane when both (cyclohexanone and cyclohexanol) are made, it can be argued that it is indeed a dehydrogenation reaction. Another reason for describing this reaction is that it is one of the few liquid phase dehydrogenation reactions and involves liquid cyclohexane being maintained in a tank in a liquified condition under carefully controlled temperature and pressure conditions. The temperature

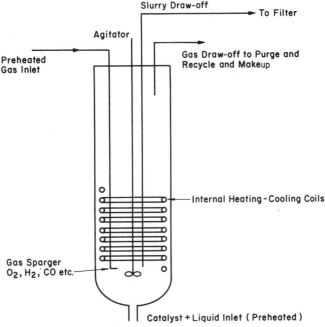

Fig. 7. A slurry converter.

is ordinarily about $150 \pm 25°C$, and the catalyst for the operation is a homogeneous catalyst such as cobalt or nickel naphthenate. The reactor is a liquid-phase tank with an air sparger and usually internal cooling coils to maintain a relatively constant temperature. This reactor is depicted in Fig. 7. Ordinarily the reaction is conducted in such a way that only a comparatively small amount of the reaction, such as 5–15%, is effected. The oxidized products are separated from the cyclohexane by distillation, and the cyclohexane is recycled. It is obvious that the reaction could also be conducted in a tray-type reactor similar to a distillation column in which the oxygen passes upstream and the liquid passes downstream in the column. This type of reactor is not shown here, but it can be visualized as a simple bubble cap distillation column with a top liquid feed and a bottom air inlet.

The combined yield of cyclohexane and cyclohexanol is approximately 85%. The by-products are those that would be expected to occur when the reaction intermediate cyclohexyl hydroperoxide decomposes and generates free radicals. These are prone to attack other ingredients in the reaction medium. This reaction is well-publicized as a route not only to cyclohexanone and cyclohexanol but also eventually to adipic acid in a separate step.

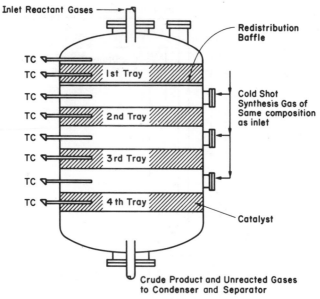

Inlet Reactant Gases

Redistribution
Baffle

TC

TC   1st Tray

TC

Cold Shot
Synthesis Gas of
Same composition
as inlet

TC

TC   2nd Tray

TC

TC   3rd Tray

TC

TC   4th Tray

Catalyst

Crude Product and Unreacted Gases
to Condenser and Separator

Fig. 8. A tray-type synthesis converter.

G. DEHYDROGENATION OF ETHYLENE PLUS ACETIC ANHYDRIDE TO VINYL ACETATE

This reaction is conducted in the vapor phase through a tubular reactor at surprisingly low temperatures (200–300°C). There is a substantial amount of exothermic heat which can be removed by a large recycle of nitrogen or another inert gas. Even ethylene in excess of that reacted can be used if the economics of recycling can be justified. In the case in which the ethylene becomes the heat sink, it is not necessary to utilize a tubular reactor, and a tray-type reactor like that illustrated in Fig. 8 is satisfactory. The catalysts employed are generally catalysts supported on silica or alumina, and the catalyst metals are generally precious metals sometimes modified with a small amount of an iron group and an alkali such as sodium or potassium.

H. DEHYDROGENATION OF METHANE TO FORMALDEHYDE AND METHANOL

The oxidative dehydrogenation of methane to formaldehyde and methanol has been extensively researched, because if methane could be oxidized directly to formaldehyde and methanol, it would eliminate the synthesis gas generation and methanol synthesis steps which are both expensive and energy-intensive. The reaction is conducted in a tubular-type reactor like those depicted in Figs. 1 and 2, and the temperature is in the range 450–600°C. The catalyst is unique in our considerations thus far in that it is a vapor phase homogeneous catalyst, nitrogen oxide. The yields are relatively high, but the conversion is extremely low, being between 2 and 4% of the methane feed. The methanol and formaldehyde are about equally divided in the product.

The process has within the last 8–10 yr received a great deal of attention in the USSR, and attempts have been made to market it in the United States, but it has not had the economic advantages necessary for its adaptation. A plant of this type was operated by Cities Service in Oklahoma for many years, but several years ago it too was shut down.

I. OXIDATIVE DEHYDROGENATION OF ETHYLBENZENE TO STYRENE

This reaction is frequently referred to in the literature, and the following is a typical example. To the author's knowledge, however, this process is not practiced commercially.

Catalytically active aluminas prepared from aluminum propoxide and aluminum sulfate and calcined at 800 to 1300°C were studied for the oxidative dehydrogenation of ethylbenzene to styrene. Higher yields of styrene were obtained when the alumina prepared from aluminum propoxide was used as a catalyst.

J. DEHYDROGENATION OF BUTYL MERCAPTAN TO THIOPHENE

This process is mentioned in this chapter because it represents a dual-functioning system. Not only is the butyl mercaptan dehydrogenated but it is also cyclized and then the thiophene is formed. The operation is conducted at 350 to 400°C in the presence of hydrogen sulfide, and the catalyst is molybdenum disulfide on tungsten disulfide supported on alumina or silica. The type of reactor used is the tubular reactor illustrated in Fig. 2.

K. MISCELLANEOUS DEHYDROGENATIONS

These dehydrogenations are mentioned only because of the uniqueness of the catalyst used. These reactions convert ethyl succinate to ethyl fumarate and acetone to methyl glyoxal; supported selenium dioxide is used as the catalyst. The support can be alumina, silica, zirconium, titanium, or cerium oxide.

CHAPTER 5

The Sasol Fischer – Tropsch Processes

MARK E. DRY

Sasol Technology (PTY) Limited
Sasolburg, South Africa

I. Introduction

Approximately 70 yr ago the feasibility of producing liquid fuels from H_2 and CO with cobalt catalysts was demonstrated in Germany [1]. Fischer and his many co-workers, in particular Tropsch and Pichler, developed the process further, and by 1936 four Fischer–Tropsch production plants were

Applied Industrial Catalysis, Volume 2

in operation [2]. At the end of World War II all synthetic fuel production in Germany ceased, but research on and development of the Fischer–Tropsch process continued, particularly in the United States. However, this too came to an end in the mid-1950s when large petroleum oil deposits were discovered in the Middle East. For full details of and references to the research on and development of the Fischer–Tropsch process during this period the reader is referred to the reviews by Storch *et al.* [3], Pichler [4], Anderson [5], and Frohning *et al.* [2].

From 1955 to 1980 the only operating Fischer–Tropsch plant was situated in Sasolburg, South Africa, and further research and development geared toward understanding and improving the process continued there. These investigations and also those carried out over the same time period at research establishments and industrial plants in other countries have been reviewed by Dry [6]. The preceding review covers the development of the various types of reactors, the theoretical and practical aspects of the Fischer–Tropsch product spectrum, the mechanism and kinetics of the reaction, and the preparation and characteristics of the catalysts used. This chapter deals mainly with the Fischer–Tropsch process as practiced commercially by Sasol using iron-based catalysts, the technological improvements made over the years, future developments in motor fuel production using combinations of Sasol and other technologies. It should be understood that a large part of the results and experience accumulated over the years is of a proprietary nature and is not available for publication.

II. Short History of Sasol

In the early 1930s the mining house Anglovaal was involved in extracting oil from shale. Because the shale deposit was associated with a coal seam, Anglovaal, in a logical extension of its interests, obtained a license in 1935 to build a Fischer–Tropsch plant [7]. However, World War II intervened, and Anglovaal resumed this endeavor only in the late 1940s. Because of the high cost of the project and because Anglovaal could not obtain a loan from the World Bank, the project was handed over to the South African government in 1950. Sasol was registered as a public company the same year, with the Industrial Development Corporation holding all the shares. Construction of the first plant, now known as Sasol One, commenced in 1952 at Sasolburg, and by 1955 the synthesis units were producing hydrocarbons.

The Sasol project was considered economically viable because of the proximity of the plant to the highest area of consumption (Johannesburg) and the low cost of the coal, $0.60 (U.S.)/ton as against $1.50 (U.S.)/barrel of crude oil. However, it was not foreseen that the huge oil deposits in the Middle East would hold down the price of oil for many years to come. As a result of initial technical problems Sasol started to show a profit only in 1960, and because of the low crude oil prices Sasol's continued growth was slow but nevertheless steady. Gasification was expanded to supply "town gas" to the Johannesburg industrial area. A nitrogen fertilizer plant, a butadiene plant, and a naphtha cracking plant to produce ethylene were also added at later stages.

With the Middle East oil crisis of 1973 and the subsequent rapid rise in the price of crude oil the profitability of the Sasol One plant rose dramatically. Because of this and because of the uncertainty of future oil supplies, it was decided in 1975 to build another, much larger, Fischer–Tropsch plant at Secunda. The Sasol Two plant came on stream in 1980. As a result of the continuing deterioration of the crude oil situation, it was decided in 1979 to build Sasol Three. This plant is virtually identical to the Sasol Two plant and came on stream in 1982. In 1979 Sasol shares were offered to the public on the Johannesburg Stock Exchange, and the offer was 30-fold oversubscribed.

III.   Synthesis Gas Production

A.   COAL GASIFICATION

At the present Sasol plants Lurgi dry ash gasifiers are used to produce synthesis gas from low-grade noncaking coal. These gasifiers cope well with this coal which contains from 20 to 40% ash. The original gasifiers at Sasol One have an internal diameter of 3.6 m and were designed to produce about 25,000 m_n^3/hr of raw gas each, where the subscript n indicates normal conditions. With time the output per gasifier was steadily improved to average $\sim$ 35,000 m_n^3/hr. Under optimum conditions 48,000 m_n^3/hr has been achieved [8]. This improvement was a result of several factors. It was realized that the ash content had an influence on the ash melting point, and this led to operation under optimum conditions near the clinkering limit. The lowering of the steam inlet temperature resulted in steam savings and higher gas production rates [9]. The gasifier is a countercurrent operation with hot ash exchanging heat with the gasification agent at the bottom and the hot product gases heating, devolatizing, and drying the coal fed in at the top (Fig. 1). For the production of 1000 m_n^3 of raw gas from coal mined at Sasol One $\sim$ 157 m_n^3 oxygen and 850 m_n^3 of steam are used. Excess steam is fed in to control the temperature in the fire zone. The normal operating

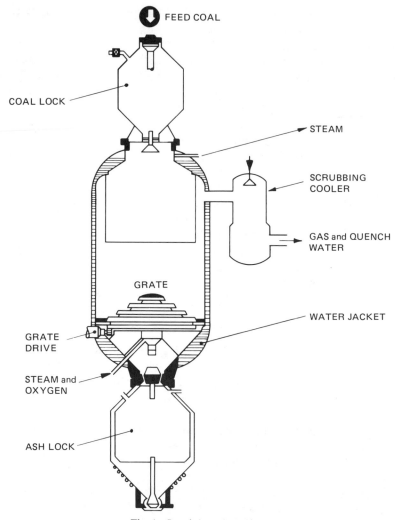

FEED COAL

COAL LOCK

STEAM

SCRUBBING
COOLER

GAS and QUENCH
WATER

GRATE

WATER JACKET

GRATE
DRIVE

STEAM and
OXYGEN

ASH LOCK

Fig. 1. Lurgi dry-ash gasifier.

pressure of the Lurgi gasifiers is ∼ 27 atm. Figure 13 is a view of a bank of Lurgi gasifiers.

The gasifiers at Sasol Two and Sasol Three are larger, with an internal diameter of 3.85 m, and are designed to produce 55,000 m_n^3/hr of raw gas. A 5.0-m-i.d. gasifier was installed at the Sasol One plant in 1980 and has been operating ever since.

B. GAS PURIFICATION AND ASSOCIATED PROCESSES

The gasifier effluent is cooled to condense out tars, oils, and excess steam. The lighter oil is hydrofined and added to the gasoline pool. At Sasol One the heavier cuts are sold as such (creosote and pitch). At Sasol Two and Sasol Three part of the pitch is recycled to the gasifiers, and the creosote oil is hydrofined to gasoline and diesel fuel. The phenols dissolved in the steam condensate are recovered by countercurrent solvent extraction (butyl acetate or diisopropyl ether) at the Lurgi Phenosolvan plant, and the NH_3 is then steam-stripped from the water. The phenols are refined and sold, and the NH_3 is converted to fertilizer. The remaining water is biologically treated and reused at the complex. (Sasol Two and Sasol Three are zero-effluent plants.)

The raw synthesis gas, which has the approximate molar composition 9% CH_4, 29% CO_2, 0.5% H_2S, 1% $(A + N_2)$, and 60% $(H_2 + CO)$, is further purified in the Lurgi Rectisol process. The gas is washed in stages with methanol down to $-55°C$. The composition of the purified gas is about 13% CH_4, 1% $(A + N_2)$, 1% CO_2, and 85% $(H_2 + CO)$. The total sulfur level of the gas is typically 0.03 mg/m_n^3. The composition of the sulfur-containing gases in the raw gas is about 97% H_2S, 2% CH_3SH, and 1% $(COS + CS_2)$, whereas in the purified gas it is about 20% H_2S, 50% CH_3SH, and the balance $COS + CS_2$. Not only does the Rectisol process decrease the sulfur to the low levels required, but it also removes the remaining contaminants such as tar naphtha vapor, ammonia, and cyanide.

IV. Fischer–Tropsch Reactors

A. COMMERCIAL UNITS

1. Introduction

Sasol One uses two types of reactors. The fixed bed reactors (Fig. 2) produce mainly heavy liquid hydrocarbons and waxes, and the transported fluidized bed reactors (Fig. 3) make predominantly gaseous hydrocarbons and gasoline. The fixed bed reactors were developed jointly by Lurgi and Ruhrchemie, and they performed with little trouble when commissioned in 1955. The circulating fluid bed reactors were scaled up directly from a 10-cm-i.d. pilot unit by Kellogg and were thus the first such units to be built. They performed poorly for several years, and only after numerous process

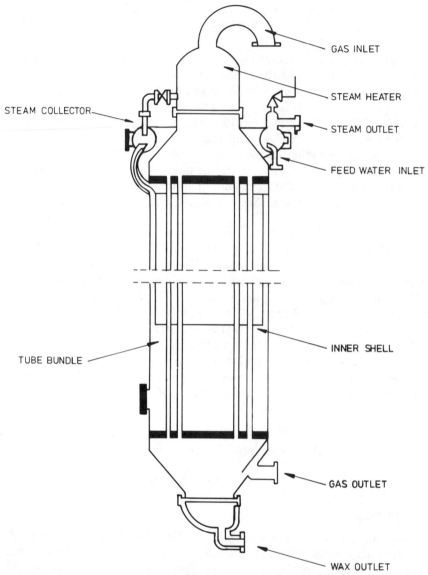

GAS INLET

STEAM HEATER

STEAM COLLECTOR

STEAM OUTLET

FEED WATER INLET

TUBE BUNDLE

INNER SHELL

GAS OUTLET

WAX OUTLET

Fig. 2. Sasol commercial fixed bed Fischer–Tropsch reactor. Only a few of the 2050 catalyst tubes are shown.

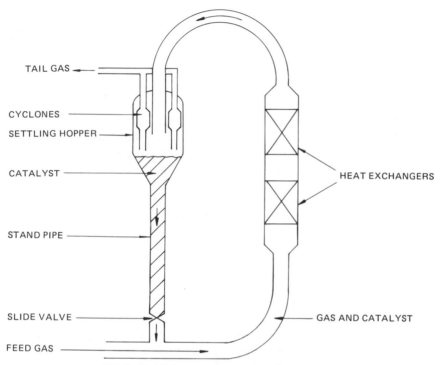

Fig. 3. Synthol reactor. Transported fluidized bed.

and mechanical modifications had been made, as well as changes in the formulation of the catalyst used, did these units operate satisfactorily. The units are now known as Sasol Synthol reactors.

The Synthol reactors have a much higher gas throughput than the fixed bed reactors. For this reason it was decided to build only Synthol reactors at Sasol Two and Sasol Three. The throughputs of the individual reactors were increased about threefold by increasing the reactor diameters, as well as by raising the operating pressure.

2. Fixed Bed Reactors (Arge)

Each Sasol One reactor consists of 2050 tubes of 50-mm i.d. and 12 m in length into which the catalyst is packed. Each tube contains ~ 20 liters of catalyst. The tubes are surrounded by water. By controlling the steam pressure above the water the desired reactor temperature is maintained. The normal operating pressure is ~ 27 atm, and the temperature can vary from 220 to 250°C. The Fischer–Tropsch reaction is highly exothermic (~ 36 kcal per reacted carbon atom), and to ensure a high rate of heat exchange between the catalyst particles and the tube walls a high gas linear velocity is essential. To obtain a high degree of conversion as well as a high gas linear velocity, part of the tail gas is recycled. Typically the recycle/fresh feed volume ratio is ~ 2. The synthesis gas enters at the top of the reactor where it is preheated and then flows through the reactor tubes. A large fraction of the hydrocarbon product is in the liquid phase inside the reactor. The catalyst's activity as well as its selectivity toward producing wax declines with age. By progressively raising the reactor temperature the conversion can be maintained constant for about a year. However, because the main objective is the production of large amounts of high-quality wax, this operating procedure is not followed and the frequency of catalyst change is entirely dictated by the wax demand.

The production capacity of the reactors has been increased considerably over the design value by improved process conditions, the development of more active catalysts, and shortened conditioning periods and catalyst turnaround times.

3. Synthol Reactors

The Synthol reactors can be described as circulating or transported fluidized beds. The overall height of the reactors is ~ 50 m. As depicted in Fig. 3, the gas (fresh feed plus recycle) is introduced into the bottom of the reactor where it meets a stream of hot catalyst flowing down the standpipe. This preheats the gas to its ignition temperature. Gas plus catalyst then flows up through the right-hand-side reaction zones. The two banks of heat exchangers inside the reactor remove a large portion of the reaction heat, the balance being absorbed by the recycle and product gases. The catalyst disengages from the gas in the wider settling hopper and flows down the standpipe to continue the cycle. The rate of catalyst flow is controlled by the slide valve at the bottom of the standpipe. The unreacted gas, together with the hydrocarbon product vapors, leaves the reactor via cyclones which remove the entrained finer catalyst particles and return them to the hopper. The reactor exit temperature is typically ~ 340°C. It is important that the process conditions and also the catalyst formulation are such that the production of heavy hydrocarbons is limited, because these would condense on the catalyst and result in defluidization of the bed. Because the iron catalyst used has a high density, it is intrinsically more difficult to fluidize than, for instance, the silica–alumina catalysts used in circulating bed catalytic crackers. To ensure the necessary fluidization quality, the particle size and the distribution of sizes must be closely controlled so that the flow down the standpipe as well as up the reactor is satisfactory.

The run lengths depend on whether or not catalyst on-line removal and addition are practiced. With on-line catalyst removal and addition the runs are much longer and are effectively determined by scheduled maintenance requirements.

The new Synthol reactors at Sasol Two and Sasol Three not only have larger diameters, but several improvements have been incorporated, e.g., improved heat exchangers and settling hopper configuration.

B. EXPERIMENTAL REACTORS

1. Fixed Fluidized Reactor

In a fixed fluidized bed (FFB) reactor the gas passes upward through the catalyst bed at velocities in the region of 10 to 60 cm/sec (calculated on a empty reactor basis under actual synthesis conditions). The catalyst bed, though expanded, is not transported and remains in a fixed position (Fig. 4). Sasol has operated 5-cm-i.d. FFB pilot reactors for many years. In spite of higher gas-to-catalyst loadings the conversions in these reactors are higher than those of the Synthol commercial reactors. The large demonstration FFB unit operated in Brownsville, Texas, in the early 1950s was initially plagued by low conversion attributed to poor catalyst fluidization [10]. Nevertheless, believing the FFB reactor to have potential, Sasol investigated the fluidization characteristics of its heavy iron catalyst in large Plexiglas models. Together with the Badger Company effective gas distribution nozzles were developed. It was also found that the addition of charcoal powder markedly improved the quality of fluidization [11]. A high-pressure demonstration unit was subsequently designed, which was scheduled to come on line in early 1983.

In the Sasol pilot units pressures up to ~ 80 atm have been investigated. The percentage of conversion was found to be independent of the pressure. The fresh feed and recycle flows are always increased in proportion to the increase in pressure. Thus, when going from 20 to 80 atm, the actual hydrocarbon production rate also must increase fourfold. No reaction heat exchange problems were encountered in the 5-cm-i.d. pilot reactor. Thus FFB reactors have a potential for increased synthesis gas throughputs.

2. Slurry Bed Reactors

In the slurry bed the finely divided catalyst is suspended in a heavy oil (preferably a product of the Fischer–Tropsch reaction itself), and the gas bubbles upward through the slurry. This type of reaction was developed by Kölbel and is claimed to be more versatile than other types of reactors [2]. For instance, it can be fed a low H_2/CO ratio gas at high temperatures, which

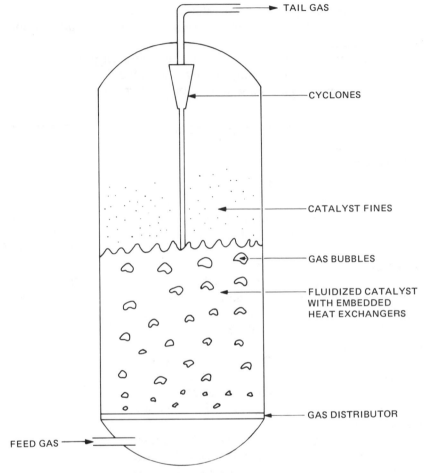

TAIL GAS

CYCLONES

CATALYST FINES

GAS BUBBLES

FLUIDIZED CATALYST
WITH EMBEDDED
HEAT EXCHANGERS

GAS DISTRIBUTOR

FEED GAS

Fig. 4. Fixed fluidized bed reactor.

would result in bed plugging (due to carbon laydown) in the case of a fixed
bed reactor, or at low temperatures, which would result in defluidization
(due to wax laydown) in the case of a "dry" fluidized bed reactor. Sasol has
compared all three systems (fixed, fluidized, and slurry) in 5-cm-i.d. pilot
reactors [6]. In the low-temperature, wax-producing mode, there was little
difference between fixed and slurry bed reactors. In the high-temperature,
gasoline-producing mode, the fluidized bed had a much higher conversion
than the slurry bed reactor. Nevertheless, Sasol is continuing its investigations of the slurry bed system mainly because of the likelihood of lower
construction costs relative to those of multitube fixed bed reactors. In a

slurry bed reactor geared toward wax production it is necessary to separate the finely divided catalyst from the product wax, and this technique needs to be fully developed. (The fixed bed reactor of course does not have this problem.)

3. High-Pressure Fixed Bed Tubular Reactors

Operation at higher pressures in these reactors has also been investigated at the Sasol pilot plant. When the pressure was doubled, from say 30 to 60 atm, the fresh and recycle flows were also doubled, and so the actual gas linear velocity through the reactor was kept constant at the required high level (Section IV.A.2). As found in the FFB studies, the percentage of conversion remained unchanged. Thus doubling the pressure means doubling the reactor's production capacity. Furthermore, neither selectivity nor catalyst life appears to be adversely affected. Thus fixed bed tubular reactors of the type presently used commercially at Sasol One also have a potential for much higher gas throughputs.

V.   Iron Catalysts

To date Sasol has exclusively used iron catalysts in its commercial reactors. Not only are they much cheaper to manufacture than their cobalt and ruthenium equivalents, but the products are more olefinic. As will be explained in Section VII, olefins play a key role in the workup process geared toward maximizing gasoline and/or diesel fuel production.

A.   PREPARATION AND CHARACTERISTICS

1. Fixed Bed Catalysts

The catalyst is prepared by precipitation techniques, supported by silica, and extruded. The catalyst originally used at Sasol was manufactured in Germany by Ruhrchemie. The preparation of this catalyst has been described by Frohning et al. [2]. The successive preparation steps were as follows. A hot solution of iron and copper nitrate was poured into a hot solution of sodium carbonate, with vigorous stirring, to lower the pH to ~ 7. The gelatinous precipitate was washed thoroughly to remove the sodium nitrate. The precipitate was then reslurried with water, and the required amount of potassium water glass added to give ~ 25 g $SiO_2/100$ g Fe). Nitric acid was added to adjust the potassium level to the desired value (~ 5 g

K₂O/100 g Fe). The slurry was filtered again, and the cake extruded and dried.

After using this catalyst for several years it was decided to manufacture it locally at Sasolburg. As a result of Sasol investigations, the catalyst was progressively improved and the production techniques modified and streamlined. The levels of the promoters, K_2O and SiO_2, were optimized, and other promoters were added to improve the conversion stability. The shape and size of the extrudates were also modified.

The influence of various preparative factors, such as type of chemicals, solution concentrations, temperature, pH, and time of addition, on the surface area and pore size distribution of the final catalyst was investigated [6].

To obtain high activity, two factors are of prime importance, the total internal surface area of the catalyst and the external area of the extrudates. The latter factor indicates that the reaction is diffusion-controlled. The smaller the extrudate size the higher the activity, but the differential pressure drop over the reactor is also higher, hence the higher the gas compression costs. Because of these factors a compromise must be made as far as extrudate size and shape are concerned. The internal surface area is controlled mainly by altering the amount of silica carrier incorporated. Table I illustrates the influence of the SiO_2 on the area of the catalyst. Although the total area progressively increases with the SiO_2 content, the amount of iron metal area (measured by CO chemisorption after reduction) decreases again beyond a certain SiO_2 level.

The catalyst is normally activated by reduction with hydrogen prior to loading it into the reactor. The presence of a small amount of copper in the catalyst enchances the rate of reduction, and this makes it possible to reduce

TABLE I

**The Influence of SiO₂ on the Surface
Area of Precipitated Catalysts**

SiO₂ content (g SiO₂/100 g Fe)	Total surface area (m²/g)	
	Unreduced catalyst	Partially reduced catalyst
0	275	35
8	345	190
19	375	250
25	390	270
29	370	265
50	405	280

the catalyst at a temperature close to that used during the subsequent synthesis. The lower the reduction temperature the less the extent of sintering, hence the formation of high-area, i.e., high-activity, catalysts. As a result of the reduction process, the extrudates shrink in volume by ~ 25%. Prereduction before loading into the fixed bed reactors therefore allows a greater volume of catalyst to be loaded. Sasol [12] has found that the required preshrinkage can also be achieved by simply soaking the unreduced catalyst in hot wax for several minutes. This allows prereduction to be carried out in the synthesis reactors themselves.

The reduction rate is very high initially but tapers off rapidly. This slow second stage is ascribed to a large part of the iron oxide being chemically complexed with the silica support [6]. The surface area of the reduced catalyst is lower than that of the unreduced catalyst (Table I). During Fischer–Tropsch synthesis the reduction process continues with the simultaneous conversion of the metallic iron to Hägg carbide (Fe_5C_2). The surface area continues to decrease and is accompanied by a shift in the pore size distribution toward larger pores [6].

2. Fluidized Bed Catalysts

As the catalyst in the Synthol reactors is turbulently transported at velocities of several meters per second, it is important that the particles have sufficient mechanical strength. At Sasol the catalyst is prepared by electrically fusing suitable iron oxides together with the required promoters. The fused ingots are then crushed to a fine powder having a specific particle size distribution to ensure satisfactory flow properties (Section IV.A.3). Sasol has investigated and used several iron sources, e.g., mill scale from steel works and high-grade iron ores such as magnetite and hematite. At the fusion temperature and in contact with air the stable iron oxide phase is magnetite, Fe_3O_4. From studies on the influence of various promoters on the unit cell size of the magnetite crystal lattice, as well as from direct microscopic investigation of highly polished sections of the fused catalyst, it was deduced that promoters or impurities such as Al_2O_3, MgO, TiO_2, and Cr_2O_3 form true solid solutions with the magnetite phase, whereas others such as SiO_2 and K_2O are present as minute occlusions of silicates dispersed throughout the fused ingots [6].

The fused and milled catalyst has a very low surface area (<1 m^2/g), but when it is reduced with hydrogen, its area increases to up to 30 m^2/g, depending on the reduction temperature and on the type and amount of promoters added. Figure 5 illustrates the influence of several additives on the surface area of the reduced catalyst [13]. The more basic the promoter cation (the lower the ratio of the ionic charge to the ionic radius), the smaller

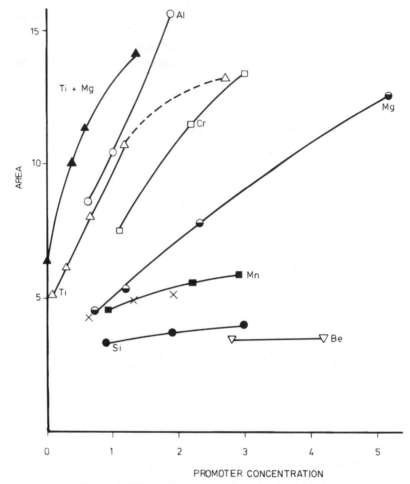

Fig. 5. The surface area of fused magnetite catalysts after full reduction. The area (square meters per gram of unreduced sample) is shown as a function of the promoter content (gram atoms of promoter cation per 100 g atom Fe).

the promotional effect. Strong bases such as Na_2O and K_2O in fact result in a decrease in surface area [6].

There is an inverse relation between the rate of reduction and the final surface area of the reduced catalyst. Thus a promoter like Al_2O_3 decreases the rate of reduction but increases the area. The presence of water vapor has a strong depressing effect on the rate of reduction. Because water is an inevitable product of the reduction with hydrogen, the use of high hydrogen linear velocities throughout the bed is recommended, as this increases the H_2/H_2O molar ratio and so increases the rate of reduction [6].

When synthesis gas is introduced to a reduced iron catalyst, the metallic phase is rapidly converted to the carbide phase (Hägg carbide, Fe_5C_2). At high pressures, or after a certain time on stream, a higher carbide (Eckstrom Adcock carbide, Fe_7C_3) is also formed [6]. As the synthesis progresses, some of the carbide phase is converted to oxide. The fact that the oxide level in larger catalyst particles is higher than in smaller particles indicates that oxidation occurs in the core of the particles. This is only to be expected when it is borne in mind that as the synthesis gas diffuses into the catalyst particle the H_2 and CO react to form hydrocarbons and H_2O; i.e., the deeper the penetration, the higher the H_2O/H_2 ratio becomes and hence the higher the oxidation potential. Because of this phenomenon, it is in fact not necessary to reduce the magnetite fully in the first place, because the cores of the particles in any event are reoxidized during synthesis. Catalysts promoted with K_2O carburize more rapidly and resist reoxidation better than catalysts not containing K_2O [6].

Although the addition of structural promoters such as Al_2O_3 and MgO results in higher area catalysts, they have little effect on one of the key characteristics of Fischer–Tropsch catalysts, namely, hydrocarbon product selectivity. This is dominantly controlled by a chemical promoter such as K_2O, as discussed in Section VI.B.

B. ACTIVITY DECLINE

Although under Synthol operating conditions a considerable amount of carbon is deposited on the catalyst, this does not appear to lower the activity, as discussed further in Section V.C. Under normal Fischer–Tropsch operating conditions conversion of the active carbide phase to the catalytically inactive oxide phase does not contribute much to activity decline because, as discussed in Section V.A, this oxidation occurs in the cores of the catalyst particles which because of their relative inaccessibility in any event do not contribute much to the overall activity of the particles. The main factors leading to activity decline of iron catalysts in the Fischer–Tropsch process are loss of active sites by sintering, fouling by products, and poisoning by sulfur compounds.

1. Sintering

At the high temperatures at which Synthol catalysts are used carbon deposition occurs. This carbon has a high surface area, hence the total

Brunauer–Emmett–Teller area increases with age. Because of this it is difficult to decide to what extent sintering of the catalyst contributes to the activity decline. At the lower temperatures at which fixed bed reactors are operated, little or no carbon deposition occurs, and here the situation is clearer. A freshly conditioned catalyst typically has an area of ~ 200 m^2/g, but after say 100 days its area is only ~ 50 m^2/g. The increased sharpness of the x-ray diffraction pattern of a used catalyst confirms that crystallite growth has occurred. When a catalyst is prepared with a lower SiO$_2$ content, it is found that the rate of activity decline is higher. This is in keeping with the accepted role of supports, namely, stabilizing the presence of small iron carbide crystallites.

Water vapor is known to enhance the rate of sintering of catalysts. As the synthesis gas moves down a fixed bed of catalyst, the gas is converted and the partial pressure of the product water must necessarily increase. Hence an increase in the sintering rate along the length of the reactor can be expected. Actual measurements have confirmed that both the intrinsic activity and the area of the catalyst, sampled at different positions down the reactor, decrease. There is thus clear evidence that for fixed bed iron catalysts crystallite sintering contributes to activity decline.

2.   Fouling by Products

Two types of fouling should be distinguished. High-molecular-weight products form in the narrow pores of the catalyst, and because of their bulk this wax impedes the diffusion of reactants through the pores to the active surface sites. This amounts to physical fouling, and it is present in both fixed and fluidized beds but particularly in fixed bed operations geared toward the production of high-molecular-weight waxes. It has been observed that, when the synthesis in a fixed bed is interrupted and the catalyst then washed with a solvent such as diesel fuel, the subsequent synthesis activity increases dramatically. This phenomenon lasts only a few minutes, however, because fresh wax is again formed and rapidly refills the catalyst pores.

At the higher temperatures used in fluidized bed reactors dienes and aromatics are formed, and these compounds are known coke precursors. These coking reactions represent chemical fouling. If used Synthol catalyst is successively extracted with heptane, benzene, and pyridine, the extracted material's H/C ratio decreases from 1.7, 0.9 to 0.8, respectively. If the catalyst is then treated with hydrogen at $\sim 400°$C, oils are evolved. This indicates that there are pyridine-insoluble deposits on the catalyst, which can be hydrocracked to oils. These deposits presumably block off active surface sites. If an aged catalyst of this type is re-reduced with hydrogen *in situ*, its activity returns to near the original value.

TABLE II

The Deactivating Effect of Sulfur on Fluidized Iron Catalysts

Synthesis gas sulfur concentration (ppm)	Activity decline (conversion percentage per day)
0.3	0.25
2.0	2.0
20	33

3. Poisoning by Sulfur Compounds

It is well known that sulfur compounds poison iron catalysts. In the Fischer–Tropsch synthesis it appears that there is no safe sulfur level, because it has been found at Sasol that in the case of fixed bed reactors there is a clearly discernible difference in the rate of activity decline when the sulfur content of the synthesis gas is lowered from 0.14 to 0.01 ppm. In a fixed bed the sulfur is rapidly adsorbed in the top layers of the bed, which completely inactivates that portion of the catalyst. It is thus found that moving down the bed the catalyst's activity at first increases (as a result of decreasing sulfur poisoning) and then decreases (as a result of H_2O sintering — see Section 5.B.1).

In a fluidized bed, because of complete mixing of the catalyst, all of the catalyst is equally deactivated by sulfur. Table II demonstrates the influence of different levels of sulfur in the synthesis gas on the rate of decline in catalyst activity.

C. CARBON DEPOSITION

Iron catalysts cannot be used at high temperatures in fixed bed reactors because of carbon deposition. On carbiding, the catalyst particles swell, and they continue to do so with further carbon deposition until they disintegrate. The fine powder formed completely plugs the catalyst bed. For this reason the operating temperatures in fixed beds are normally kept below 260°C.

In fluidized bed synthesis the iron particles also swell and disintegrate, but here at least bed plugging cannot occur and the process can continue effectively. The main drawback of catalyst disintegration is that the fines produced can be carried out of the reactor, even though the cyclones are very efficient, and foul downstream equipment. There is no clear evidence that carbon deposition (as distinct from coke deposition) results in the deactivation of iron catalysts. As the reaction rate is diffusion-controlled (i.e., is affected by particle size), it is possible that particle disintegration compensates for any concomitant deactivation process.

Dry [6] has reviewed the carbon deposition studies done at Sasol in predominantly carbon monoxide atmospheres, as well as in typical synthesis gas atmospheres. In both atmospheres the presence of strong alkalis (e.g., K_2O) enhances the rate of carbon deposition. If the catalyst also contains compounds with which the alkali can react, such as SiO_2 and Al_2O_3, the basicity of the catalyst is effectively lowered and the rate of carbon deposition is diminished.

Under synthesis conditions, it has been found that, for fluidized beds operating in the vicinity of 300°C, the rate of carbon deposition is proportional to $p_{CO}/p_{H_2}^2$ at the reactor entrance (p_{CO} and p_{H_2} are the partial pressures of CO and H_2, respectively). This relation holds over a wide range of synthesis gas composition, pressure, and fresh feed/recycle feed ratios [6]. Figure 6 illustrates the relation for two different iron catalysts. Table III gives details of the gas partial pressures for the two cases.

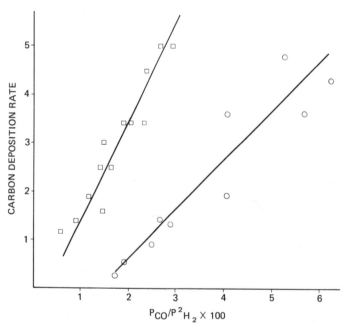

Fig. 6. The carbon deposition rate as a function of the ratio $p_{CO}/p_{H_2}^2$. The plots ○ and □ are, respectively, for catalysts A and B in Table III.

TABLE III

The Relation between $p_{CO}/p_{H_2}^2$ and the Carbon Deposition Rate

Set	Entrance partial pressure (bars)			$p_{CO}/p_{H_2}^2 \times 100$	Carbon deposition rate[a]
	H_2	CO	CO_2		
A	11.2	2.2	2.2	1.7	0.24
	9.6	1.8	1.9	1.9	0.53
	8.3	1.7	1.8	2.5	0.86
	7.8	1.8	0.7	2.9	1.3
	7.2	1.4	1.5	2.7	1.4
	7.5	2.3	3.9	4.1	1.9
	6.6	1.8	1.0	4.1	3.6
	6.9	2.7	6.8	5.7	3.6
	6.3	2.5	2.7	6.3	4.3
	5.5	1.6	1.2	5.3	4.8
B	32	6.4	6.4	0.6	1.2
	24	5.1	4.6	0.9	1.4
	16.4	4.0	4.1	1.5	1.6
	17.0	3.6	1.3	1.2	1.9
	15.6	3.4	3.7	1.4	2.5
	12.5	2.6	2.5	1.7	2.5
	13.4	2.7	1.1	1.5	3.0
	9.2	1.6	0.6	1.9	3.4
	11.5	2.8	3.4	2.1	3.4
	8.6	1.8	0.7	2.4	3.4
	8.5	1.7	1.5	2.4	4.5
	8.1	1.8	1.7	2.7	5.0
	8.8	2.3	1.6	3.0	5.0

[a] Mass of carbon per mass of iron per unit time.

An interesting result of the parameter $p_{CO}/p_{H_2}^2$ is that, using the same feed gas, as the total synthesis pressure is increased, the numerical value of $p_{CO}/p_{H_2}^2$ decreases and hence the rate of carbon deposition decreases. This occurs in spite of the fact that more Fischer–Tropsch product is being produced per mass of catalyst as the pressure is increased (see Section VIII for the kinetics). Similarly, when the reactor production rate is increased by increasing the fresh feed flow and decreasing the recycle flow, the rate of carbon deposition decreases because the value of $p_{CO}/p_{H_2}^2$ is decreased by decreasing the recycle ratio [6]. Thus increased Fischer–Tropsch production with a fixed mass of iron catalyst need not be accompanied by higher carbon deposition rates.

VI. Fischer – Tropsch Product Selectivity

A. COMMERCIAL PLANTS

The typical start-of-run selectivities obtained in the Sasol fixed and fluidized commercial reactors is given in Table IV. As the catalyst ages, the hard wax selectivity in the case of the fixed bed reactors decreases and the selectivities at the opposite end of the spectrum (e.g., C_1 to C_4 hydrocarbons) increase. The same phenomenon, but to a lesser degree, occurs in the case of the fluidized catalyst. At fixed operating temperatures the overall activity of the catalysts also decreases with age. It has been suggested [6] that the decline in both selectivity and activity are linked. It is well known that K_2O promotes both activity and the selectivity toward higher molecular weight products (Section VI.B.2). As it is likely that the most active sites (those associated with the K_2O promoter) will also deactivate sooner, the shift to lower selectivity associated with declining activity is to be expected.

Table V compares the properties of the unrefined gasoline and diesel cuts produced in the two types of Sasol reactors. As can be seen from both Tables IV and V, the Synthol products are more olefinic than the fixed bed products. Probably because of the higher operating temperature the Synthol products are more branched and also contain more ring compounds (see Table VIII), whereas the fixed bed products are completely free of aromatics.

TABLE IV

Selectivity of Sasol Commercial Operations[a]

Product	Fixed bed at 220°C (%)	Synthol at 325°C (%)
CH_4	2.0	10
C_2H_4	0.1	4
C_2H_6	1.8	4
C_3H_6	2.7	12
C_3H_8	1.7	2
C_4H_8	3.1	9
C_4H_{10}	1.9	2
C_5 to C_{11} (gasoline)	18	40
C_{12} to C_{18} (diesel fuel)	14	7
C_{19} to C_{23}	7	⎫
C_{24} to C_{35} (medium wax)	20	⎬ 4
>C_{35} (hard wax)	25	⎭
Water-soluble nonacid chemicals	3.0	5
Water-soluble acids	0.2	1

[a] Carbon atom basis.

TABLE V

Selected Properties of Primary Unrefined Straight-Run Gasoline and Diesel Cuts[a]

Product cut	Property	Fixed bed	Synthol
Gasoline, C_5 to C_{11}	Olefins	32	57
	Paraffins (total)	60	14
	n-Paraffins	57	8
	Aromatics	0	7
	Ring compounds (total)	0	15
	Alcohols	7	6
	Ketones	0.6	6
	Acids	0.4	2
	RON (Pb-free)	~ 35	~ 65
Diesel, C_{12} to C_{18}	Olefins	25	73
	Paraffins (total)	65	10
	n-Paraffins	61	6
	Aromatics	0	10
	Alcohols	6	4
	Ketones	<1	2
	Acids	0.05	1
	Cetane number	75	55

[a] Mass percentage of cut, except for RON and cetane number.

Ketone formation is also associated with high-temperature operations [6]. As can be seen, the Synthol products have a much higher ketone content than the fixed bed products.

Table VI gives the composition of the oxygenated products present in the reaction water phase for the two types of reactors. Note that the low-temperature process produces much more methanol relative to the high-temperature Synthol process. Again it can be seen that Synthol reactors produce relatively more ketones.

B.   CONTROL OF CARBON NUMBER DISTRIBUTION

The selectivities given in Table IV are for the reactors as currently operated, but if desired the selectivities of both types of reactors could be changed considerably. The variables used to achieve this are the composition and formulation of the catalyst, the synthesis temperature, the fresh feed gas composition, the recycle ratio, and the total pressure. By manipulation of these parameters Sasol has varied the CH_4 selectivity from 2 to 80% and, at the other extreme, the hard wax (cut $>500°C$) has been varied from 0 to over 50%.

TABLE VI

Composition of Water-Soluble Products from Sasol Commercial Reactors

Compound	Fixed bed reactors (wt%)	Synthol reactors (wt%)
Nonacid		
CH_3CHO ⎫		3
C_2H_5CHO ⎬	~2	1
C_3H_7CHO ⎭		0.5
CH_3COCH_3	2	10
$C_2H_5COCH_3$	1	3
$C_2H_5COC_2H_5 + C_3H_7COCH_3$	—	1
CH_3OH	24	1
C_2H_5OH	50	55
$n\text{-}C_3H_7OH$	11	13
$iso\text{-}C_3H_7OH$	—	3
$n\text{-}C_4H_9OH$	6	4
$iso\text{-}C_4H_9OH$	—	3
$2\text{-}C_4H_9OH$	—	1
$C_5H_{11}OH$	4	1
Acid		
Acetic	—	70
Propionic	—	16
Butyric	—	9

1. Interrelation between Products

In the experiments mentioned previously in which wide ranges of selectivities were covered, it was evident from the results that a clear relationship existed between the selectivities toward the individual carbon number products irrespective of how the different selectivity changes were brought about. These experimental results are illustrated in Figs. 7 and 8 where the selectivities of various product cuts are plotted against the CH_4 and hard wax selectivities, respectively. These distributions are mathematically predictable if a growth mechanism, such as the addition of one carbon entity at a time to growing hydrocarbon chains or the polymerization of simple surface entities, is assumed. This subject has been reviewed elsewhere [6]. Figure 9 illustrates the results of a simple mathematical treatment of hydrocarbon chain growth. A surprisingly good agreement is observed between the calculated and the observed selectivity correlations. A notable exception is the C_2 selectivity, which is lower than predicted. The various possible reasons for this have been discussed [6].

The overall conclusion from these investigations is that, utilizing present technology, the classic Fischer–Tropsch process will always produce a wide

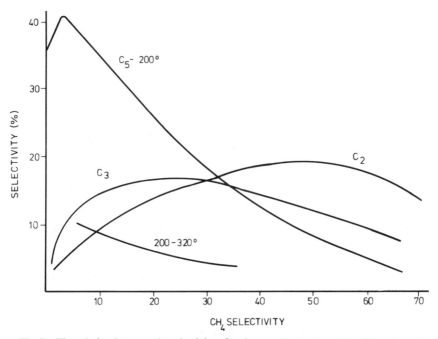

Fig. 7. The relation between the selectivity of various product cuts and the CH$_4$ selectivity (selectivities are on carbon atom basis).

range of hydrocarbons. It remains a challenge to research scientists to determine if the selectivity toward a single cut, such as diesel fuel, can be markedly increased. At the moment only the two extremes, CH$_4$ and wax, can be obtained in very high amounts. By utilizing further processes downstream the maximization of a particular product can be achieved, and this is discussed in Section VII.

2. The Influence of Promoters on Selectivity

Catalysts consisting of only iron are unsatisfactory, as they produce mainly paraffinic hydrocarbon gases. The addition of promoters is essential to obtain a more olefinic product and to shift the selectivity toward the desired higher molecular weight hydrocarbons. The key promoters are the strong group I alkali metals, Na, K, Rb, and Cs. Lithium is ineffective, and because of their high cost Rb and Cs are not used in practice. The influence of all other promoters can in fact be interpreted in terms of how they affect the key K$_2$O component. Thus the presence of SiO$_2$ in a catalyst results in the formation of potassium silicate which is less basic than the K$_2$O alone. This catalyst then behaves like a catalyst with a lower K$_2$O content. This also

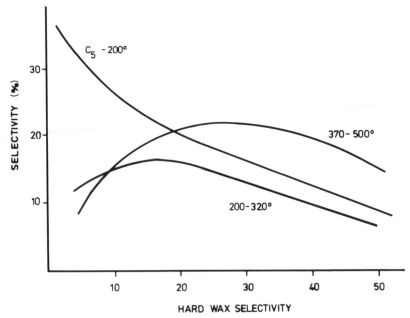

Fig. 8. The relation between various product cuts and the hard wax selectivity in fixed bed Fischer–Tropsch synthesis.

applies to catalysts containing an acidic component, such as Al_2O_3 and TiO_2. The same reasoning explains why many catalysts prepared by impregnating iron into preformed supports are not satisfactory. Such supports are commonly acidic in nature, e.g., Al_2O_3 and zeolites, and because the support inevitably is the major component of the total catalyst, very high K_2O levels are required to obtain the desired basicity. At very high K_2O levels, however, the catalyst's activity is usually found to very poor.

Table VII illustrates the influence of the K_2O level on the performance of the two types of commercial catalysts used by Sasol, namely, fused magnetite for the Synthol reactors and precipitated supported Fe_2O_3 for the fixed bed reactors. Because of the intercorrelation between the products (Section VI.B.1), it is only necessary to give either the CH_4 or the hard wax selectivity to define the carbon number distribution of the products. (In the case of the fixed bed process reference is made to the hard wax rather than to the CH_4 selectivity, because the CH_4 selectivity is very low and therefore the measurement less accurate.) As can be seen from Table VII, increasing the K_2O in both catalysts results in a strong shift in selectivity toward higher molecular weight products. The olefinity of the products increases, and more oxygenated products are also formed, as illustrated by the light-acid selectiv-

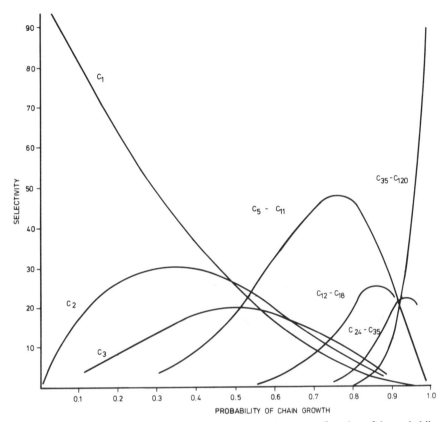

Fig. 9. The calculated selectivities of the various product cuts as a function of the probability of chain growth of the adsorbed complex on the catalyst surface.

ities. In the case of the Synthol catalyst the activity increased with increasing K_2O content, whereas for the fixed bed catalyst the opposite was observed. This is not unique to these particular iron catalysts but simply reflects previous observations that the activity of iron catalysts first increases and then decreases with increasing K_2O content [6].

3.   The Influence of Temperature

Increasing the synthesis temperature shifts the selectivity spectrum toward lower molecular weight products, as shown in Table VIII. As mentioned in Section VI.A, the formation of aromatics and ketones occurs only at higher temperatuers. As can be seen in Table VIII, the aromaticity of the gasoline cut increases as the temperature increases. Similarly the ketone

TABLE VII

Influence of the K_2O Level on Catalyst Performance

Process	K_2O level[a]	Selectivity[b] (%)		Olefinity, C_3H_6/C_3H_8 ratio	Acid selectivity[c] (%)	Activity[a]
		CH_4	Hard wax			
Synthol	0	41	—	0.2	0.05	63
	1.75	20	—	3.0	0.2	90
	2.1	15	—	5.7	0.3	90
	2.65	13	—	7.6	0.5	92
	3.4	12	—	11	0.7	94
	3.95	10	—	12	0.8	93
Fixed bed	12	—	18	—	0.2	112
	16	—	20	—	0.3	109
	21	—	30	—	0.3	85
	24	—	38	—	0.5	83
	32	—	44	—	0.7	75

[a] Relative quantities.
[b] Carbon atom basis.
[c] Acids in the product water phase. Carbon atom basis.

TABLE VIII

Influence of Temperature on Selectivity

Process	Temperature (°C)	Selectivity (%)[a]		Aromatic content of 80–160°C cut (%)	Water-soluble ketone selectivity
		CH_4	Hard wax		
Fixed bed	213	—	47	0	<0.1
	227	—	34	0	<0.1
	237	—	24	0	<0.1
	247	—	17	0	<0.1
Synthol	310	10	—	4	0.4
	330	14	—	8	0.8
	350	17	—	10	1.1
	360	20	—	13	1.3
	370	23	—	18	1.2
	380	28	—	26	0.8

[a] Carbon atom basis.

selectivity increases up to $\sim 360\,^\circ$C and then decreases again. (This observation is further discussed elsewhere [6].)

4. The Influence of Gas Composition and Pressure

The actual partial pressures of the gases in contact with the catalyst inside the reactor are dependent on the compositions and flows of fresh feed gas and of tail gas recycled to the reactor, as well as the total pressure. Rather than deal with these variables individually it is more meaningful to correlate the selectivity directly with the partial pressures inside the reactor.

If the Fischer–Tropsch reaction mechanism is as described in Section VIII, then it could be argued that one of the factors controlling the probability of chain termination is the amount of chemisorbed hydrogen on the catalyst surface. Thus the product spectrum should be related to the hydrogen partial pressure in the gas phase. However, hydrogen chemisorbs rather weakly on iron catalysts relative to CO, CO_2, and H_2O [6]. This means that the amount of hydrogen chemisorbed would not only be proportional to its own partial pressure but would also be inversely proportional to the partial pressures of CO, CO_2, and H_2O. Based on this line of reasoning the carbon number selectivity spectrum could be proportional to a ratio such as

$$ap_{H_2}^w/(bp_{CO}^x + cp_{CO_2}^y + dp_{H_2O}^z), \tag{1}$$

where the constants a, b, c, d, w, x, y, and z can have different values depending on the catalyst's characteristics and the synthesis conditions. A simplified form of this ratio is p_{H_2}/p_{CO}. The H_2/CO ratio is very commonly used when correlating selectivity with gas composition.

Table IX gives the partial pressures of H_2, CO, and CO_2 at the reactor entrances for two sets of experiments. More details and more experimental results are given in the review by Dry [6]. As can be seen from Table IX, the simple H_2/CO ratio does in fact correlate well with the hard wax selectivity in the case of the fixed bed process. The hard wax selectivity decreases as the H_2/CO ratio increases. For the fluidized bed, however, there is no clear relation between the CH_4 selectivity and the H_2/CO ratio. This difference is corroborated by the observed effect of total pressure on the selectivity. As can be seen in Table X, the actual H_2/CO ratio did not change much in each set of experiments. Whereas the hard wax selectivity for the fixed bed reactor did not change, the CH_4 selectivity for the fluidized bed reactor changed significantly. As can be seen from Tables IX and X, there is, however, a relation between the value of the ratio $p_{H_2}^{0.5}/(p_{CO} + p_{CO_2})$ at the reactor entrance and the CH_4 selectivity. This ratio is another version of Eq. (1).

Studies at Sasol showed that with Synthol catalysts at $\sim 320\,^\circ$C the selectivity spectra were independent of the percentage of conversion for each

TABLE IX

The Relation between Reactor Entry Partial Pressures and Product Selectivity

| | Partial pressure (bars) | | | H_2/CO ratio | $p_{H_2}^{0.5}/$ $(p_{CO} + p_{CO_2})$ | Selectivity (%) | |
Process	H_2	CO	CO_2			CH_4	Hard wax
Fixed	10.8	10.8	1.5	1.0	—	—	48
bed	10.0	5.4	1.4	1.9	—	—	37
	12.1	5.6	2.0	2.2	—	—	32
	13.3	3.6	1.5	3.7	—	—	22
	16.6	2.8	3.8	7.3	—	—	13
Fluidized	7.52	1.72	4.28	4.4	0.46	8	—
bed	7.24	2.48	1.86	2.9	0.62	12	—
	5.66	1.31	1.52	4.3	0.84	17	—
	7.38	1.66	0.62	4.4	1.19	22	—
	10.14	1.31	0.28	7.7	2.0	27	—

fresh feed composition studied. These studies were carried out in a high-internal-recycle laboratory reactor of the Berty [14] type. In these experiments the ratio $p_{H_2}^{0.5}/(p_{CO} + p_{CO_2})$ was not constant at the different conversion levels. However, the ratio $p_{H_2}^{0.5}/(p_{CO} + p_{CO_2} + p_{H_2O})$ was found to be constant. Because in high-temperature reactors there is always an excess of H_2, the value of $p_{H_2}^{0.5}$ does not change much with conversion. The constancy of the ratio $p_{H_2}^{0.5}/(p_{CO} + p_{CO_2} + p_{H_2O})$ is due to the fact that in the Fischer–Tropsch

TABLE X

The Influence of Total Pressure on Product Selectivity[a]

| | Total pressure (bars) | Partial pressure (bars)[b] | | | H_2/CO ratio | $p_{H_2}^{0.5}/$ $(p_{CO} + p_{CO_2})$ | Selectivity | |
Process		H_2	CO	CO_2			CH_4	Hard wax
Fixed	11.4	5.91	2.35	0.66	2.5	—	—	18
bed	27.6	14.07	5.87	1.39	2.4	—	—	18
Fluidized	10.8	4.3	1.0	0.9	4.3	1.10	18	—
bed	40.8	14.2	3.6	2.8	3.9	0.59	11	—
	75.8	30.7	6.7	4.6	4.6	0.49	6	—

[a] In each set of experiments the temperature, fresh feed gas composition, and recycle ratios were kept constant. As the pressures were increased, all the gas flows were increased in proportion so as to maintain the same total gas linear velocity.

[b] The partial pressures are those at the reactor entrance.

reaction each molecule of CO is effectively replaced by a molecule of H_2O.

$$2H_2 + CO \rightarrow (CH_2) + H_2O \tag{2}$$

This keeps the value of the denominator more or less constant.

At the entrance of an integral reactor the value of p_{H_2O} is low, and so the values of $p_{H_2}^{0.5}/(p_{CO} + p_{CO_2})$ and $p_{H_2}^{0.5}/p_{CO} + p_{CO_2} + p_{H_2O})$ are almost the same. This explains why in Table IX and X the value of $p_{H_2}^{0.5}/(p_{CO} + p_{CO_2})$ at the reactor entrance appears to correlate satisfactorily with the CH_4 selectivity.

In summary, it appears that for low-temperature fixed bed operations selectivity is dependent on the simple H_2/CO ratio, whereas for high-temperature operations over iron catalysts the selectivity is controlled by the more complex ratio $p_{H_2}^{0.5}/(p_{CO} + p_{CO_2} + p_{H_2O})$. Further studies and evaluations are needed to quantify properly the influence of gas composition on product selectivity. (In an earlier publication [15] data were presented indicating that in high-temperature reactors the selectivity was simply inversely proportional to the entry CO_2 partial pressure. Further investigations, however, showed that this was an inadequate correlation.)

VII.   Process Schemes

A.   SASOL COMMERCIAL PLANTS

The description of Sasol process flow schemes will be confined to the Fischer–Tropsch sections and the downstream product workup units. The upstream processes have already been described in Section III.B. (These include the recovery of phenols and ammonia from the raw synthesis gas and the workup of the naphtha, tar oil, and pitch produced in the Lurgi gasifiers.)

The details of the two types of Fischer–Tropsch synthesis reactors and the way in which they are operated have been dealt with in Section VI.A.

The Sasol One complex also incorporates plants that produce styrene, butadiene, NH_3 (synthesized from N_2 and H_2), HNO_3, and NH_4NO_3. Since these products are not directly associated with the Fischer–Tropsch process, however, they will not be dealt with in this chapter.

1.   Sasol One at Sasolburg

Figure 10 depicts the Sasol One flow scheme. Purified synthesis gas is fed into the two types of Fischer–Tropsch reactors. The fixed bed reactors are geared toward high wax production, and the Synthol reactors toward gasoline production. Figure 11 is a photograph of the fixed bed Fischer–Tropsch

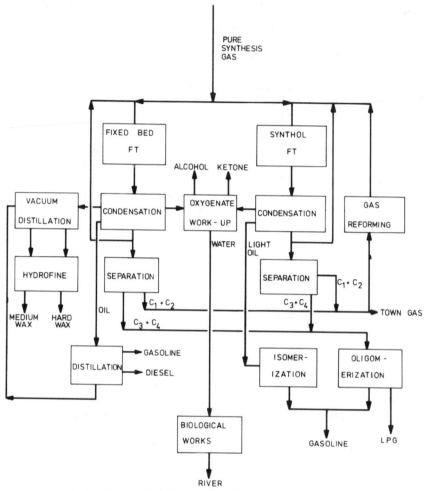

Fig. 10. Process flow diagram of the Sasol One plant at Sasolburg.

reactors. The total reactor effluents are progressively cooled to ambient temperature to condense the water and the liquid hydrocarbon products. A large part of the residual gas is recycled back to the reactors. Typically the total feed gas to the Fischer–Tropsch units is composed of 1 part fresh synthesis gas and 2 parts recycle gas (after liquid product removal).

The condensed water phase contains about 2–6% dissolved lower molecular weight oxygenates (e.g., alcohols and ketones). Table VI gives the typical compositions of the oxygenates. The water is steam-stripped in a column to remove the oxygenates overhead, leaving only the carboxylic

Fig. 11. The fixed bed Fischer–Tropsch reactors at Sasol One.

acids in the bottoms. The alcohols and ketones are fractionated into the individual components, purified, and sold as solvents. Aldehydes are hydrogenated (over nickel catalysts at low temperatures) to alcohols which are then recycled to the front end of the oxygenate separating unit. The residual water is biologically treated to destroy the dissolved carboxylic acids, and the purified water is returned to the nearby Vaal River.

The tail gas containing the uncondensed hydrocarbons is passed through oil absorption towers which remove the C_3 and heavier components. The C_3 and C_4 hydrocarbons are fed to the "cat poly" reactors. In these reactors the olefins are oligomerized to gasoline over a phosphoric acid–kieselguhr catalyst at $\sim 190°C$ and 38 atm. The C_3 and C_4 paraffins pass through unreacted and are sold as light petroleum gas (LPG).

The Synthol light oil, which is $\sim 75\%$ olefinic, is vaporized and passed over an acidic zeolite catalyst at $\sim 400°C$ and ~ 1 atm. This decarboxylates acids, dehydrates alcohols to olefins, and isomerizes the olefin double bonds from the alpha to an internal position. Some skeletal isomerization also occurs. This process increases the research octane number (RON) of the gasoline cut from ~ 65 to ~ 86 (Pb-free). This gasoline is combined with the "cat poly" product to give a gasoline of RON 90 (Pb-free). For over a decade $\sim 15\%$ Fischer–Tropsch ethanol was also added to the gasoline. The oil from the fixed bed process is fractionated into diesel fuel (cetane number ~ 75) and gasoline (octane number ~ 35). This gasoline can be upgraded by catalytic isomerization, resulting in a product of RON ~ 65.

The waxes from the fixed bed Fischer–Tropsch units are vacuum-distilled into medium wax (370–500°C) and hard wax (>500°C). These cuts are individually hydrofined (olefin saturation and oxygen removal) over nickel catalysts.

The remaining tail gas consists of unreacted CO and H_2, CO_2, ethylene, ethane, and methane. This gas is diluted with pure synthesis gas (~ 0.016 GJ/m_n^3 or ~ 400 Btu/scf) to give a town gas (~ 0.020 GJ/m_n^3 or ~ 500 Btu/scf) which is sold to neighboring industries. As this gas is completely sulfur-free, it is of prime quality. Any excess tail gas is reformed over nickel catalysts at ~ 1000°C with steam and oxygen. For CH_4 the approximate equation is

$$1.0CH_4 + 0.24H_2O + 0.62O_2 \rightarrow 0.52CO + 0.48CO_2 + 2.24 H_2 \qquad (3)$$

The reformer product gas is water-scrubbed to remove most of the CO_2, and the gas is then recycled to the Synthol reactors to be converted to hydrocarbons. The reforming reaction lowers the overall thermal efficiency of the process and so is used only when there is an excess of tail gas.

2. Sasol Two and Sasol Three at Secunda

These two plants are virtually identical and so do not require separate descriptions. Figure 12 shows the Sasol Two flow scheme. Only the new large, high-capacity Synthol reactors are used at these plants (Section IV.A.3). There are several major differences in the product separation and workup between the older Sasol One and the new Sasol Two and Sasol Three plants. As before, the reaction water and liquid oils are condensed out. Whereas the tail gas at Sasol One is then passed through an oil absorption tower, the tail gas at Sasol Two is first passed through a Benfield unit to scrub out the CO_2, and it then goes to a cryogenic unit that separates the gas into CH_4-rich, H_2-rich, C_2, and C_3–C_4 streams. Although this separation technique is more expensive, it makes possible recovery of the high-priced ethane and ethylene components. The C_2 hydrocarbons go to the ethylene plant where the ethane is steam-cracked to ethylene. (At the Sasol One plant the ethylene is sold together with the CH_4 as town gas.) The CH_4-rich stream from the cryogenic unit is sent to the reformer where the CH_4 is converted to synthesis gas, as at Sasol One, and recycled to the Synthol units. Since the feed to the reformer at Sasol Two has a much higher CH_4 concentration than at the Sasol One plant, the Sasol Two process is more efficient. The H_2-rich stream from the cryogenic unit is recycled back to the Synthol units. Pure hydrogen, required in the various hydrotreating units, is extracted from the H_2-rich stream by pressure swing absorber (PSA) units.

The C_3–C_4 stream is oligomerized over a phosphoric acid–kieselguhr

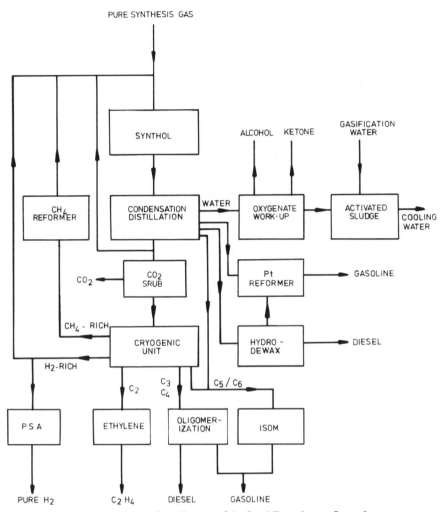

Fig. 12. Process flow diagram of the Sasol Two plant at Secunda.

catalyst as is done at Sasol One, but diesel fuel production is maximized by recycling a portion of the gasoline product. A diesel fuel selectivity of ~ 75% can be achieved. The Fischer–Tropsch oil boiling above ~ 190°C is hydro-treated, and the heavier portion then further treated in a selective hydrode-waxing operation employing a special Mobil zeolite catalyst. The diesel fuel produced from the heavy-oil feed has a cetane number of ~ 55. For diesel fuel production there is considerable flexibility at the plant. By altering the operating conditions of the hydrodewaxing and the olefin oligomerization

Fig. 13. Lurgi gasifiers.

units and also by adjusting the cut-point temperatures, the overall gasoline/diesel fuel ratio can be varied from $\sim 10:1$ to $\sim 1:1$.

Whereas the Sasol One gasoline can simply be blended into the gasoline pool produced from petroleum crude oil, the amount of gasoline produced at the Sasol Two and Sasol Three plants is so large that it is essential for the gasoline to "stand on its own feet." To ensure this the C_7 to 190°C straight-run Fischer–Tropsch oil is first hydrofined to saturate olefins and remove all oxygenates and then reformed over a $Pt–Al_2O_3$ catalyst. A United Oil Products moving bed continuous-regeneration unit is employed. The $C_5–C_6$ cut is isomerized over a zeolite catalyst to improve its octane rating.

The final gasoline and diesel fuels produced meet all the specifications for modern motor fuels, and furthermore they are completely compatible with the normal products derived from petroleum crude oils.

Figure 13 is a view of a bank of Lurgi gasifiers, Fig. 14 is a view of four Synthol reactors, and Fig. 15 shows the alcohol and ketone refining section at Secunda.

B. DIESEL FUEL-PRODUCING SCHEMES

Fischer–Tropsch synthesis directly produces predominantly straight-chain products. For this reason it has a unique advantage over other direct or indirect processes that convert coal into motor fuels. Thus direct coal

Fig. 14. Synthol reactors at Secunda.

hydrogenation methods such as the H–coal process and Exxon's donor solvent process produce highly aromatic feedstocks which are excellent gasoline precursors but require severe hydrogenation to upgrade them to acceptable diesel fuels. The Mobil indirect route first produces methanol from synthesis gas and then converts it over a special zeolite catalyst to a high-quality, highly aromatic gasoline, but no diesel fuel is produced.

The straight-run diesel fuel produced in low-temperature Fischer–Tropsch operations (using either fixed or slurry bed reactors) has a cetane number of ~ 75, and the diesel fuel obtained by selectively hydrocracking the waxes produced has a cetane number of ~ 70. This diesel fuel is completely free of aromatics, naphthenes, sulfur, and nitrogen compounds, and thus it is an attractive fuel because there is an ever-increasing demand for less noxious engine exhaust emissions. An advantage of the high cetane number of this diesel fuel is that other lower quality diesel fuels can be blended into the diesel pool. For instance, the highly branched diesel fuel produced by oligomerizing C_3 to C_6 olefins over acidic catalysts such as phosphoric acid–kieselguhr or amorphous silica–alumina have cetane numbers of only ~ 30. Such diesel fuels require boosting by the addition of a high-quality diesel fuel. Such blends still maintain the advantage of being aromatics-, sulfur-, and nitrogen-free.

Figure 16 depicts a process scheme in which high diesel fuel production is

Fig. 15. The alcohol and ketone refining section at Secunda.

the objective. A low-temperature fixed or slurry bed Fischer–Tropsch synthesis is used. The operating conditions are geared toward producing a high wax selectivity and yield. Under relatively mild conditions the wax is selectively catalytically hydrocracked to yield predominantly diesel fuel. The product boiling above 385°C is recycled to extinction. In prolonged pilot plant tests carried out at Sasol, diesel/naphtha ratios of ~ 11 were achieved in the wax hydrocracking process. Although the diesel fuel produced in this operation is more branched than the straight-run Fischer–Tropsch diesel fuel, it still has a very high cetane number (~ 70).

As can be seen from Fig. 16, 77% of the final product is diesel fuel. The cetane number of the final diesel pool is ~ 65, and it also meets all other diesel fuel specifications. The 10% naphtha is an excellent feedstock for cracking to ethylene because the naphtha is free of ring compounds. The residual C_3 and C_4 products can also be fed into the ethylene cracker.

When high-temperature fluidized bed reactors are used instead of low-temperature fixed bed reactors in the Fischer–Tropsch section, the overall diesel fuel yield is not as favorable (57% as against 77% for the fixed bed route). The two systems are compared in Table XI. The calculation basis is the same for both schemes. Only 7% of the fixed bed route diesel fuel is produced by olefin oligomerization, whereas for the fluidized bed route it is 54%. In the fluidized bed route it is therefore essential that the oligomeriza-

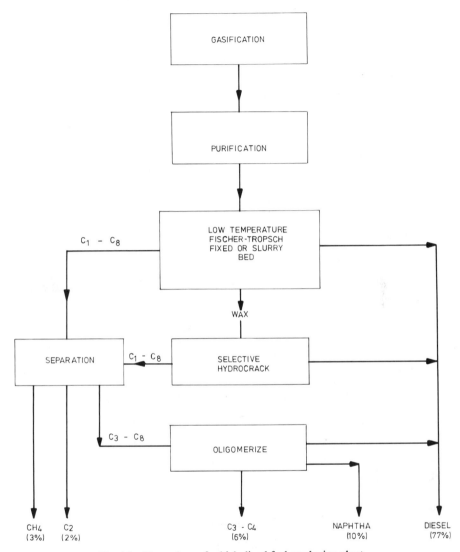

Fig. 16.   Flow scheme for high diesel fuel producing plant.

tion process produce good quality diesel fuel. Oligomerization over phosphoric acid and similar catalysts does not produce satisfactory diesel fuel. Either thermal noncatalytic or catalytic oligomerization over special zeolites can be used. Pilot plant investigations at Sasol using both of the two latter techniques have shown that satisfactory diesel fuel can be made (cetane number ~ 55). The quantity and quality of the diesel fuel made using the

TABLE XI

Diesel Fuel Production from Low-Temperature Fixed Bed and High-Temperature
Fluidized Bed Fischer–Tropsch Operations

Product selectivity after workup	Fixed bed scheme (%)	Fluidized bed scheme (%)
CH_4	3	11
C_2	2	8
$C_3 + C_4$	6	4
Light naphtha (C_5 to 110°C)	10	14
Diesel fuel (110–385°C)	77	57
Balance (water-soluble oxygenates)	2	6
Origin of diesel		
Straight-run	28	38
Olefin oligomerization	7	54
Wax hydrocracking	65	8

fixed bed Fischer–Tropsch route is, however, clearly superior to that produced by the fluidized bed route.

C. LIGHT-HYDROCARBON PRODUCTION

1. C_2 to C_4

There is considerable interest in selectively producing C_2 to C_4 olefins for the petrochemical industry. Because of the interrelationship that exists between the different products (Section VI.B.1), it does not seem to be possible to produce more than ~ 18 wt% C_2, $\sim 17\%$ C_3, or $\sim 13\%$ C_4 hydrocarbons using iron Fischer–Tropsch catalysts. Several patents have claimed low methane selectivities coupled with high C_2 to C_4 olefin selectivities [see the review by Dry (6)]. However, apparently none of these processes have to date been proven on a large scale.

Dow Chemical Company has developed a K_2O-promoted molybdenum catalyst that yields $\sim 70\%$ C_2 to C_5 [16]. The C_2 selectivity peaks at $\sim 30\%$ (as against Sasol's 18%).

Both the Dow and the Sasol processes also produce large amounts of CH_4 as well, contrary to the previously mentioned patent claims. With a high-temperature fluidized bed reactor and a suitably promoted iron catalyst, the Sasol system can produce the following product spectrum: 35% CH_4, 40% C_2 to C_4, and 24% light oil (carbon atom basis). This can be achieved at a high conversion ($\sim 90\%$) and a high fresh feed volume space velocity

($\sim$ 1000/hr). To operate such a system, however, it is essential that there be a ready market for the CH_4 (e.g., pipeline gas) and the light oil (e.g., feedstock for gasoline production or for cracking to ethylene).

2. Methane

When only methane is required, the recommended route is one of the proven methanation processes [17] using nickel catalysts. However, when lighter hydrocarbons are also desired by-products, a Sasol system can be considered. Lurgi gasifiers, which produce a large amount of CH_4, are then the preferred coal gasification process. With a high-temperature fluidized bed with a low-basicity iron catalyst, the following final product spectrum can be achieved; 81% CH_4, 8% C_2, 7% ($C_3 + C_4$), and 4% light naphtha (carbon atom basis). Increasing the catalyst's basicity decreases the CH_4 production and increases the naphtha production, with C_2 products peaking at $\sim$ 18%.

D. COMBINATING THE FISCHER–TROPSCH PROCESS WITH OTHER FUEL-PRODUCING PROCESSES

As discussed in Section VII.B, the low-temperature Fischer–Tropsch process can be utilized to produce large quantities of high-quality diesel fuel. Direct coal liquefaction processes (e.g., H–coal and Exxon donor solvent) produce gasolines that contain more than 50% aromatics, and thus some dilution with nonaromatic material is required. The diesel fuel cut is also highly aromatic, and even with complete ring saturation the cetane number of the product is low. Combining direct coal liquefaction and the indirect Fischer–Tropsch process has an obvious advantage. The completely paraffinic gasoline produced in the Fischer–Tropsch operation can be blended with the highly aromatic gasoline of the coal liquefaction process, and the high-quality diesel fuel of the former will upgrade the diesel fuel of the latter.

If all the coal is being gasified to produce synthesis gas, then two combinations present themselves. In one, the Mobil process, which first produces methanol and then converts it to highly aromatic gasoline, can be coupled with the low-temperature Fischer–Tropsch process which produces highly paraffinic diesel fuel. In the other, the high-temperature fluidized Fischer–Tropsch process can be coupled with the low-temperature process. The former produces large amounts of low-molecular-weight olefins which can be oligomerized to highly branched gasoline, and the latter produces predominantly straight-chain high-quality diesel fuel.

VIII. Fischer – Tropsch Kinetics and Mechanism

To date neither the kinetics nor the mechanism of the Fischer – Tropsch process has been satisfactorily clarified, and the true mechanism in particular remains a contentious matter. A recent review has dealt with these subjects [6], and so only summarized versions will be presented here and only the Sasol iron-based catalysts will be discussed.

A. KINETICS

Under typical commercial operating conditions it has been found that, for the fixed bed process in which an extruded catalyst is used, the overall activity increases as catalyst particle size decreases. This also applies to the fine iron catalyst (less than 100 μm) used in the fluidized bed process. Additional evidence for this in the case of the fluidized catalyst is the observation that the finer particles have a higher deposited carbon content than the larger particles. This indicates that the smaller particles have done more "work," i.e., were more active, than the larger particles. In fixed bed pilot plant units operating under high-feed, low-conversion conditions (i.e., near differential) it was found for both the precipitated and the fused catalyst types that the activity was linearly proportional to the external area of the catalyst particles. These observations indicate that diffusional resistances were present. In practice there is always some high-molecular-weight hydrocarbon product (waxes) on the catalyst, and diffusion of the reactant gases through the wax film must retard the overall kinetics. This probability is confirmed by the observation that, when the catalyst is washed with a light liquid solvent to strip it of waxes, there is a marked increase in conversion for a short time (Section V.B.2).

The observation that at lower levels K_2O promotion increases activity but at higher levels decreases activity (Section VII.B.2) is probably also associated with the amount and type of wax on the catalyst. Because K_2O definitely promotes the formation of high-molecular-weight waxes, it can be expected at high K_2O levels that diffusional resistances will progressively increase and so retard the overall rate of reaction.

The measured activation energies of Sasol catalysts have been found to cover the range 13 – 24 kcal/mol. Because of the presence of diffusional resistances these are only apparent activation energies, and their values depend on catalyst type, particle size, and synthesis conditions.

The influences of the various reactants and products on the Fischer – Tropsch reaction over iron catalysts have been extensively investigated for

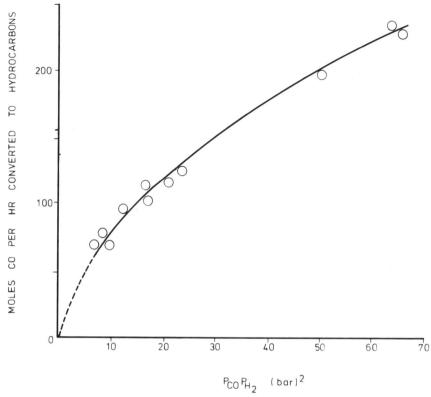

Fig. 17. The relation between the product $p_{CO}p_{H_2}$ at the reactor entrance and the amount of CO converted to hydrocarbons. The linear gas velocity and the catalyst charge were constant for all the experiments.

both fixed bed and fluidized bed operations. Thus the partial pressures of H_2, CO, CO_2, and H_2O have been varied over wide ranges, and both light (gasoline) and heavy (wax) hydrocarbon products have been deliberately recycled to the reactor to study the effects on the rate at which CO was converted to hydrocarbons. It was found that the hydrocarbons and also CO_2 had little influence on the rate. Water vapor had a strong negative influence. In integral reactors there was a good correlation between the product $p_{CO}p_{H_2}$ and the activity (Fig. 17), whereas in differential reactors the rate was proportional to p_{H_2} only. The influence of the partial pressures of H_2, CO, and H_2O is demonstrated by a few examples in Table XII. Qualitatively all the foregoing observations can be accommodated by the simple

TABLE XII

The Influence of Partial Pressures at the Reactor Entrance on the Rate of Hydrocarbon Production

| Variable | Fixed bed precipitated catalysts at 223°C | | | | | Fluidized bed fused catalyst at 330°C | | | | | |
| | Partial pressure (bars) | | | | | Partial pressure | | | | | |
	H_2	CO	CO_2	H_2O	Activity[a]	H_2	CO	CO_2	H_2O	Total pressure (bars)	Activity[a]
p_{H_2}	12.2	5.7	0.2	0.0	97	7.2	1.7	1.0	0.0	18	89
	15.2	5.6	0.2	0.0	109	10.1	1.6	0.8	0.0		100
	15.2	5.6	0.2	0.0	109	11.6	1.5	0.1	0.0	13.8	41
p_{CO}	—	—	—	—	—	11.7	3.0	0.1	0.0	15.8	66
	15.7	8.0	0.2	0.0	134	11.9	4.6	0.1	0.0	18.8	99
	8.4	4.3	0.1	0.0	69	9.8	1.8	0.7	0.0		107
p_{H_2O}	—	—	—	—	—	9.6	1.8	0.9	0.7	20.8	105
	8.4	4.1	0.1	0.35	51	10.0	1.8	1.2	1.4		98
						10.3	1.8	1.6	2.8		83

[a] Gram atoms of carbon in hydrocarbon produced per unit time.

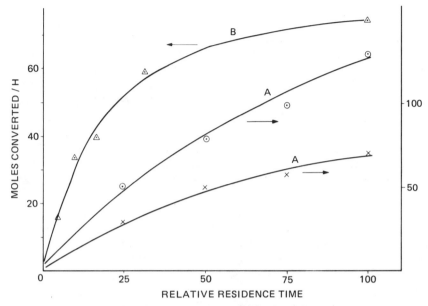

Fig. 18. Reaction profiles in fixed bed (A) and fluidized bed (B) Fischer–Tropsch operations. The symbols show the actual measured amounts of moles converted, and the solid lines are the profiles as calculated from the equation. $r = mp_{CO}p_{H_2}/(p_{CO} + ap_{H_2O})$

rate equation

$$r = mp_{CO}p_{H_2}/(p_{CO} + ap_{H_2O}), \tag{4}$$

where m is a temperature-dependent constant. This equation also quantitatively describes the actual reaction profiles observed in fixed and in fluidized reactors. Figure 18 demonstrates the fits obtained between the observed and calculated reaction profiles.

In the kinetic simulation model it is also necessary to incorporate the water gas shift reaction.

$$CO + H_2O \rightarrow CO_2 + H_2 \tag{5}$$

At higher temperatures this reaction is rapid over an iron catalyst, and it is in equilibrium all along the length of the catalyst bed. However, at lower temperatures the reaction is much slower and does not reach equilibrium. In this case the simple equation $r = kp_{CO}$ satisfactorily reflects the rate of the forward reaction. This observation is supported by the findings of others [18].

B. MECHANISM

Many apparently different mechanisms have been suggested in the past, and their relative merits are still being debated [6]. Some researchers believe that, upon chemisorption on the metal catalyst surface, the CO molecule first dissociates into surface carbon and oxygen atoms [19]. The carbon is then hydrogenated to CH_2 units which link up to form hydrocarbon chains. Others believe that the link between the carbon and oxygen atoms is maintained and that the CO is hydrogenated to surface alcohol-type species which are involved in the chain propagation process. Because CO insertion reactions have been well established in homogeneous catalysis, it has been assumed by many that this reaction is also involved in the chain growth mechanism.

With iron-based catalysts the hydrocarbons are commonly highly olefinic (in Synthol products the average olefinity is $\sim 75\%$). If the partial pressures of the olefins, paraffins, and hydrogen in the reactor are considered, it becomes apparent that the olefin/paraffin ratio is far higher than thermodynamics predicts. A logical conclusion thus appears to be that olefins are primary products of the surface reaction. Similarly the alcohol concentrations are much higher than expected [6]. With iron catalysts a very high alcohol selectivity can be achieved, for instance, by using a high-basicity catalyst at low conversions and at high pressures. Again it is tempting to interpret this as indicating that alcohols are primary products also. Considerations such as these makes it easier to accept the mechanism involving oxygenated surface complexes rather than the one involving complete CO dissociation followed by hydrogenation of the atomic carbon to CH_2 species.

However, it is the author's opinion that these differences in the basic mechanisms proposed are more apparent than real. By way of illustrating this consider the mechanistic scheme illustrated in Fig. 19. Because oxygen has a high affinity for a metal like iron, it seems very probable than in the chemisorbed CO surface complex there is a bond not only between the carbon and the active catalyst site (S) but also between the oxygen and the catalyst (complex I in Fig. 19). If the bond between C and O is very weak, then complete dissociation will occur as per step b in the scheme. The atomic carbon can then be hydrogenated (step d) to CH_2, complex III. If, however, the C—O bond remains intact, complex I can be hydrogenated (step c) to complex II which can be further hydrogenated (step g) to methanol or to complex III (step e). As illustrated in Fig. 19, complexes II, and III represent the two extremes, yet complex III is only the hydrogenated product of complex II. Whether either of these two complexes exists as such is also debatable. The true complex could be a hybrid of these two, or even of

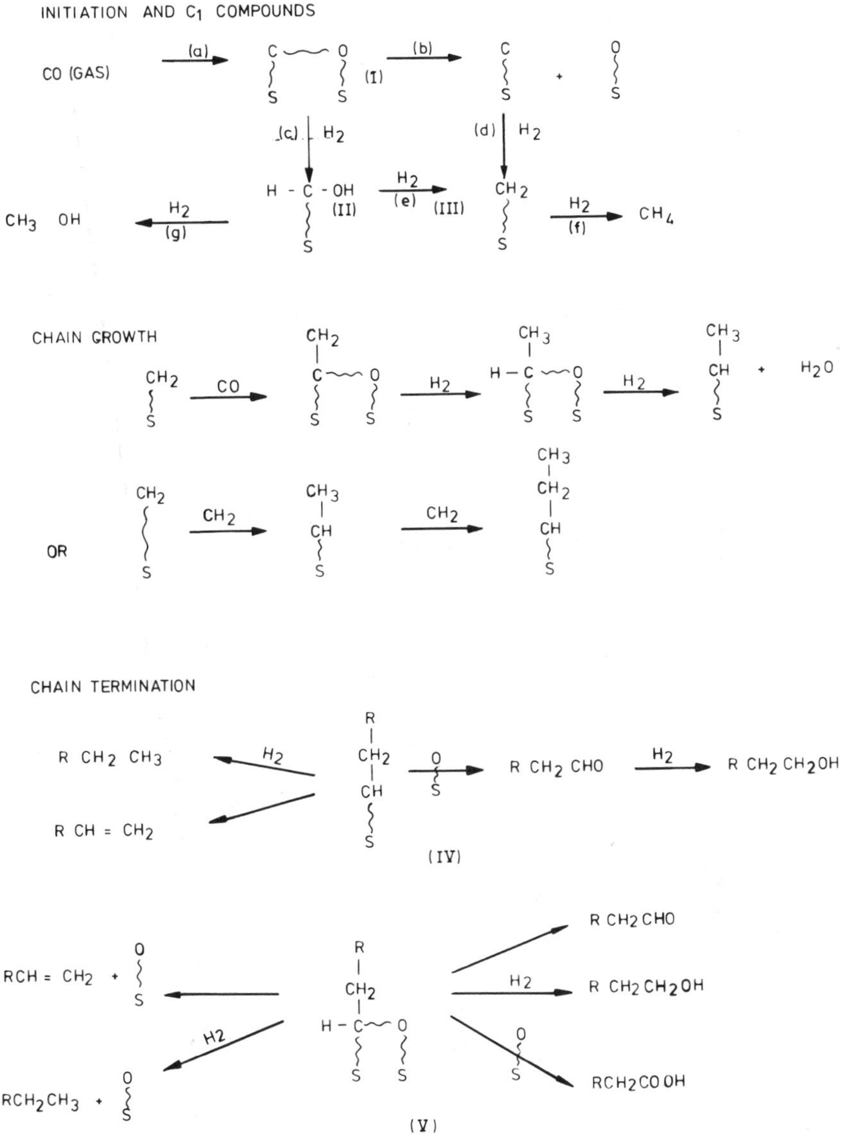

INITIATION AND C_1 COMPOUNDS

CO (GAS) $\xrightarrow{(a)}$ C⌇O (I) $\xrightarrow{(b)}$ C ⌇S + O ⌇S

$(c) \downarrow H_2$ $(d) \downarrow H_2$

CH_3 OH $\xleftarrow[(g)]{H_2}$ H–C–OH (II) $\xrightarrow[(e)]{H_2}$ CH_2 (III) $\xrightarrow[(f)]{H_2}$ CH_4

CHAIN GROWTH

CH_2 $\xrightarrow{CO}$ $\overset{CH_2}{C}$⌇O $\xrightarrow{H_2}$ $\overset{CH_3}{H–C}$⌇O $\xrightarrow{H_2}$ $\overset{CH_3}{CH}$ + H_2O

OR

CH_2 $\xrightarrow{CH_2}$ $\overset{CH_3}{CH}$ $\xrightarrow{CH_2}$ $\overset{CH_3}{\underset{CH}{CH_2}}$

CHAIN TERMINATION

R CH_2 CH_3 $\xleftarrow{H_2}$

R CH = CH_2 $\xleftarrow{}$ $\overset{R}{\underset{CH}{CH_2}}$ (IV) $\xrightarrow{}$ R CH_2 CHO $\xrightarrow{H_2}$ R CH_2 CH_2OH

RCH = CH_2 + O ⌇S $\xleftarrow{}$ $\overset{R}{\underset{H–C}{CH_2}}$ (V)

RCH = CH_2 ... $\xrightarrow{}$ R CH_2 CHO

$\xrightarrow{H_2}$ R CH_2 CH_2OH

RCH$_2$CH$_3$ + O ⌇S $\xleftarrow{H2}$

$\xrightarrow{}$ RCH$_2$CO OH

Fig. 19. Possible surface reaction scheme for the Fischer–Tropsch synthesis on iron and other catalysts. S, Active surface catalyst site; ⌇ , denotes a bond of indefinite order.

other complexes. The objective is, however, to suggest that such complexes are not really different but that they are closely interrelated. If it is maintained for the moment that both complexes II and III can coexist, their relative statistical abundance will depend on the nature of the catalyst surface (e.g., whether Ni, Ru, Fe, or K_2O-promoted Fe) as well as the process conditions (temperature, gas composition, etc).

Chain growth is depicted to occur by the insertion of units of either CO or CH_2 between the carbon – catalyst bond. Which of these two possible insertions dominates is dependent on the relative abundance of complexes I and III. Thus it does not seem unreasonable to assume that complexes of types IV and V can both exist on the catalyst surface. Note that CO insertion converts complex IV to complex V.

Chain termination is depicted as formation of the simple desorption, hydrogenation, or oxidation products of complexes IV and V. All the normally observed products of Fischer – Tropsch synthesis can be accounted for in this way. It can be seen that many of the chain termination steps involve hydrogenation, whereas the absence of absorbed hydrogen only increases the probability of chain growth (by insertion of CO or of CH_2). Thus the probability of chain termination (and hence the resultant carbon number distribution of products) can be expected to be related to coverage of the surface by chemisorbed hydrogen. The relation between the carbon number distribution and a ratio of the type

$$ap_{H_2}^w/(bp_{CO}^x + cp_{CO_2}^y + dp_{H_2O}^z)$$

follows (Section VI.B).

Another advantage of the generalized scheme illustrated in Fig. 19 is that it accounts for the products of other related reactions such as methanol synthesis, hydroformulation (oxo synthesis), and the synthesis of polyhydric alcohols [6]. In these cases the bond between the C and the O is never broken.

Deluzarche et al. [20] and Sapienza et al. [21] have proposed mechanisms in which only M—O—C bonds are present. The former envisages chain growth by CO insertion to give M—O—CO—R intermediate complexes, and the latter suggests chain growth by the interaction of neighboring M—O=CH_2 units, etc. The concept of M—O linkages is also present in Fig. 19.

It is natural for surface scientists to want to present mechanistic concepts in terms of recognized chemical structures. However, because the prime role of catalysts is to "loosen" the existing bonds of reacting entities, it seems almost inevitable that the intermediate complexes will have ill-defined labile structures. If this concept is accepted, it means that arguments about closely related mechanisms are not really constructive.

References

1. BASF, German Patent No. 293,787, 1913.
2. C. D. Frohning, H. Kölbel, M. Ralek, W. Rottig, F. Schur and H. Schultz, *In* "Chemierohstoffe aus Kohle" (J. Falbe, ed.). Thieme Verlag, Stuttgart, 1977.
3. H. H. Storch, N. Golumbic, and R. B. Anderson, *In* "The Fischer–Tropsch and Related Synthesis." Wiley, New York, 1951.
4. H. Pichler, *In* "Advances in Catalysis" (W. G. Frankenburg, V. I. Komarewsky, and E. K. Rideal, eds.), Vol. 4. Academic Press, New York, 1952.
5. R. B. Anderson, *In* "Catalysis" (P. H. Emmett, ed.), Vol. 4. Reinhold, New York, 1956.
6. M. E. Dry, *In* "Catalysis Science and Technology" (J. R. Anderson and M. Boudart, eds.), Vol 1. Springer-Verlag, Berlin, 1981.
7. A. H. Stander, *Financial Times Symposium, London,* September, 1975.
8. J. C. Hoogendoorn *Symposium on Clean Fuels from Coal.* Institute of Gas Technology. Chicago, September, 1973.
9. J. P. Ingram, *School in Coal Processing at R.A.U. South African Coal Processing Society,* September, 1978.
10. P. C. Keith, *Oil Gas J.,* 1946, **45,** 102.
11. D. H. Jones, T. Shingles, F. T. Kelly, and M. E. Dry, U.S. Patent No. 4,225,531, 1980.
12. R. de Haan, B. N. S. Britz, C. S. Botha, and M. E. Dry, S.A. Patent Application No. 81 0071.
13. M. E. Dry, J. A. K. du Plessis, G. M. Leuteritz, *J. Catal.,* 1966, **6,** 194.
14. J. M. Berty, *Chem. Eng. Prog.,* 1974, **70**(5), 78.
15. M. E. Dry, *Ind. Eng. Chem. Prod. Res. Dev.,* 1976, **15**(4) 282.
16. C. B. Murchison and D. A. Murdick, *Hydrocarbon Process.* **60**(January), 159, 1981.
17. K. H. Eisenlohr, F. W. Moeller, and M. E. Dry, *In,* "Methanation of Synthesis Gas" (L. Seglin, ed.) Adv. Chem. Ser. 146 Am. Chem. Soc., 1975.
18. D. S. Newsome, *Catal. Rev. Sci. Eng.,* 1980, **21**(2) 275.
19. P. Biloen, J. N. Helle, and W. M. H. Sachtler, *J. Catal.,* 1979, **58,** 95.
20. A. Deluzarche, J. P. Hindermann, R. Kieffer, A. Muth, M. Papadopoulos and C. Tanielian, *Tetrahedron Lett.,* 1977, **9,** 797.
21. R. Sapienza, W. Slegeir, and T. O'Hare, *A I Ch E meeting Detroit,* August, 1981.

CHAPTER 6

Methanol Synthesis

F. Marschner   F. W. Moeller

Lurgi Kohle und Mineraloeltechnik GmbH
Frankfurt, Federal Republic of Germany

I.   Introduction

Methanol has become one of the major chemical raw materials (world production in 1980 was approximately 12 million tons), ranging third in volume behind ammonia and ethylene. Therefore the history of methanol is certainly of interest.

Methanol was first obtained in 1661 by P. Boyle when he rectified the neutral components of wood vinegar over lime. He called the liquid he obtained in this way "adiaphorous spiritus lignorum." This substance was rediscovered by Taylor in 1822, and its chemical individuality was first established by Justus v. Liebig and, at the same time, by J. B. Dumas and A.

Pictet. The latter two investigators also stated the correct formula and introduced the name "methyl" in 1835–1836.

After 1830, so-called wood alcohol or wood spirit was increasingly produced through dry distillation of wood. Wood distillation remained the only major source of methanol until 1923.

The first attempts to synthesize methanol were made at the beginning of the twentieth century after metals and metal oxides had been discovered to act as catalysts in building up chemical compounds from their constituents, e.g., ammonia from nitrogen and hydrogen, and after the physicochemical basis had been established and suitable equipment had become available to apply high pressures and temperatures. At this time the findings from research on ammonia synthesis (F. Haber, W. Nernst, and others) were consistently being applied to the synthesis of methanol.

A. Mittasch and co-workers obtained oxygen-containing compounds— methanol among others—from carbon monoxide and hydrogen over iron catalysts as early as 1913. This synthesis was described by M. Patart in 1921. Badische Anilin- & Soda-Fabrik (BASF) was the first to achieve a breakthrough to industrial-scale synthesis on the basis of tests performed between 1920 and 1923 by a team headed by M. Pier. These workers used zinc oxides activated by means of chromic acid as catalysts, thereby introducing the basic type of catalyst.

By means of these catalysts, methanol was produced on an industrial scale from water gas at Leuna-Werke in 1924.

Synthesis plants employing zinc oxide–chromium oxide catalysts operate at 250 to 350 bars and at temperatures between 320 and 400°C and thus must be considered high-pressure methanol synthesis plants. Until well into the 1960s this technology formed the basis of synthetic methanol production at practically all plants.

When largely sulfur-free synthesis gases came to be used about 1970 in conjunction with highly active copper catalysts—which, in principle, had also been known since the 1920s—the stringent reaction conditions could be greatly relaxed.

The resulting low-pressure methanol synthesis technology operates at approximately 50 to 100 bars and at temperatures between 230 and 280°C and, owing to its far greater economic merits, has completely superseded high-pressure technology.

Synthesis technologies known as medium-pressure synthesis, which operate at pressures and corresponding temperatures between the low-pressure and the high-pressure range and also employ copper catalysts, have been of little significance so far.

The continuously rising demand for methanol as a chemical raw material and the prospect of methanol playing a very significant role as a source of energy in the future (e.g., methanol as a motor fuel additive or the conversion of methanol to gasoline) together with the fact that raw materials are becoming scarcer and more expensive compels us to make the most economical use of materials and to obtain the highest possible yield of the desired products.

Attention is being paid to economical integration of the overall process route starting from such raw materials as natural gas, refinery gas, certain naphtha fractions, residue oils, and even coal. It is in this light that proposals for recovering part of the heat produced in nuclear reactors by means of suitable heat carriers and using it as process heat to produce synthesis gas from coal to arrive at an optimum overall process must be reviewed. The investigation of the structure and function of catalysts by means of the latest physical methods is a field not yet fully developed.

Catalyst activity, selectivity, stability, and price exert an influence upon the economics of a methanol process that will become even greater in the future.

II. Physical Properties

Methanol (CH_3OH, MW 32.042) is a neutral, colorless liquid and has a smell similar to that of ethanol. Methanol can be mixed with alcohols, esters, ketones, hydrogen chlorides, and aromatic hydrocarbons, as well as with water.

The most important physical data for methanol [1] are summarized in Table I. Table II gives the vapor pressure of methanol, and Table III the density of aqueous methanol solution at 25°C. Table IV shows the specific heat of methanol vapors, and Table V the freezing point of aqueous solution. The viscosities of methanol–water mixtures are summarized in Table VI, and binary azeotrope mixtures of methanol in Table VII.

III. Basic Process Theory

A. CHEMICAL REACTIONS IN METHANOL SYNTHESIS

In methanol synthesis based on carbon oxides (CO and CO_2) and hydrogen, equilibrium reactions are involved that are exothermic in the direction of methanol formation.

$$CO + 2H_2 \rightleftharpoons CH_3OH, \qquad \Delta H_{298\,K} = -90.84 \quad kJ/mol \qquad (1)$$

$$CO_2 + 3H_2 \rightleftharpoons CH_3OH + H_2O, \qquad \Delta H_{298\,K} = -49.57 \quad kJ/mol \qquad (2)$$

TABLE I

Physical Data for Methanol

Boiling point, 1013 mbars	64.509°C
Melting point	−97.88°C
Density	
d_4^0	0.81009
d_4^{25}	0.78687
d_4^{45}	0.76761
Specific heat, liquid	
−125.30°C	1.6953 J/g K
−51.47°C	2.2325 J/g K
−6.15°C	2.3698 J/g K
+0.42°C	2.3920 J/g K
+19.85°C	2.4979 J/g K
Combustion heat, 25°C	22.6926 kJ/g
Evaporation heat	
20°C	38.48 kJ/mol
64.7°C	35.346 kJ/mol
Melt heat	99.23 J/g
Free energy of formation, gaseous, $\Delta F_{298.16}^\circ$	−162.03 kJ
Formation enthalpy from the elements	
Liquid, 298.16 K	−238.65 kJ/mol
Gaseous	−201.39 kJ/mol
Entropy, liquid	126.9 J/mol K
Entropy, gaseous, 298.16 K	241.46 J/mol K
Critical data	
Temperature	239.43°C
Pressure	80.92 bars
Density	0.272 g/cm³
Flash point according to Abel-Pensky, DIN 51 755	+6.5°C
Ignition temperature, DIN 51 794	470°C
Explosion limit in air	6.72–36.5 vol%
Dielectricity constant, liquid	
20°C	33.79
40°C	28.3 (2 bars)
Refractive index	
n_D^{20}	1.3285
n_D^{45}	1.31876
n_D^{52}	1.326
Surface tension against air at 15°C	22.99 mN/m

As the exothermic reactions in Eqs. (1) and (2) also proceed with volume contraction, one obtains maximum methanol yields at low temperatures and high pressures.

A further reaction that should be mentioned is the shift conversion reaction

$$CO + H_2O \rightleftharpoons CO_2 + H_2 \qquad \Delta H_{298\,K} = -41.27 \quad kJ/mol \qquad (3)$$

TABLE II

**Temperature Dependence of the
Vapor Pressure of Methanol**

Temperature (°C)	Pressure (mbars)	Temperature (°C)	Pressure (mbars)
−16	13.3	40	352
−6	26.9	50	552
40	39.7	60	841
10	73.2	70	1246
20	128.6	80	1801
30	216.7	90	2546

which in the thermodynamic calculation of equilibrium conditions does not represent a reaction independent of reactions (1) and (2). Methanol formation according to Eqs. (1) and (2) is thermodynamically little favored by the application of high-CO and H_2-rich synthesis gases. There are numerous

TABLE III

Density of Aqueous Methanol solutions at 25°C (2)[a]

Methanol (mass %)	Methanol (vol%)	Relative density d_4^{25}
0	0.00	0.99708
5	6.28	0.9887
10	12.46	0.9804
15	18.55	0.9726
20	24.53	0.9649
25	30.42	0.9572
30	36.20	0.9492
35	41.84	0.9405
40	47.37	0.9316
45	52.74	0.9220
50	57.98	0.9122
55	63.05	0.9019
60	67.96	0.8910
65	72.64	0.8792
70	77.19	0.8675
75	81.54	0.8553
80	85.66	0.8424
85	89.60	0.8293
90	93.33	0.8158
95	96.79	0.8015
100	100.00	0.7867

[a] In [2] data for 30°C and 40°C are also given. For further data on the density of aqueous methanol solutions see [3, 4].

TABLE IV

**Temperature Dependency of the
Specific Heat of Methanol
Vapors at Atmospheric
Pressure [5]**

Temperature (°C)	Specific heat (J/mol K)
74.2	86.92
83.4	66.66
100.2	56.23
125.8	53.01
128.0	54.05
158.3	55.81
169.0	56.06
184.2	57.3
204.6	57.28
211.9	58.61
225.8	60.17
248.2	61.59
282.8	63.98
308.2	66.40
312.2	66.86

secondary reactions, the formation tendency of which is far higher. These secondary reactions are undesirable, as they mean a loss of synthesis gas and in addition make methanol distillation more expensive.

Formation of the following by-products is possible: hydrocarbons, higher alcohols, dimethyl ether, esters, and ketones and aldehydes.

TABLE V

**Freezing Point of Watery
Solutions [6]**

Methanol (mass %)	Freezing point (°C)
0	0
10	−6.5
20	−15.0
30	−26.0
40	−39.7
50	−55.4
60	−75.7
64	−83.4

TABLE VI

Viscosity of Methanol–Water Mixtures at Various Temperatures in mPas [4]

Temp. (°C)	Methanol[a]					
	99.95 (99.91)	90.01 (83.51)	79.98 (69.20)	70.00 (56.75)	60.07 (45.82)	50.04 (36.03)
10	0.6795	1.023	1.394	1.760	2.103	2.388
0	0.7966	1.250	1.772	2.320	2.865	3.349
−10	0.9425	1.556	2.302	3.144	4.043	4.912
−20	1.134	1.971	3.074	4.420	5.961	7.587
−30	1.380	2.546	4.231	6.478	9.261	12.46
−40	1.715	3.384	6.061	9.991	15.34	22.10
−50	2.184	4.665	9.124	16.46	27.55	42.74
−60	2.868	6.705	14.65	29.56	54.43	—[b]
−70	3.899	10.22	25.68	58.96	121.8	—
−80	5.596	16.90	50.24	135.3	—	—
−90	8.624	31.32	115.3	376.4	—	—
−100	—	68.36	333.3	—	—	—
−110	—	189.7	—	—	—	—

[a] The nonparenthetical values give the mass % methanol, the parenthetical values give the mol% methanol.

[b] The dash indicates that no data were available.

The formation of methane and higher hydrocarbons during methanol synthesis based on the equations

$$CO + 3H_2 \rightleftharpoons CH_4 + H_2O \tag{4}$$

$$CO_2 + 4H_2 \rightleftharpoons CH_4 + 2H_2O \tag{5}$$

$$nCO + (2n - 1)H_2 \rightleftharpoons C_nH_{2n+2} + nH_2O \tag{6}$$

can then always be limited to a minimum if the impurities in the iron, cobalt, and nickel in the base materials of the catalyst are kept to a minimum during catalyst production. Low reaction temperatures are also advantageous in reducing methane formation.

Carbon formation due to the Boudouard reaction

$$2CO \rightleftharpoons CO_2 + C \tag{7}$$

is of no importance in methanol synthesis despite a very high formation tendency so long as certain catalyst-specific maximum temperatures are not exceeded.

The formation of higher alcohols such as ethanol, propanol, and butanol according to

$$nCO + 2nH_2 \rightleftharpoons C_nH_{2n+1}OH + (n - 1)H_2O \tag{8}$$

TABLE VII

Binary Azeotropic Mixtures of Methanol [7]

Components	Boiling point of components (°C)	Boiling point of azeotrope (°C)	Methanol content in azeotrope (mass %)
Acetone	56.2	55.7	12
Carbon disulfide	46.5	37.65	14
Carbon tetrachloride	76.5	55.7	20.66
Chloroform	61.7	54.0	12.6
Trichlorethylene	87.1	59.4	38.0
Methylene dichloride	41.6	37.8	7.3
Ethyl nitrate	87.68	61.77	57.0
Acetonitrile	81.68	63.65	19
Acrylonitrile	77.3	61.4	61.3
Ethyl formate	54.10	50.95	16
Methyl acetate	57.2	54.0	19.5
Ethyl acetate	77.1	62.25	44
Isopropyl acetate	88.7	64.5	80.0
n-Pentane	36.07	30.4	15.5
Thiophene	84.0	59.55	55
Benzene	80.1	57.5	39.1
Toluene	110.6	63.8	69

can be repressed if one excludes promoters such as alkali and alkaline earth dioxide from the catalyst.

Dehydrogenating methanol to dimethyl ether is then possible when Al_2O_3 is used as a promoter and stabilizer in the methanol catalyst.

$$2CH_3OH \rightleftharpoons CH_3OCH_3 + H_2O \qquad (9)$$

However, by taking certain precautions a Cu–Zn catalyst with a relatively high percentage of Al_2O_3 can be manufactured without causing critical values (<0.2 wt%) with respect to dimethyl ether formation.

The formation of acids or esters as well as ketones and aldehydes by an undesirable secondary reaction is normally extremely low. The formation of methyl formate

$$CH_3OH + CO \rightleftharpoons HCOOCH_3 \qquad (10)$$

under certain reaction conditions of methanol synthesis alone is worth mentioning. Thus, Kotowski [8] reported on the formation of 1.5–2.5 wt% methyl formate in raw methanol during a test involving methanol synthesis with a $CuO-ZnO-Al_2O_3$ catalyst.

This high methyl formate content must be viewed as an extreme value in commercial methanol synthesis units utilizing Cu–Zn catalysts. The selectivity was improved, and therefore a methyl formate content approximately 0.15 wt% in crude methanol must be expected.

B.   THERMODYNAMIC EQUILIBRIUM IN
METHANOL SYNTHESIS

The maximum obtainable conversions of CO and CO_2 to methanol according to Eqs. (1) and (2) are limited by the position of the thermodynamic equilibrium.

The equilibrium constants K_f [Eq.(1)] and K_f [Eq.(3)] for the CO conversion reaction [Eq.(1)] and the shift conversion reaction [Eq.(3)] based on fugacities are sufficient for complete determination of the simultaneous equilibrium.

The following definition equations apply.

$$K_f[\text{Eq.(1)}] = \frac{f_{CH_3OH}}{f_{CO}f_{H_2}^2} = \frac{p_{CH_3OH}}{p_{CO}p_{H_2}^2}\frac{\gamma_{CH_3OH}}{\gamma_{CO}\gamma_{H_2}^2}, \tag{11}$$

$$K_f[\text{Eq.(3)}] = \frac{f_{CO_2}f_{H_2}}{f_{CO}f_{H_2O}} = \frac{p_{CO_2}p_{H_2}}{p_{CO}p_{H_2O}}\frac{\gamma_{CO_2}\gamma_{H_2}}{\gamma_{CO}\gamma_{H_2O}}. \tag{12}$$

However, in practice the equilibrium constants

$$K_p[\text{Eq.(1)}] = \frac{K_f[\text{Eq.(1)}]}{K_\gamma[\text{Eq.(1)}]} = \frac{p_{CH_3OH}}{p_{CO}p_{H_2}^2} \tag{13}$$

and

$$K_p[\text{Eq.(3)}] = \frac{K_f[\text{Eq.(3)}]}{K_\gamma[\text{Eq.(3)}]} = \frac{p_{CO_2}p_{H_2}}{p_{CO}p_{H_2O}}. \tag{14}$$

based on partial pressures are largely used for calculations, and therefore the quotients $K_f[\text{Eq.(1)}]/K_\gamma[\text{Eq.(1)}]$ and $K_f[\text{Eq.(3)}]/K_\gamma[\text{Eq.(3)}]$ must be determined.

The equilibrium constant K_f based on fugacities can be calculated from the change in free reaction enthalpy

$$K_f = \exp(-\Delta G/RT) \tag{15}$$

and is dependent only on the temperature.

Various authors [9–11] have shown the temperature dependency of the equilibrium constant $K_f[\text{Eq.(1)}]$ for methanol formation from CO in the form of a numerical equation.

This equation, according to Thomas and Portalski [9], is

$$\log K_f[\text{Eq.(1)}] = 10.20 + (3921/T) - 7.971 \log T \\ + 2.499 \times 10^{-3} \times T - 2.953 \times 10^{-7} \times T^2. \tag{16}$$

The low deviations in the equilibrium constants of other authors [10, 11] are caused by the differing coefficients in the determination equation for the temperature dependency of the specific heat of the individual components.

With respect to the shift conversion reaction [Eq.(3)], the temperature dependency on the equilibrium constant K_f[Eq.(3)] was also shown in a numerical equation by Bisset [12]:

$$\ln K_f[\text{Eq.(3)}] = (5639.5/T) + 1.077 \times \ln T + 5.44 \times 10^{-4}T \\ - 1.125 \times 10^{-7}T^2 - (49{,}170/T^2) - 13.148. \tag{17}$$

Good agreement with further data on the equilibrium constants of the conversion reaction in the literature [13, 14] is to be noted.

The correction K_y based on the fugacity coefficients must be determined taking into account the real behavior of the gases. The fugacity coefficients of the individual components (γ_i), dependent upon the pressure and temperature, can be calculated using the equation

$$RT \ln \gamma_i = \int_0^P V' - (RT/p)\, dp. \tag{18}$$

Here the equation can be solved by graphic integration of experimental pVT data or by analytic integration of the pVT dependency of a suitable state equation.

Using graphic integration Newton [15] determined the coefficients γ_i for a large number of gases from experimental pVT data and compiled a general diagram for the relation.

$$\gamma_i = f(P/P_c, T/T_c).$$

More accurate results for the fugacity coefficients are obtained when known state equations (Benedict – Webb – Rubin, Beattie – Bridgeman, or Redlich – Kwong) are taken as a basis.

On technically performing methanol synthesis accurate matching of temperature, pressure, concentration, and catalyst activity is absolutely necessary to obtain maximum yields and optimum economics in view of the conversion limitation imposed by the equilibrium.

Table VIII shows the equilibrium conversion of carbon monoxide and carbon dioxide reached dictated by pressure and temperature. A conditioned coal gasification gas with 4 vol% CO_2, 26 vol% CO, 60 vol% H_2, and 10 vol% inert gases has been taken as a basis.

For the sake of simplicity the recycling of product gas has not been taken into account. It is clearly recognizable that as the temperature rises the carbon monoxide conversion drops considerably, whereas the carbon dioxide conversion increases.

In the case of synthesis gases rich in carbon monoxide the CO_2 conversion only participates to a low degree in methanol production. Because CO_2 reaction is not kinetically favored, the assessment can almost exclusively be limited to carbon monoxide conversion.

TABLE VIII

CO$_2$ and CO Conversion at the Equilibrium Dependent upon Pressure and Temperature

Temperature (°C)	CO Conversion			CO$_2$ Conversion		
	50 bars	100 bars	300 bars	50 bars	100 bars	300 bars
250	0.524	0.769	0.951	0.035	0.052	0.189
300	0.174	0.440	0.825	0.064	0.081	0.187
350	0.027	0.145	0.600	0.100	0.127	0.223
400	0.015	0.017	0.310	0.168	0.186	0.260

Critical observation of the carbon monoxide conversion up to attainment of the equilibrium shows that:

(1) When a highly active catalyst is available that ensures an extensive establishment of the equilibrium conversion for the carbon monoxide at low temperatures of about 260°C, methanol synthesis even at low pressures of about 50 to 100 bars is technically optimum.

(2) In the case of a less active catalyst appropriate higher reaction temperatures of about 380°C and higher reaction pressures of about 300 bar are necessary for the methanol synthesis (termed high-pressure process).

(3) As the carbon monoxide conversion is limited by the equilibrium to 30 to 50% dependent upon reaction conditions, the product gas from the synthesis must be recycled to the synthesis after condensing out methanol and water to ensure optimum gas utilization.

C. REACTION MECHANISM AND KINETICS

Very different theories are put forward in the literature on the reaction mechanism in methanol synthesis on copper–zinc catalysts. A successive hydrogenation of carbon monoxide as well as a reaction procedure based on carbon monoxide with a formate intermediate stage which is subsequently hydrogenated and dehydrogenated are discussed [16].

Contrary to this, various Russian authors [17, 18] favor another reaction mechanism for methanol synthesis on Cu–ZnO catalysts. Here, methanol formation is dominated by hydrogenation of CO$_2$ formed during reaction by the water gas reaction from CO.

$$CO + H_2O \rightleftharpoons CO_2 + H_2 \xrightarrow{+2H_2} CH_3OH + H_2O$$

The known influence of the CO$_2$ on the methanol formation rate is given as proof for this reaction theory. It is known [18, 19] that optimum methanol yields can be obtained at a CO_2 concentration of about 5 vol% in the reactor inlet gas. If the CO_2 contents are lower or higher, the methanol formation rate on the Cu–ZnO catalyst drops. Methanol formation is not noted if the synthesis gas is free of CO_2 and H_2O. Although the reaction mechanism in methanol synthesis has not been determined, there has been no shortage of attempts to draw up kinetic equations for the methanol formation rate.

From numerous data, Natta [20] compiled the following reaction kinetic conditional equation for the methanol formation rate r_{CH_3OH}, where a trimolecular reaction takes place on the surface between a CO molecule and two adsorbed H_2 molecules

$$r_{CH_3OH} = \frac{\gamma_{CO}\gamma_{H_2}^2 - \gamma_{CH_3OH}/K_p}{(A + B\gamma_{CO} + C\gamma_{H_2} + D\gamma_{CH_3OH})^3},$$

$\gamma_i = f_i p_i$ is the fugacity of the individual reactants.

This classic conditional equation has been altered or extended by other authors [21-23].

Another rate expression was proposed by Uchida and Ogino [24] assuming that the desorption of methanol was the rate-limiting step:

$$r_{CH_3OH} = k(p_{H_2}^2 p_{CO})^{0.7}.$$

D. CATALYSTS

The requirement for commercial methanol synthesis at relatively low operating pressures of 50 to 100 bars is a catalyst featuring a very high activity at relatively low temperatures. It has been known for a considerable amount of time that catalysts containing copper have a much higher activity than zinc–chrome catalysts. These catalysts containing copper were tested with respect to their application for commercial methanol synthesis in the 1930s and in the technical literature are occasionally called "Blasiak Catalysts" [25].

Various of the first works on methanol synthesis with catalysts containing copper were published in 1928–1938. These concern the system Cu–Zn and Cu–Mg.

Commercial application of these catalysts at that time was not successful because satisfactory catalyst lives were not obtained. The very high sensitivity of copper catalysts to catalyst poisons such as sulfur, chlorine, phosphorus, and carbonyl compounds, as well as a certain temperature sensitivity, were mainly responsible for this.

The zinc–chrome oxide catalysts for high-pressure synthesis exhibit relatively good resistance to catalyst poisons such as sulfur with 30 ppm H_2S and more in the feed gas to the synthesis still permissible. Contrary to this, the catalytic poisons in the form of H_2S must be limited to <1 ppm using a copper catalyst. Only the development of new gas purification processes allowed the content of sulfur compounds in synthesis to be reduced to 0.1 ppm, resulting in catalyst lives of more than 3 yr.

A further reason for the rapid aging of copper catalysts at that time must be viewed as the recrystallization of the copper due to excessively high temperatures.

The copper in the catalyst is metal under the conditions of methanol synthesis. The Tamman temperature of the copper, i.e., the temperature at which a noteworthy mobility is first noted in the crystal lattice is about 190°C. For this reason, the upper temperature limit for copper catalysts is about 270°C. At higher temperatures, rapid aging takes place as a result of recrystallization of the copper. The first reported attempt to apply copper catalysts in a commercial methanol synthesis unit was undertaken at the Polish Chemical Works in Oswiecim [8].

During the years after 1959 a new phase in the development of catalysts containing copper was introduced and permitted economic commercial methanol synthesis at the lowest possible pressures and correspondingly low temperatures.

At almost the same time ICI in Billingham [26] and Lurgi in Frankfurt [27] used copper catalysts in various commercial low-pressure methanol synthesis units with success.

As shown in the accompanying table, attempts were made to optimize the copper–zinc catalyst system with respect to its activity and thermostability by adding promoters. This listing cannot claim to be complete.

Catalyst manufacturer	Catalyst system	Reference
ICI	Cu–Zn–Cr	[28]
	Cu–Zn–Al	[29]
Lurgi	Cu–Zn–Cr	[30]
	Cu–Zn–Mn–V	[31]
BASF	Cu–Zn–Mn–Cr	[32]
	Cu–Zn–Mn–Al–Cr	
	Cu–Zn–Mn–Al	
	Cu–Zn–Al	
CCI	Cu–Zn–Al	[33]
Shell	Cu–Zn–Ag	[34]
Mitsubishi	Cu–Zn–Cr	[35]

Varying the composition alone does not produce an optimum catalyst, and the method of manufacture can also be of decisive importance. Copper catalysts are generally manufactured by precipitating the metal components together out of a solution diluted to a greater or lesser degree. Maintenance of a definite precipitation temperature and a certain pH value is extremely important. After drying and calcining the filter cake, the catalyst material obtained is pressed to the desired size.

Prior to going into operation the catalyst must be reduced; i.e., the copper oxide must be converted to the metallic, catalytically active form. As the reduction of copper oxide is highly exothermic and catalyst overheating must be avoided to obtain the highest possible activity, reduction is carried out at low hydrogen concentrations in an inert gas. A subsequent forming phase is not necessary.

IV. Feedstocks for Methanol Production

Generally, all naturally occurring carboniferous fuels can be used as feedstocks to produce suitable synthesis gases. Depending on the form in which the fuel occurs, a larger or smaller number of preparatory steps may, however, be required to obtain the desired mixture of CO, CO_2, and H_2 with as low an inert gas content as possible and without exceeding the maximum allowable degree of contamination by sulfur, chlorine, and other compounds that may be detrimental to the catalyst.

Whereas the first industrial-scale methanol synthesis plants used synthesis gas from coke or coal [36], development of the steam-reforming technology led to the use of naphtha and natural gas about the middle of the 1950s. However, the growing scarcity and rising prices of these feedstocks have made it necessary to use also heavy residues from oil fractionation; for this purpose, suitable wash processes have been developed to obtain synthesis gas of high purity, which is absolutely mandatory in using highly sensitive copper catalysts [37].

If the shortage of natural gas and oil products becomes even more pronounced, coal will also have to be used again to produce methanol synthesis gas — a technology for which modern and cost-effective processes are already available or are currently being developed [38]. The following text describes the production of methanol synthesis gas from a few selected raw materials on the basis of a low-pressure methanol synthesis process using copper catalysts.

A. NATURAL GAS

Figure 1 shows the production of methanol synthesis gas from natural gas. Depending on the content and kind of sulfur compounds the gas is first

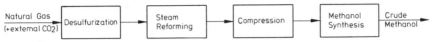

Fig. 1. Methanol from natural gas.

desulfurized to less than 0.5 ppm total sulfur by means of zinc oxide or other suitable materials, such as molecular sieves, with a hydrogenation stage preceding the sulfur removal if necessary.

Thereafter the desulfurized natural gas is steam-reformed to obtain methanol synthesis gas which is compressed and passed directly to the methanol synthesis unit.

Because synthesis gas from natural gas normally contains excess hydrogen, CO_2 can be added from outside until a $(H_2 - CO_2)/(CO + CO_2)$ ratio of approximately 2 is reached.

Without external CO_2 the synthesis gas contains approximately 8% CO_2, 16% CO, 73% H_2, and approximately 3% CH_4; after the addition of external CO_2 the percentages are approximately 10% CO_2, 18% CO, 69% H_2, and 3% CH_4.

The steam-reforming process employs nickel catalysts and operates at pressures of approximately 15 to 20 bars and temperatures of 850 to 880°C.

In the case of large methanol plants—approximately 1500 tons/day or more—based on natural gas, it is normally reasonable to use a combination of steam reforming and catalytic autothermal reforming with oxygen. Figure 2 shows such a process arrangement. Only about 50% of the natural gas feed is conventionally steam-reformed at pressures of about 35 bars and temperatures of about 780°C. Thereafter, the remaining natural gas is added, and this gas mixture is autothermally reformed by means of oxygen over a nickel catalyst at approximately 950°C to obtain the desired methanol synthesis gas. With this arrangement the number of tubes in the steam reformer can be greatly reduced and the synthesis gas pressure increased. Less energy is therefore required to compress the gas to the necessary synthesis pressure, and the synthesis gas has a composition that, being nearly stoichiometric, is more favorable for methanol synthesis: CO_2 8%, CO

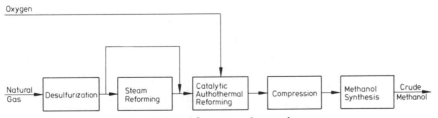

Fig. 2. Methanol from natural gas and oxygen.

21.5%, H_2 67.5%, $(CH_4 + N_2 + A_y) \sim 3\%$. The synthesis loop is thus also smaller than in the case of pure steam reforming.

B. NAPHTHA

In view of the general scarcity of naphtha, methanol plants based on conventional naphtha may be expected to be built in the future only in exceptional cases.

However, it may be an economically promising route to use raffinate in combination with refinery gases [39].

Figure 3 shows such a process arrangement where treated naphtha and refinery gases are jointly processed in a hydrogenation stage (converting sulfur compounds to H_2S and saturating the olefins to paraffins).

Gas generation is particularly efficient if the steam reformer is preceded by a rich-gas reactor (as in the BASF–Lurgi Recatro process). The hydrocarbons are first reformed with a relatively small quantity of steam in the rich-gas reactor containing a high-nickel catalyst. The resulting gas mixture is rich in methane and is then treated in the actual steam reformer filled with the normal catalyst to produce methanol synthesis gas. This ensures a CO_2/CO ratio favorable for methanol synthesis and allows the catalyst volume in the methanol synthesis reactor to be kept low [40]. Moreover, high catalyst cycle times are achieved in the steam reformer.

After the addition of external CO_2 the synthesis gas has approximately the following composition: CO_2 11%, CO 16.5%, H_2 67.5%, and about 3% CH_4 and 2% N_2. The pressures employed in this gas generation process range between 15 and 20 bars.

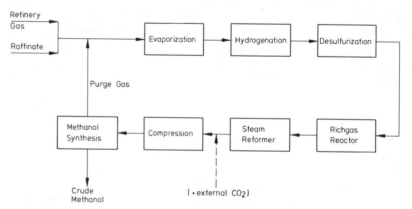

Fig. 3. Methanol from refinery gas and raffinate.

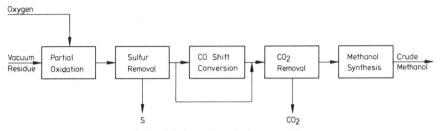

Fig. 4. Methanol from vacuum residue.

C.   HEAVY RESIDUE OIL

A raw material such as heavy residue oil cannot be converted to synthesis gas by steam reforming but must be partially oxidized with oxygen. Suitable processes developed, for instance, by Shell and by Texaco, are being employed at several commercial-scale methanol plants. Partial oxidation takes place without a catalyst at pressures between 40 and 90 bars and a temperature of about 1400°C.

Figure 4 shows such a process arrangement. Special attention should be paid to the gas purification stage where a scrubber using cold methanol, the so-called Rectisol wash unit, offers particular advantages. In addition to approximately 1% sulfur compounds and other contaminants, the raw gas from the partial oxidation stage has approximately the following composition: CO_2 5%, CO 47%, H_2 46%, $(CH_4 + N_2 + A_y)$ 1%. Therefore, a certain percentage of the CO has to be converted and excess carbon has to be washed out in the form of CO_2. A particularly efficient process arrangement for a synthesis gas plant based on heavy residue oil is obtained by coupling methanol synthesis with ammonia synthesis [37].

The synthesis gas composition is nearly ideal for methanol synthesis: CO_2 3%, CO 29%, H_2 67%, $(CH_4 + N_2 + A_y)$ 1%. It is particularly worth mentioning that the synthesis gas can be passed on to the methanol synthesis unit without further compression; i.e., this process arrangement can operate without the normally required synthesis gas compressor [41].

D.   COAL

A larger number of process steps is necessarily required to produce methanol synthesis gas from coal, since the raw gases contain a larger number of undesirable by-products, such as sulfur compounds, tar, and

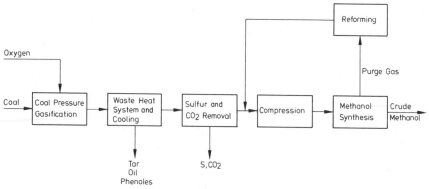

Fig. 5. Methanol from coal.

phenols. Figure 5 shows methanol synthesis gas production from coal by the Lurgi pressure gasification process. After the coal has been gasified by means of oxygen and steam at a pressure of approximately 30 bars, a first-purification stage is required to remove such volatile components as tar, oils, and phenols. This is followed by a fine-cleaning stage for which cold methanol is employed as described in Section IV.C.

A typical clean gas which is then compressed and passed to the synthesis unit contains $CO_2 < 1\%$, CO 30%, H_2 55%, and $(CH_4 + N_2 + A_y) \sim 14\%$.

Depending on the origin and age of the coal, the methanol synthesis gas may contain as much as 15 vol% methane. The purge gas from the methanol synthesis unit should therefore be steam-reformed and recycled to the inlet of the methanol synthesis unit. This ensures an optimum synthesis gas composition for methanol synthesis [42]. Another economically favorable route, however, is also the coproduction of methanol and synthetic natural gas through methanation of the purge gas from the methanol synthesis unit [43].

In addition to the Lurgi fixed bed process for the pressure gasification of coal, there are a number of other processes, such as the Winkler fluidized bed process, dust gasification processes developed by Koppers–Totzek and by Texaco, and others [38].

Recent developments aim at using, for instance, helium at a temperature of approximately 950°C from a high-temperature nuclear reactor for the endothermal coal gasification reaction (nuclear process heat).

This ensures that the coal is exclusively used to produce the synthesis gas components rather than burning some of it to generate the necessary process heat.

V. Processes for Methanol

With regard to the basic arrangement, currently employed methanol processes differ only slightly from each other and have practically all been derived from the original high-pressure methanol synthesis technology. Since the formation of methanol from carbon oxides and hydrogen takes place in an exothermal reaction, removing the reaction heat and/or limitating the catalyst temperature are of utmost importance. The methods by which this problem is solved differ among the various known methanol processes either in the type of reactor construction or in the way in which the heat of reaction is discharged. The most widely used methanol processes are considered in the following discussion.

A. CLASSICAL METHANOL PROCESSES

The classical methanol processes operate at pressures of approximately 300 bars, using chromium oxide–zinc oxide catalysts for the catalytic conversion of CO and CO_2 with H_2 at temperatures of 320 to 400°C to produce methanol. Figure 6 shows a typical arrangement of a high-pressure methanol synthesis unit.

The compressed synthesis gas is mixed with hydrogen-rich recycle gas and

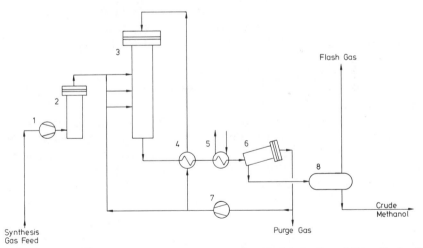

Fig. 6. High-pressure methanol synthesis by Union Rheinische Braunkohlen-Kraftstoff AG. (1) Synthesis gas compressor, (2) activated carbon filter, (3) methanol reactor, (4) interchanger, (5) cooler, (6) high-pressure separator, (7) recycle compressor, (8) let-down vessel.

then passed through the reactor which in most cases contains several catalyst beds. The temperatures at the inlets to the various catalyst beds range between 320 and 350°C, whereas the outlet temperatures — because of the exothermic nature of the methanol reaction — are about 380–400°C. Between the catalyst beds the temperature is reduced by adding cold gas (cold gas quenching). Temperatures above 400°C may lead to the dreaded highly exothermic methane formation, causing temperatures to run away and destroy both the catalyst and the vessels.

Methane is catalytically formed from CO and H_2 in the presence of active iron which is produced when iron pentacarbonyl is decomposed under reaction conditions. This iron pentacarbonyl is formed from carbon monoxide corroding the surfaces of loop components, such as heat exchangers and coolers, at temperatures between 120 and 210°C if normal carbon steel is used without a special lining or coating instead of high-alloy steel.

A special type of high-pressure methanol process has been developed by Union Rheinische Braukohlen-Kraftstoff AG, Wesseling, on which Fig. 6 is based. In this process, the partial pressure of the carbon monoxide in the loop is kept at less than 18 bars, corresponding to 6 vol% CO. This allows major components of the synthesis unit to be manufactured from carbon steel or steel that is resistant to hydrogen under pressure, because with normal steels the formation of iron pentacarbonyl was found to be negligible even in the above-mentioned critical temperature range. The CO-rich synthesis gas is fed directly between the catalyst beds so that yields of as much as 1.6 kg/liter of catalyst per hour are obtained [36].

Other high-pressure methanol processes employ reactors in which the catalyst is indirectly cooled by means of cold recycle gas passing through internal heat exchangers or by means of water.

In addition to water, high-pressure methanol synthesis produces a considerable yield of by-products which may range between 5 and 10 wt% depending on the process.

The disadvantages of the high-pressure process are the expensive high-pressure components, the high energy costs for synthesis gas compression, the relatively high maintenance costs, and the above-mentioned considerable yield of undesirable by-products such as dimethyl ether, hydrocarbons, and higher alcohols. Although some high-pressure methanol synthesis plants are still operating, they continue to be replaced by modern low-pressure plants.

B. MODERN METHANOL PROCESSES

Synthesis gas of very high purity can be obtained with the application of special gas purification processes by which catalyst poisons are almost

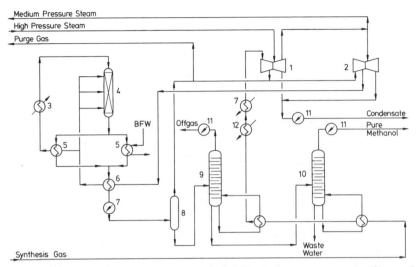

Fig. 7. ICI low-pressure methanol process. (1) Synthesis gas compressor, (2) recycle compressor, (3) start-up heater. (4) methanol reactor, (5) BFW preheater, (6) heat exchanger, (7) cooler, (8) separator, (9) low-boiler column, (10) pure methanol column, (11) condensator, (12) BFW degasser.

completely removed. This has paved the way for the use of highly active copper catalysts, leading in the 1960s to the breakthrough of the low-pressure methanol processes used today.

These processes employ pressures between 50 and 100 bars depending on the size of the plant. The copper catalysts operate at temperatures between 230 and 300°C, higher temperatures leading to recrystallization of the catalyst and thus to a decrease in activity.

The basic arrangement of a low-pressure synthesis unit is very similar to that for high-pressure synthesis.

Since — as already pointed out — copper catalysts are very sensitive to high temperatures, it is of utmost importance to observe the allowable catalyst temperatures. Therefore the currently employed low-pressure methanol processes differ essentially in reactor design and in the method by which the heat of reaction is removed and further used. Well-known processes include the ICI process (60%) [44], the Lurgi process (30%) [37], the Topsøe process (formerly Nissui-Topsøe and hitherto negligible) [45], and the process developed by Mitsubishi Gas Chemical Company, Inc. (about 6%) [46].

Figure 7 shows the ICI low-pressure methanol process which operates at pressures between 50 and 100 bars depending on the plant size. This process uses a quench-type reactor; i.e., a preheated mixture of recycle gas and fresh

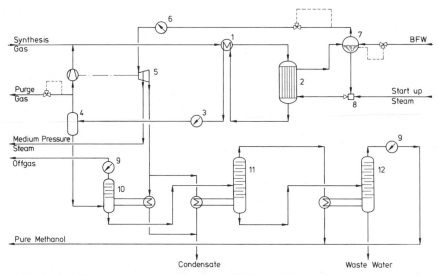

Fig. 8. Lurgi low-pressure methanol process. (1) Heat exchanger, (2) methanol reactor, (3) cooler, (4) separator, (5) recycle compressor, (6) steam superheater, (7) steam drum, (8) steam injector, (9) condenser, (10) low-boiler column, (11) pure methanol column I, (12) pure methanol column II.

gas is injected at different levels to limit the catalyst temperature. In most cases, an economizer is provided downstream of the reactor, through which a major part of the reaction heat can be turned to good use, increasing the overall efficiency of the process. This efficiency is further increased by having part of the sensible and latent heat of the synthesis gas removed in the reboilers of the methanol distillation unit instead of cooling down the synthesis gas completely by water or air in the steam-reforming unit.

The raw methanol is processed in two distillation columns arranged in series. The first column serves to boil off the light-end contaminants such as dimethyl ether and methyl formate overhead, and the second column is used to remove all contaminants with boiling points higher than that of methanol, such as higher alcohols, hydrocarbons, and reaction water.

Figure 8 shows the Lurgi low-pressure methanol process which operates at 40 to 100 bars and employs a tube reactor which is cooled by means of boiling water.

With this process the maximum catalyst temperature is only about 10°C above that of the boiling water even if the recycle gas/fresh gas ratios are as low as 2.5–3.5, so that catalyst cycle times of more than 5 yr are obtained. Up to 1.4 tons of steam at 40 to 50 bars per ton of methanol can be produced depending on the preheating temperature of the boiler feedwater. The

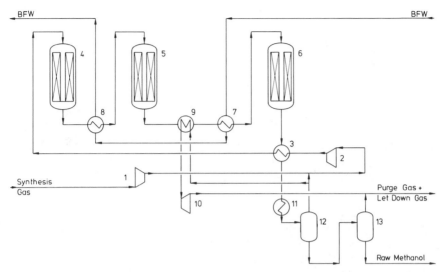

Fig. 9. Haldor Topsøe low-pressure methanol process. (1) Synthesis gas compressor, (2) recycle compressor, (3) heat exchanger, (4–6) methanol reactors, (7 and 8) BFW preheaters, (9) purge gas preheater, (10) purge gas expander, (11) cooler, (12) separator, (13) flash drum.

extremely uniform catalyst temperature ensures that hardly any by-products are formed. The steam generated in the methanol reactor may profitably be used to drive the recycle compressor, and the spent steam from the turbine can be used for raw methanol distillation. Temperature control in the reactor is extremely simple. Steam is saved in the methanol distillation stage by passing the overhead vapors from the first pure methanol column to the reboilers of the second column. For this purpose the pure methanol column is subdivided into two sections, the first section operating at a correspondingly higher pressure.

Figure 9 shows the loop of the Topsøe methanol process. A characteristic feature of this process is the three reactors connected in series with interposed boiler feedwater heating to reduce the temperature. Normally, the reaction takes place at pressures between 50 and 90 bars and temperatures of 200 to 310°C. The catalyst is arranged within the reactor in the form of an upright annular layer. The radial flow of the gases processed with this arrangement leads to a slight pressure drop which saves energy in the recycle compressor. Moreover, a catalyst with relatively small dimensions can be used so that the space–time yield rises and the necessary catalyst volume can be reduced. To make further use of the available energy, a purge gas expander in the purge gas flow is proposed for plants based on natural gas. The design of the distillation stage is similar to that shown in Fig. 7.

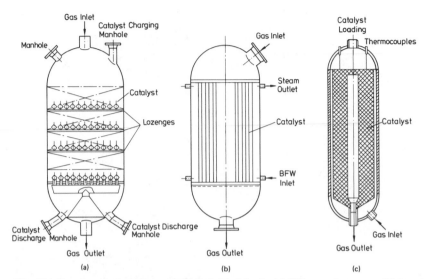

Fig. 10. Reactors for low-pressure methanol synthesis. (a) ICI quench reactor, (b) Lurgi tubular reactor, (c) Haldor Topsøe one-bed radial flow reactor.

The MGC methanol process of Mitsubishi Gas Chemical Company, Inc., also uses a multibed reactor and low-pressure steam generation in external heat exchangers between individual beds. The literature indicates flexible reaction conditions between pressures of 50 to 200 bars and temperatures of 220 to 300°C.

Figure 10 shows three types of reactors that differ in the arrangement of the catalysts and the type of cooling employed.

The catalyst volume and hence also the reactor output can be increased in the case of the ICI reactor by increasing the reactor diameter, in the case of the Lurgi reactor by increasing the number of tubes filled with catalyst, and in the case of the Topsøe reactor by extending the upright, radial annular layer. Space–time yields of more than 1 kg of methanol per liter of catalyst per hour can thus be obtained.

Low-pressure methanol plants can today be constructed as single-train units with turbo compressors for outputs ranging between 150 and 3000 tons/day.

Whereas earlier high-pressure synthesis plants operating with natural gas as a feedstock had an energy demand of approximately 9 MMkcal/1000 kg of methanol, modern low-pressure methanol processes need only 7–7.5 MMkcal/1000 kg of methanol; or, with heavy residue oils as a feedstock, approximately, 8–8.5 MMkcal/1000 kg of methanol; or, with coal as a feedstock, approximately 10–11 MMkcal/1000 kg of methanol.

C. RECENT DEVELOPMENTS

New catalysts may crucially influence current methanol processes, particularly in regard to reactor size, recycle ratio, and reaction conditions.

There are, for instance, reports of a methanol synthesis catalyst consisting of thorium–copper intermetallics [47] from which, at 60 bars and 280°C, an inlet standard hourly space velocity of 31,000, and a 16:1 hydrogen–carbon monoxide synthesis gas, 44 mol% of methanol (excluding hydrogen) was found in the off-gas from a methanol synthesis microreactor, compared with 6.5 mol% found when a commercial $Cu–ZnO–Al_2O_3$ catalyst was used.

Other work deals with zinc-promoted Raney copper catalysts for the synthesis of methanol [48]. These catalysts are based on the caustic extraction of aluminum–copper–zinc alloys. In addition to the high activity and selectivity, there is also a tendency to form dimethyl ether, which may turn out to be an advantage if the methanol—dimethyl ether mixture used to produce gasoline according to the Mobil process [49].

Simultaneous production of methanol and dimethyl ether is advocated also by [50] who recommends the use of a fixed bed consisting of a mixture of methanol synthesis catalyst and dehydration catalyst.

Energy-saving combinations involving synthesis gas production and methanol synthesis could also be devised if special raw materials, such as off-gases from the steel industry, which essentially contain carbon monoxide, are used [51].

Other work, aiming at applications in the rather distant future, deals with the production of methanol from CO_2 and H_2, assuming that CO_2 may, for instance, be obtained through absorption from the air and hydrogen is made available through electrolysis of water and that nuclear fusion energy can be turned to advantage for the overall process [52]. This would, however, require the development of special catalysts with sufficiently high activity for the conversion of CO_2 and H_2 to methanol.

VI. Methanol Quality Specification

The quality requirements placed on pure commercial methanol are generally very high, and in some cases dictated by further processing. The specifications of the individual manufacturers generally differ very little.

The U.S. Federal Specification O-M-232F 1975 for methanol has established itself throughout the world.

Table IX contains the quality requirements for synthetic pure methanol, a distinction being made between grades A and AA.

TABLE IX

Requirements for Grades A and AA

Characteristic	Grade A	Grade AA
Acetone and aldehydes, maximum (%)	0.003	0.003
Acetone, maximum (%)	—	0.002
Ethanol, maximum (%)	—	0.001
Acidity, maximum as acetic acid, (%)	0.003	0.003
Appearance and hydrocarbons	Free of opalescence, suspended matter, and sediment	Free of opalescence, suspended matter, and sediment
Carbonizable substances		
Color	Not darker than Color Standard No. 30 of ASTM D1209, Platinum–Cobalt Scale	Not darker than Color Standard No. 30 of ASTM D1209, Platinum–Cobalt Scale
Color	Not darker than Color Standard No. 5 of ASTM D1209, Platinum–Cobalt Scale	Not darker than Color Standard No. 5 of ASTM D1209, Platinum–Cobalt Scale
Distillation range	Not more than 1°C and including 64.6°C ± 0.10°C at 760 mm	Not more than 1°C and including 64.6°C ± 0.10°C at 760 mm
Specific gravity, maximum	0.7928 at 20/20°C	0.7928 at 20/20°C
Methanol, minimum (wt%)	99.85	99.85
Nonvolatile content, maximum (g/100 ml)	0.0010	0.0010
Odor	Characteristic, nonresidual	Characteristic, nonresidual
Permanganate	No discharge of color in 30 min	No discharge of color in 30 min
Water, maximum (%)	0.15	0.10

The federal specification also contains the test methods to be applied, which are largely available as American Society for Testing and Materials (ASTM) Standards:

(1) D1078—distillation range for volatile organic liquids;
(2) D1209—color of clear liquids (platinum–cobalt scale);
(3) D1296—odor of volatile solvents and dilutants;

(4) D1353—nonvolatile matter in volatile solvents for use in paint, varnish, lacquer, and related products;

(5) D1363—permanganate time of acetone and methanol;

(6) D1364—water in volatile solvents (Fischer reagent titration method);

(7) D1612—acetone in methyl alcohol (methanol);

(8) D1613—acidity in volatile solvents and chemical intermediates used in varnish, lacquer, and related products.

VII. Uses of Methanol

More than half the world's methanol is processed to formaldehyde which, in particular, is the basis of the manufacture of urea, phenols, and melamine resins and adhesives. Today, a multitude of other products such as methyl acrylate, methylamine, methyl halides, and acetic acid are customarily also manufactured on the basis of methanol. Methanol itself serves as an anti-frost agent, aircraft fuel, inhibitor, and solvent.

Methanol was processed in the United States, Western Europe, and Japan to the following products [53] in 1979: formaldehyde 49.2%, solvent 9.5%, methyl halides 6.8%, methyl amines 5.2%, methyl methacrylate 4.9%, dimethyl terephthalate 4.9%, acetic acid 4.2%, and miscellaneous 15.3%.

An increase in methanol consumption for the production of acetic acid using the Monsanto process based on methanol and carbon oxide is to be expected.

It must be assumed that in the future methanol will be applied not only as a chemical raw material but also as an energy carrier for combustion machines.

Low-level methanol–gasoline blending appears to be an attractive meth-anol application. Approximately 5 to 7% methanol can be added to gasoline without any modification of the internal combustion engine. Adding up to about 15% methanol requires only a few plastic items associated with the fuel supply system to be replaced. Engines running on pure methanol have already been developed. In any case, the addition of methanol increases the octane number of the mix and reduces the discharge of noxious substances such as nitrous oxides and sulfur.

In the future, methanol will be just as important as a raw material for the production of oxygenated high-octane gasoline additives such as methyl *tert*butyl ether. This compound is obtained using a simple process to convert methanol with isobutylene.

References

1. Verlag Chemie, *Ullmanns Encyklopädie der technischen Chemie,* Weinheim, 1978.
2. C. Carr, and J. A. Riddick, *Ind. Eng. Chem.,* 1951, **43,** 693.
3. M. L. McGlashan and A. G. Williamson, *J. Chem. Eng. Data,* 1976, **21**(2), 196–199.
4. T. W. Yergovich, G. W. Swift, and F. Kurata, *J. Chem. Eng. Data,* 1971, **16**(2), 222–226.
5. E. Strömsöe, H. G. Rönne, and A. L. Lydersen, *J. Chem. Eng. Data,* 1970, **15**(2), 286–290.
6. H. K. Ross, *Ind. Eng. Chem.,* 1954, **46,** 607.
7. L. H. Horsley, *"Azeotrop data."* Amer. Chem. Soc., Washington, 1952.
8. W. Kotowski, *Chem. Tech.,* 1963, **15,** 204.
9. S. Portalski and W. Thomas, *Ind. Eng. Chem.,* 1958, **50,** 967.
10. B. F. Dodge and R. H. Newton, *Ind. Am. Chem. Soc.,* 1934, **56,** 1287.
11. R. H. Ewell, *Ind. Eng. Chem.,* 1940, **32,** 147.
12. L. Bissett, *Chem. Eng.,* 1977, October, **24,** 155.
13. D. D. Wagman, J. E. Kilpatrick, W. J. Taylor, K. S. Pitzer, and F. D. Rossini, *J. Res. Natl. Bur. Stand.,* 1945, **34,** 143; 1946, **35,** 467.
14. W. Hirschwald, O. Knacke, and P. Reinitzer, *Erzmetall,* 1957, **10,** 123.
15. R. H. Newton, *Ind. Eng. Chem.,* 1935, **27,** 302.
16. H. H. Kung, *Cat. Rev. Sci. Engl,* 1980, **22,** 235.
17. A. Y. Rosovskii, *Khim. Prom.,* 1980, **11,** 652.
18. Y. B. Kagan, *et al. Kinet. Catal.,* 1976, **17,** 440.
19. I. G. Popov, Y. V. Lender, M. M. Karavaev, and V. G. Vislogusova *Khim. Prom.,* 1974, **6,** 420.
20. G. Natta, *Chem. Ind.,* 1953, **35,** 705.
21. G. Donali and G. B. Ferraris, *Eng. Chem.,* 1971, **7,** 53.
22. A. Cappelli, A. Collina, and M. Dente, *Ind. Eng. Chem. Prod. Res. Dev.,* 1972, **11,** 184.
23. G. Luft and O. Schermuly, *Ger. Chem. Eng.,* 1978, **1,** 222.
24. Y. Ogina and U. Uchida, *Bull Chem. Soc. Jpn,* 1958, **31,** 45.
25. E. Blasiak Poln. Patent No. 34000, 1947.
26. D. H. Bolton *Chem. Eng. Techn.,* 1969, **41,** 129.
27. H. E. Liebgott, W. Herbert, and G. Baron, *Brennst. Chem.,* 1972, **25,** 75.
28. ICI, U.S. Patent No. 3326956, London, 1963.
29. ICI, DOS 2302658, London, 1973.
30. Metallgesellschaft (LURGI) DBP 1300917, Frankfurt, 1967.
31. Metallgesellschaft (LURGI) DBP 1930702, Frankfurt, 1969.
32. BASF (1969), DOS 1930003, 1969; 2026165, 1970; 2026182, 1970; 2056612, 1970.
33. Catalyst and Chemicals Inc., DOS 1956007, Louisville, 1969.
34. SHELL International Research, DOS 2163317, Den Haag, 1971.
35. Mitsubishi Gas Chemicals Co., DOS 2165378, Tokyo, 1971.
36. O. Neuwirth, *Petrochemie vereinigt mit Brennstoff-Chemie,* 1976, **29**(2), 57–60.
37. E. Supp, *Nitrogen,* 1977, **109,** 36–40.
38. J. Falbe, *"Chemierohstoffe aus Kohle",* pp. 114–166. Georg-Thieme-Verlag,Stuttgart, 1977.
39. F. Marschner and H. J. Renner, *Hydrocarbon Process.,* 1982, **4,** 176–180.
40. E. Supp, *Chemtech,* 1973, July, 430–435.
41. H. Hiller and F. Marschner, *Hydrocarbon Process.* 1970, Sept., 281–285.
42. U.S. Patent No. 4087449, 1978.
43. U.S. Patent No. 3993457, 1976.

44. A. Pinto and P. L. Rogerson, *CEP,* 1977, July, 95–100.
45. Ib. Dybkjaer, *Chem. Econ. Eng. Rev.* 1981, **13**(6,149).
46. H. Takahashi and Y. Tado, *Chem. Econ. Eng. Rev.,* 1974, **6**,(11,79).
47. Elizabeth G. Baglin, G. B. Atkinson, and L. J. Nicks, *Ind. Eng. Chem. Prod. Res. Dev.,* **20,** 87–90.
48. W. L. Marsden, M. S. Wainwright, and J. B. Friedrich, *Ind. Eng. Chem. Prod. Res. Dev.,* 1980, **19,** 551–556.
49. S. E. Voltz, and J. J. Wise, "Development Studies on Conversion of Methanol and Related Oxygenates to Gasoline", 1976, Final Report, ERDA.
50. J. C. Zahner, U.S. Patent No. 4011275, 1977.
51. U.K. Patent Application No. GB2022074A, 1979.
52. Meyer Steinberg and Dang VI-Duong *"Energy conversion"* Vol. 17, pp. 97–112. Pergamon, Oxford, 1977.
53. World Bank Report, Washington D.C., 1982.

CHAPTER 7

Oxidation Catalysts for Sulfuric Acid Production

J. R. DONOVAN, R. D. STOLK, and M. L. UNLAND
Monsanto Enviro-Chem Systems, Inc.
St. Louis, Missouri

List of Symbols

BAV	Russian catalyst (barium, aluminum, vanadium)	KS	Russian catalyst (fluidized bed type)
C_p	heat capacity (cal/g mol °C)	K_p	equilibrium constant $(atm^{-1/2})$
CPG	controlled pore size glass	k	rate constant (units defined by equation form)
ΔF_f°	free energy of formation (ideal gas state)		
ΔF_T	free energy change for specified reaction	k_i	rate constant at absolute pressure P_i
ΔH_f°	heat of formation (ideal gas state)	k_0	rate constant at 1 atm absolute
ΔH_T	enthalpy change for specified reaction at $T(K)$	liter/MT	liters per metric ton of 100% H_2SO_4 per day
IK-1,2,3,4	Russian catalysts (limited data—see text)	MT/day (ST/day)	metric tons (short tons) of 100% H_2SO_4 per day

Applied Industrial Catalysis, Volume 2

MSV/A	Russian catalyst (molten sulfovanadate on aerosil)	T	absolute temperature (K)
MSV/D	Russian catalyst (molten sulfovanadate on diatomite)	TS	Russian catalyst (thermally stable)
P or P_i	total pressure (atmospheres), i signifies specific value	X	degree of conversion (fraction of one)
$p_{O_2, SO_2, \text{ or } SO_3}$	partial pressure (atmospheres) of subscript compound	x	V_2O_5 concentration in catalyst (percent)
r	net reaction rate (usually gram mol SO_2 per second per gram of catalyst) or difference between forward and reverse reaction rates	x_e	equilibrium percent conversion of SO_2 to SO_3 at specified conditions
r_f	forward reaction rate only, not including reverse reaction	y	molar ratio, potassium : vanadium
SVB	Russian catalyst (sulfovanadium — phosphorus	α	particular crystalline form of specified compound
SVD	Russian catalyst (sulfovanadium on diatomite)	β	equilibrium approach quotient defined by Eq. (11) (nondimensional)
SVS	Russian catalyst (sulfovanadium on silica gel)	δ	thickness of liquid layer (Å)
SVNT	Russian catalyst (sulfovanadium — low temperature)	η	pellet effectiveness factor (nondimensional)
S_{sp}	specific surface per unit catalyst mass (meters squared per gram)		

I. The Sulfuric Acid Process

Sulfuric acid is produced from a number of different sulfur containing raw materials that are processed in various ways to generate gas streams containing SO_2 and O_2. The SO_2 is catalytically oxidized to SO_3, which is then hydrated to produce H_2SO_4.

In the eighteenth and nineteenth centuries sulfuric acid was produced by the now essentially obsolete "chamber" process, which used nitrogen oxides (in the form of nitrosyl compounds) as homogeneous liquid phase catalysts for SO_2 oxidation. The product made by this process is of rather low concentration (typically no higher than about 77–78 wt% H_2SO_4), which is not great enough for many uses.

In the second half of the nineteenth century, development of the synthetic organic chemicals industry in Europe produced a need for high concentration sulfuric acid as a sulfonation reagent. This demand fostered development of the "contact" sulfuric acid process using platinum catalysts. Unfortunately platinum catalysts are susceptible to poisoning or deactivation by small amounts of some extraneous impurities. To overcome this problem (and the high cost of platinum),Badische in Germany developed a novel vanadium based catalyst in 1914. Vanadium catalysts of this general type were subsequently developed by several U.S. companies during the 1920s and soon replaced platinum. The modern sulfuric acid process still employs vanadium catalyst, although its production and properties have been gradually modified and improved over the years.

Major raw materials for sulfuric acid production are elemental sulfur, H_2S, spent (diluted or degraded) sulfuric acid, and SO_2 gas derived from sulfide ores of nonferrous metals. In earlier years, pyrites or other iron sulfides were widely used as raw materials but such use is now rare, except in Spain, South Africa, and eastern Europe.

In all cases, combustion, roasting, or other high temperature operations are carried out to generate SO_2 gas from raw materials. When using good quality sulfur or H_2S, purification of the SO_2 gas is not normally needed. However, other raw materials typically require wet scrubbing of SO_2 gas followed by electrostatic precipitators or other high efficiency devices to remove fine particulates prior to further processing.

The clean gas stream containing sulfur dioxide is usually either dried before catalytic oxidation or, when produced from good quality elemental sulfur, the air used for combustion is dried instead. If the oxygen concentration of any SO_2 gas stream is low, additional air or oxygen is added prior to (or during) catalytic oxidation to assure that an excess exists. Some plants that burn H_2S use undried SO_2 gas for catalytic oxidation, followed by collection of the acid thus produced. Such "wet catalysis" is relatively rare because of the possibility of equipment corrosion and/or acid mist emissions when unwanted acid condensation occurs before or after the acid collection step.

After catalytic oxidation, SO_3 is absorbed and hydrated in one or more packed towers irrigated with sulfuric acid. Preferably 98–99% H_2SO_4 is used because it is the azeotrope having minimum total vapor pressure. At lower acid concentrations sulfuric acid aerosols or mists are formed by vapor phase reaction of SO_3 and H_2O, and at higher concentrations, emissions of

SO_3 (and H_2SO_4 vapor) become significant. In either event, gas leaving the absorbing tower would be a source of downstream corrosion or objectionable emissions to atmosphere. Sulfuric acid is used as the drying agent for incoming SO_2 gas or air and is normally interchanged with acid used for SO_3 absorption to maintain water balance and desired acid concentrations.

The catalytic oxidation of SO_2 to SO_3 is strongly exothermic, consequently, equilibrium becomes more unfavorable for SO_3 formation as temperature increases above $410-430°C$. Unfortunately, this is about the minimum temperature at which available catalysts will operate. Consequently, plant catalytic reactors (converters) are normally designed as multistage fixed bed units with gas cooling between each adiabatic catalyst stage (or "pass"). Typical SO_2 concentrations entering the converter are about $4-11\%$ by volume, other concentrations occasionally being handled by special process or plant modifications.

In early years the contact process frequently employed 2 or 3 catalyst passes to obtain overall conversions of about 95 to 96%. Later, 4-pass converters were used to obtain conversions of about 97 to 98% with high SO_2 concentrations (or occasionally, 3-pass converters for low SO_2 concentrations). Currently, air pollution requirements have caused adoption of a modified process called "double contact" or "double absorption," which gives overall conversions of about 99.5 to 99.8%, depending on particular circumstances.

The "double absorption" process uses intermediate removal of the reaction product (SO_3) to obtain favorable equilibria and kinetics in later stages of the catalytic reaction. Gas leaving the second or third catalyst pass is cooled, passed through an acid irrigated tower for SO_3 removal, heated, and then reacted further in one or two additional catalyst passes before entering a final gas cooler and SO_3 absorber. Most industrialized nations now have emission standards that require double absorption plants (or auxiliary tail gas scrubbers for single absorption plants).

Catalytic oxidation is normally carried out at absolute pressures of about 1.2 to 1.5 atm. Theoretically, higher pressures improve both equilibrium and reaction kinetics. However, the practical and economic difficulties of building and operating plants at higher pressures are such that only one large plant of this type has been built to date [1]. A few small catalytic units to generate SO_3 gas for sulfonations or other special uses have been operated at pressures approximating 4 to 5 atm or more.

Considerable heat is evolved by combustion of sulfur or H_2S and additional heat is generated by catalytic oxidation to SO_3 and by the reaction of SO_3 and H_2O to form H_2SO_4. Much of the heat is typically used to produce steam, a useful product that can be converted to power by a turbine if desired. In many cases, large sulfur burning plants are essentially coproducers of sulfuric acid and steam or power, both products having significant economic value. Spent acid usually is decomposed in furnaces fired by gas, oil, or other fuels (sometimes H_2S), and the high temperature gas from such furnaces can also generate steam or power.

To achieve economies of scale, plant sizes have increased dramatically since World War II. Currently plants with capacities of 2000 to 3000 ST/day (as 100% H_2SO_4) are considered relatively standard in large phosphate fertilizer complexes.

II. Current Business/End Use Situation

Sulfuric acid is by far the largest volume chemical commodity produced. It is used in a wide variety of applications, primarily as a reagent or intermediate for production of other chemicals or end products. Its major use is for production of phosphatic fertilizers (and some other fertilizers), an application that consumes about 70% of the total U.S. supply. Almost all large phosphatic fertilizer plants burn sulfur to produce sulfuric acid and steam for use on site. Consequently, the amount of "merchant" sulfuric acid available for sale is considerably less than total production.

Approximately 70–80% of U.S. acid production is derived from sulfur, 8–15% from spent acid or H_2S, and the balance from nonferrous sulfide ores or pyrites. The acid produced at smelters is sometimes called "fatal acid" because it is a by-product and will be generated almost regardless of market or price. Remotely located smelters may sometimes obtain less money for their "fatal acid" than its cost of production (at zero SO_2 value) because of freight costs to consuming locations.

Other major uses of sulfuric acid are in production of organic chemicals (particularly alcohols, phenols, and dyes), inorganic pigments, textile fibers, explosives, petroleum products, pulp and paper, detergents, inorganic chemicals (particularly alum and hydrogen fluoride) and as a leaching agent for ores, a pickling agent for metals and a component of lead storage batteries. Some of these uses are declining and others are growing, the net effect being a very small total growth rate for uses other than fertilizers.

In the period 1960–1980, U.S. consumption of sulfuric acid increased from 17,454,000 to 41,200,000 MT/yr. Over the same period, world production increased from about 55,000,000 to 142,000,000 MT/yr. Equivalent average annual growth rate is about 5%, most of this growth being attributable to increases in fertilizer production. In 1981 acid production and consumption decreased from 1980 levels because of declines in fertilizer usage and production. No appreciable change in consumption is expected in

1982 and total production in 1982 is estimated to be no more than about 70% of rated plant capacity. It appears likely that future annual growth will continue to be lower than in the past, although cyclical spurts may occur.

The United States is a net importer of sulfuric acid, most imports being smelter produced acid exported from Canada. Imports are typically about 0.5 to 1.0% of consumption. The Scandinavian countries (except for Denmark) also export some smelter-produced acid to other European countries, and Japan exports acid to other Asian countries.

Current (1982) U.S. prices for sulfuric acid are about $75–95/ST, depending on geographic location. Oleum (fuming sulfuric acid) typically sells at a premium of about $3 to 5/t (or more for concentrations above 30% free SO_3 by weight). These figures are based on equivalent 100% H_2SO_4 although actual assays may be lower or higher than 100%. The price of sulfuric acid is highly dependent on sulfur pricing, which has increased rapidly in the past few years. Undoubtedly the resulting price increases for sulfuric acid and fertilizers have contributed to static or reduced consumption.

Very little information is published on catalyst production or consumption, but rough estimates can be made from government statistics on vanadium [2]. Assuming an average 7.0–8.0% V_2O_5 in catalyst plus some yield losses during production, it appears that about 4,000,000 to 6,000,000 liters of catalyst were produced annually in the United States during the period 1975–1980. Of this total, possibly 20–30% is exported. Consumption of sulfuric acid catalyst averages approximately half for new acid plant installations and half for replacements or additions to catalyst in existing plants. An assumed annual replacement rate of about 6% or more [3] can also be used to estimate consumption, in conjunction with specific figures on new plant requirements.

Because sulfuric acid catalyst is a specialty product, not a standardized commodity, its price will vary somewhat depending on the supplier and type of catalyst. Typical 1982 prices appear to be in the approximate range of $2 to 3/liter.

III.   General Catalyst Characteristics

Vanadium catalysts for oxidation of SO_2 are produced and sold by a number of different companies throughout the world. There are differences in catalyst performance and life, physical properties, pellet dimensions, manufacturing processes, and price. However, overall or superficial chemical compositions are rather similar.

A noteworthy feature of this general catalyst is that at operating temperatures its active ingredient is a molten-salt mixture supported on and within the pores of a porous silica pellet. In other words, it is a supported liquid phase catalyst analogous to those used in the Deacon process for oxidation of HCl or supported phosphoric acid catalysts used for polymerizations [4].

The active molten-salt mixture consists primarily of vanadium oxides dissolved in alkali metal pyrosulfates (or similar compounds containing more or less SO_3). The higher molecular weight (MW) alkali elements are preferred, i.e., potassium, rubidium, or cesium. Potassium is the element used in commercial catalysts because of its abundance and relatively low price. The technical advantages of rubidium and cesium are apparently insufficient to provide commercial justification for their use. Small amounts of sodium are sometimes used in conjunction with potassium.

In spite of extensive world-wide research efforts over the past 60 years, details of catalyst reactions and even the overall kinetics are not fully understood or quantified. Both subjects are very complex and much confusion has occurred because analytical or property measurements at room temperature (where the active ingredient is solid) generally do not relate well to the molten-salt condition existing at reaction temperatures.

The catalyst is unusually rugged and resistant to extraneous materials, as compared with other heterogeneous catalysts used in gas-phase reactions. Typically, good quality vanadium catalysts will exhibit useful lifetimes of about 5 to 10 yr in high temperature ($\pm$ 600 – 650°C) first-pass service and considerably longer in subsequent lower temperature passes. This good resistance to deactivation is a major reason that vanadium catalysts have replaced platinum given that catalytic activities are rather similar.

It is common practice to operate a catalytic converter until extraneous dust deposited on the first-pass catalyst causes a substantial pressure drop increase. The catalyst is then removed, mechanically sieved to eliminate dust, and recharged to the converter. Undoubtedly this practice has some benefit in maintaining catalyst activity by abrading off contaminated surface layers from the pellets. Sieving losses will vary with catalyst origin or composition, plant operating conditions, sieving procedure, rate of sieving, and ambient conditions. Even under the best conditions, sieving losses are significant and require additions of new catalyst to keep catalyst bed volumes constant. Such additions also help to maintain average catalyst activity at satisfactory levels. Dust fouling usually is not a serious problem in catalyst passes after the first one.

A significant effort has been devoted over the years to defining optimum pellet shapes and sizes. In general, a considerable portion of the overall reaction occurs at or near pellet external surfaces directly exposed to flowing gases. Hence, small sized pellets or those with special shapes of high surface area are more effective per unit volume than large sized pellets or those with low ratios of surface to volume. Pellet sizes typically range from 4 to 10 or 12 mm in diameter.

In earlier years most catalysts were produced as cylindrical or spherical pellets or as solid shapes approximating these forms. Some producers offered large sized pellets for first- and second-pass service and smaller pellets for subsequent passes. In this way an efficient balance of reaction rates and pressure drops was achieved. Ring-shaped pellets have been offered by several catalyst manufacturers and are beginning to replace solid pellets. The ring pellets offer substantial gas pressure drop reductions as compared with solid pellets, thus significantly reducing energy consumption as well as increasing production capacity in existing plants. Bayer in Germany has also developed small granular particles for fluidized bed operation, (patent numbers* 3,793,230; 3,880,985; 3,933,991; and Re 29,145 also see [5]) as have some U.S.S.R. workers [6–9].

In addition to offering different pellet sizes or shapes for different converter passes, some manufacturers offer different catalyst formulations. For example, one formulation is used for low to moderate SO_3 concentrations at high temperatures (i.e., the first one or two passes) and one for service in highly converted gases (SO_3 rich or low SO_2) at relatively low temperatures (i.e., the last few passes). Monsanto Company introduced this concept commercially in 1963 by offering Type 11 catalyst for service in the last few converter passes to supplement Type 210, the original catalyst used in all passes.

In general, changing the catalyst formulation or production process to enhance one property usually causes another property or properties to shift to less desirable values. In other words, it is very difficult to get something for nothing. The most important catalyst properties from the standpoint of an acid plant operator are activity, pressure drop, useful life, and hardness (as reflected in low sieving or handling losses). Obviously, the property called "life" is based on resistance to high temperatures and/or extraneous substances. The optimum balance of catalyst properties from an economic standpoint typically results in compromises or "trade-offs" between individual properties.

Traditionally vanadium catalysts are sold and used on a volumetric basis, employing liters as units. This practice is useful because catalyst bulk densities can vary widely, depending on pellet geometry and on the amount of contained sulfur oxides (largely SO_3). The quantity of SO_3 in a given catalyst varies depending on temperature, pressure, SO_3 concentration of the gaseous environment, as well as on conditions during cool-down to ambient temperature. Most catalyst producers sell the product as a sulfated

---

* For a complete citation of these patents, refer to the Appendix.

material, but some have offered low sulfur products that must be sulfated *in situ* in an acid plant prior to normal use. This practice is not desirable because catalyst can be damaged if the sulfating operation is not carefully controlled. In addition, unsulfated catalyst is typically rather soft or weak and can easily be crushed or damaged by careless handling.

Color of the catalyst can vary widely depending on the conditions to which it is exposed and on oxidation state (valence) of the vanadium in it. Unsulfated catalyst is normally almost white, whereas oxidized catalyst (V^{5+}) that has been sulfated will be yellow or various light brown or red tints. Reduced catalyst (V^{4+}) will be green, light gray, or blue. The catalyst is somewhat hygroscopic and exposing it to moist atmospheres generally causes it to turn green and to become soft. Normal color and hardness are usually restored by careful heating.

Because the active ingredient is a molten salt, it will readily migrate into or equilibrate with any porous solid or fine dust in contact with the catalyst at operating temperatures. Undoubtedly interfacial tensions and capillary action are important in this effect. Misleading conclusions about catalyst attrition or breakage are sometimes drawn from chemical analyses of dust because molten-salt migration is ignored. Typical properties of catalysts supplied by a major manufacturer are given in Table I. Vasilev *et al.* [10] also present data on various catalysts available outside Russia.

The amount of catalyst employed in any given acid plant will vary with plant capacity and specific design of the plant. It is usually reported as liters per daily short ton (or metric ton) of capacity (as 100% H_2SO_4). Typical catalyst "loadings" range from about 150 to 225 liters/ST day. In general, sulfur burning plants need less catalyst than plants utilizing byproduct SO_2 from smelters because sulfur burner gases have higher oxygen concentrations at the same SO_2 concentration. The magnitude and duration of conversion guarantees and SO_2 concentrations will also affect catalyst loadings.

Refer to Cuchetto *et al.* [11], Duecker and West [3], Fariss [12] and Table

TABLE I

Approximate Physical Properties of Sulfuric Acid Catalyst

Property	Solid pellets	Ring pellets
Porosity (pore volume) (%)	50–60	Same
Single pellet density (g/cc)	0.9–1.1	—
Bulk density (kg/liter)	0.56–0.60	0.48–0.52
Pellet specific heat (cal/g °C)	0.20–0.24	Same
Pellet conductivity (J/sec m °C)	0.45–0.53[a]	Same

[a] From Balakrishnan and Pei [134]

TABLE II

Thermodynamic Properties of SO_2, SO_3, and $SO_2 + \frac{1}{2}O_2 \rightarrow SO_3$ [a]

Temperature (K)	Sulfur dioxide			Sulfur trioxide			$SO_2 + \frac{1}{2}O_2 \rightarrow SO_3$			
	$\Delta Hf°$ (kcal/mol)	$\Delta Ff°$ (kcal/mol)	C_p (cal/mol °C)	$\Delta Hf°$ (kcal/mol)	$\Delta Ff°$ (kcal/mol)	C_p (cal/mol °C)	ΔHT (kcal/mol)	ΔFT (kcal/mol)	$\log K_p$	K_p (atm$^{-\frac{1}{2}}$)
600	−86.58	−76.08	11.71	−110.00	−86.02	16.90	−23.42	−9.94	3.621	4180.0
700	−86.59	−74.32	12.17	−109.86	−82.04	17.86	−23.27	−7.72	2.410	237.0
800	−86.59	−72.57	12.53	−109.67	−78.08	18.61	−23.08	−5.51	1.505	32.0
900	−86.57	−70.81	12.82	−109.44	−74.15	19.23	−22.87	−3.34	0.811	6.47
1000	−86.55	−69.06	13.03	−109.16	−70.24	19.76	−22.61	−1.18	0.258	1.81

[a] Evans and Wagman [14]

V for information on how a total catalyst charge is typically subdivided between passes. The Fariss paper shows significantly less catalyst in the first pass than appears prudent, but illustrates the general principle of increasing catalyst volume in each successive pass as gas progresses through the catalytic converter.

IV. Reaction Equilibria and Kinetics

The thermodynamics of the exothermic reaction

$$SO_2(g) + \tfrac{1}{2}O_2(g) \rightleftarrows SO_3(g) \tag{1}$$

have been the subject of several papers [13–20] dating at least back to Bodenstein and Pohl [13]. However, early workers did not have the advantage of modern spectroscopic data such as that used by Evans and Wagman [14]. Table II contains thermodynamic properties for sulfur dioxide and sulfur trioxide from Evans and Wagman as well as values of the equilibrium constant defined by

$$K_p = p_{SO_3}/p_{SO_2}p_{O_2}^{1/2} \quad \text{atm}^{-1/2}. \tag{2}$$

Temperature dependence of the equilibrium constant is most simply expressed by an equation of the form

$$\log K_p = (A/T) - B, \tag{3}$$

although more complicated correlations have been published.

As shown in Table III, a least squares fit of equilibrium data from various sources to Eq. (3) gives varying results. These discrepancies become highly significant if one attempts to extrapolate from measured ranges to high

TABLE III

Equilibrium Data Fit to Log $K_p = (A/T) - B$

Source	Number of points	Temperature ranges (K)	A	B	K_p (425°C)	$X_e{}^a$
Ross [18]	18	600–1000	5139.4	4.8812	302.2	98.71
Evans [14]	14	300–1500	5093.6	4.8176	300.8	98.70
Mikhailov [17]	10	673–898	5077.8	4.8220	282.6	98.61
Evans [14]	5	600–1000	5049.3	4.7992	271.1	98.56
Boreskov[b]	10	673–898	4914.3	4.6571	241.0	98.39

[a] X_e is the equilibrium percent conversion of SO_2 for initial gas of composition 10% SO_2, 11% O_2, 79% N_2 at 1 atm and 425°C.

[b] From Mikhailov [17], presumably from experimental data by Kapustinsky and Shamovsky [15].

pressures or temperatures, or to develop kinetic models incorporating a numerical value for the approach to equilibrium. Work in Monsanto laboratories indicates that the top two rows of equilibrium data in Table III probably should not be used.

In addition to thermodynamics, the kinetics of catalyzed SO_2 oxidation are also important in reactor design and operation and have been widely studied. Unfortunately, there is very little agreement on this subject and a generally accepted model or equation cannot be found in the literature. Reviews by Kenney [21] and by Urbanek and Trela [22] discuss this topic in detail.

Kinetic studies are usually carried out for one of two reasons: to better understand the mechanism of the reaction and/or to obtain data for designing plants or analyzing their operation. The conditions of the experiment and the expression of results will often differ depending on the objective. Obtaining mechanistic information from kinetic studies is especially difficult for the oxidation of SO_2 over potassium – vanadium catalysts (silica supported liquid films) because of their sensitivity to temperature and gas composition. Urbanek and Trela [22] point out that "so far no ultimate and unambiguous scheme has been developed for the mechanism of the chemical reactions proceeding in a vanadium catalyst pellet."

As examples of popular mechanisms, Mars and Maessen [23] proposed a two-step mechanism:

$$SO_2 + 2V^{5+} + O^{2-} \rightleftarrows SO_3 + 2V^{4+} \tag{4}$$
$$\tfrac{1}{2}O_2 + 2V^{4+} \rightleftarrows O^{2-} + 2V^{5+} \tag{5}$$

whereas Glueck and Kenney [24] presented evidence for a three-step mechanism:

$$V_2O_5 \cdot SO_3 + SO_2 \rightleftarrows (VOSO_4)_2 \tag{6}$$
$$(VOSO_4)_2 \rightleftarrows V_2O_4 \cdot SO_2 + SO_3 + \tfrac{1}{2}O_2 \tag{7}$$
$$V_2O_4 \cdot SO_2 + \tfrac{1}{2}O_2 \rightleftarrows V_2O_5 \cdot SO_3 \tag{8}$$

Studies of the chemistry of the molten salt by electrochemical [25, 26] and other techniques [27 – 35] are beginning to add support to some mechanistic ideas. One of the more interesting studies of the $K_2S_2O_7-K_2SO_4-V_2O_5$ molten salt is by Hansen [29] who suggests that the equilibrium

$$VO(SO_4)_3{}^{3-} + SO_4{}^{2-} \rightleftarrows VO_2(SO_4)_2{}^{3-} + S_2O_7{}^{2-} \tag{9}$$

where only $VO_2(SO_4)_2{}^{3-}$ is catalytically active, can explain a number of empirically observed properties of industrial catalyis. Whether the Lux-Flood [36, 37] acid – base concepts advocated by Hansen are useful or not in this system, the proposed active species seem chemically more "reasonable" than the ones in Eqs. (4)–(8).

Other useful reviews of various rate expressions are given by Neth [38],

Boreskov [39], Weychert [40], and Livbjerg [41]. Most of the rate expressions found in these review articles can be put in the form:

$$r = r_f(1 - \beta^s), \tag{10}$$

where β is defined by

$$\beta = (p_{SO_3}/K_p)/(p_{SO_2}p_{O_2}^{1/2}) \tag{11}$$

and r_f is the forward reaction rate.

Several authors attempt to determine s in Eq. (10) from the experimental data and thereby learn something of the mechanism. According to Livbjerg [41] if $s = 1$, SO_2 (or SO_3) is involved in the rate determining step whereas if $s = 2$, O_2 is involved in the rate determining step. There is no agreement on the correct value of s and it is doubtful that one value is best for all cases.

The majority of forward reaction rate expressions r_f found in the literature are of the form:

$$r_f = k\, p_{SO_2}^l p_{O_2}^m p_{SO_3}^n \tag{12}$$

or

$$r_f = k\, p_{SO_2}^l p_{O_2}^m p_{SO_3}^n /(1 + A p_{SO_2}^d + B p_{O_2}^e + C p_{SO_3}^f), \tag{13}$$

where Eqs. (12) and (13) are the power law and the Langmuir–Hinshelwood form, respectively.

There is no agreement on the parameters in Eqs. (12) and (13) except that in the majority of studies l and m are between 0.5 and 1.0 and n is usually 0 to -1. In Eq. (13), d, e, and f are usually 1.0 and the parameters A, B, and C may or may not be temperature dependent. The temperature dependence of the reaction rate constant k is very complicated, and in one study on a commercial catalyst [41] the apparent activation energy varied from 65 to 16 kcal/mol in the range 416–484°C. Breaks in Arrhenius plots of reaction rate logarithm versus $1/T$ are found in most studies if the experiment covers a large temperature range.

The description of global (or observed) reaction rates close to equilibrium was considered by Happel [42] who contends on the basis of tracer studies [43–45] that "although chemisorption of oxygen may be the rate controlling step far from equilibrium, as equilibrium is approached . . . , the rate of sulfur dioxide absorption may also contribute in reducing the overall rate." From this one might expect that a rate expression suitable for last pass conditions (high conversions near equilibrium) might be very different from an expression useful for the early stages of conversion.

From Eq. (1) it is apparent that an increase in total pressure should favor the forward reaction, and in 1972 Ugine Kuhlman commissioned a plant based on a high-pressure (5 atm) process [1]. Of several pressure kinetics papers [46–50] the study by Mukhlenov *et al.* [49] at pressures up to 10 atm seems most useful. Best agreement for their data was obtained by combining the kinetic expression of Ivanov [51] with a decrease in rate constant at increasing pressure according to

$$k_0 = k_i P_i^{1/2}, \tag{14}$$

where k_0 is the rate constant at ~ 1 atm and k_i is the rate constant at pressure P_i. More detailed studies of reaction rates close to equilibrium under pressures up to 10 atm are needed.

Ideally, kinetic equations should help the engineer decide how much catalyst to place in each stage of a given converter to achieve optimum cost effectiveness. In practice, many of the equations in the literature are of little or no value in this respect because they do not accurately describe catalyst activity under all conditions of interest, particularly where appreciable preconversion of SO_2 to SO_3 has occurred. Also, observed rates are not always governed by intrinsic kinetics alone. The effectiveness of catalyst pellets can be greatly altered by one of several external or internal (inside the pellet) mass and heat transport processes. Urbanek [22], Kovenklioglu [52] and Livbjerg [41] have discussed the problem of catalyst effectiveness in detail. The order of importance of possible diffusion effects is *internal mass transport* > *external mass* and *heat transport* > *internal heat transport*.

The engineer deals with this problem by applying appropriate correction factors (called effectiveness factors η) to the intrinsic rate expression. That is,

$$r_{obs} = \eta r. \tag{15}$$

In practice heat transport effects are usually ignored. Satterfield [53] discusses in detail the calculation of η for mass transport processes and Rony [54, 55] also considers this topic. The problem of an appropriate expression for r remains. The Harris and Norman [56] model:

$$r = P_{SO_2} P_{O_2}^{1/2} (1 - \beta)/(A + BP_{SO_2} + CP_{SO_3})^2, \tag{16}$$

which is based on commercial catalyst data, has been used for predicting converter performance via Eq. (15) with relatively satisfactory results. In Eq. (16), A, B, and C are temperature dependent parameters. For example, for Monsanto Type 210 catalyst:

$$\begin{aligned} \ln A &= -6.80 + (4960/T), \\ \ln B &= 10.32 - (7350/T), \\ \ln C &= -7.58 + (6370/T). \end{aligned} \tag{17}$$

Equation (16) was also useful in correlating data from several authors.

As a further illustration of the complexity of the kinetics, an entirely empirical model described by Fariss [12] uses 21 arbitrary constants. For proprietary reasons, specific values of the constants are not given.

As mentioned previously, in a commercial converter the catalyst is separated into several adiabatic layers or stages with interstage cooling of the gas before introduction to the next stage. The need for such an arrangement is the equilibrium limit on conversion plus the inability of available catalysts to function continuously for more than short periods at temperatures below 400°C. Some increase in conversion can be achieved by the use of interpass absorption (of SO_3) but this typically increases plant cost. A catalyst with good activity at $\sim 350°C$ would eliminate the need for interpass absorption and save capital but, so far, a satisfactory "low temperature" catalyst has not been discovered.

Catalyst bed depths (or contact times) and inlet temperatures are the main independent variables for plant design or operation. Some published studies comparing kinetic calculations with plant data are reported by Chartrand and Crowe [57] and Cuccetto et al. [11].

V.   Catalyst Formulation and Preparation

Details of catalyst preparation are generally considered highly proprietary information by each manufacturer. However, sulfuric acid catalyst has been produced for many years and its patent literature is very extensive and informative. Over 75 U.S. patents have been issued specifically on the preparation of catalyst for SO_2 oxidation plus many more in other countries. A list of some U.S. patents is provided in the appendix to supplement an earlier list published in 1959 [3].

The chemistry of the active part of the catalyst has been reported for both acidic and basic pH during manufacture as well as wet and dry systems [58, 59]. Raw materials generally include V_2O_5 or another vanadium salt (NH_4VO_3 or $VOSO_4$) and potassium (and sodium) as hydroxides, carbonates, or sulfates (patent number* 3,186,794). Frequently sodium or potassium silicate or colloidal silica is added as a binder [60]. These materials are then combined with a powdered siliceous material such as diatomaceous earth, amorphous silica, or silica gel, formed into pellets and sulfated. The final catalyst typically contains 6–9% V_2O_5, 6–12% K_2O (plus Na_2O), and 60–75% SiO_2 [59, 61]. Other elements in Group 1A of the Periodic Table can also provide a promoting effect for vanadium. These promoter effects have been reported in patents (patent numbers* 3,448,061; 3,789,019;

---

* For a complete citation of these patents, refer to the Appendix.

3,987,153; and 4,193,894) and by Topsoe and Nielsen [62], Tandy [63], and in the Russian literature [64–69]. Other minor constituents of the catalyst generally enter as impurities in the particular silica used as the support, although specific additives are sometimes proposed [70–73].

One of the more common production techniques consists of mixing the catalyst components either in liquid or solid state (or a combination) to form a "pastelike" mass. Pellets of desired sizes and shapes are then formed by extrusion at ambient temperature. Other methods of compacting powders or pastes have been used such as disk granulators or tablet machines [74, 75]. Detailed formulations have appeared in numerous patents [see the appendix plus references 58, 59, 66, 76–83].

Once the solid catalyst has been formed, it is dried to provide sufficient strength for handling. This is usually followed by calcining in SO_2 or SO_3 to convert the alkali materials into pyrosulfates while simultaneously increasing the mechanical strength substantially. As an alternate, sulfuric acid solutions or sulfates are sometimes used as starting materials. For a given composition of elemental constituents, the final properties of the catalyst depend heavily on the way the preparative steps were carried out, which, of course, relates to the knowledge and experience of the catalyst manufacturer.

A wide range of supports have been used including zeolites, carborundum, pumice, titanium dioxide, aluminum oxide, aluminum silicates, silica gel, Marshalite, and several other forms of silica [60, 76, 84–88]. Silica in the form of diatomaceous earth is the most widely used support material. Selection of the specific diatomaceous earth type (marine or fresh water, natural, calcined, or flux calcined) and grade (usually associated with particle size and processing conditions) has been the subject of research effort and patents (patent numbers* 4,127,509; 4,158,048; 4,206,086; 4,213,882; and 4,284,530). Some workers focus on optimized pore size distribution [76, 89, 90] whereas others aim at obtaining high thermal stability (patent numbers* 4,127,509 and 4,158,048).

Other work aimed at defining optimum pore structures for silica gel or other silicas is reported in References 60, 91–96. These workers believe that optimum pore radius is about 1000 Å and that pores below 500-Å radius are smaller than the mean free paths of reactant and product molecules.

Much of the research work on this catalyst has been an attempt to optimize chemistry of the catalytically active salts versus physical characteristics of the chosen support. In addition to improving activity, other important objectives include reducing the pressure loss caused by collection of dust in the first layer of the catalyst, increasing mechanical strength to

---

* For a complete citation of these patents, refer to the Appendix.

reduce losses during screening, and achieving stability at high temperature without significant loss of activity [97].

One of the interesting stories in catalyst formulations has been the development of sulfuric acid catalysts in the Soviet Union, under the general leadership of Academician G. K. Boreskov. Russian investigators have introduced a series of catalyst formulations for high and low temperature application as well as fixed and fluid bed reactors. One of the first formulations produced was designated BAV (barium, aluminum, vanadium) having the following composition (before calcining):

$$V_2O_5 \cdot 12SiO_2 \cdot 0.5Al_2O_3 \cdot 2K_2O \cdot 3BaO \cdot 2KCl \qquad (18)$$

It is prepared by reacting an alkali solution of potash water glass (K_2SiO_3) at 70°C with a hydrochloric acid solution of $AlCl_3$ and an aqueous solution of $BaCl_2$ to form a precipitate from the combined solutions. The precipitate is collected on a filter and then pressed hydraulically to obtain a cake with 40 to 45% water content. The cake is formed into rings, pellets, or granules and dried at 60 (initially) to 115°C. It is then contacted with SO_2, which forms polyvanadates or some other form of V_2O_5 with the release of chlorine [98, 99].

Another formulation designated SVD (sulfovanadium–diatomite) catalyst is prepared using vanadium pentoxide, gypsum, potassium bisulfate, and diatomite (or infusorial earth). The composition can be described as follows:

$$V_2O_5 \cdot 2.7K_2O \cdot 0.6SO_3 \cdot 0.75CaO \cdot 25SiO_2 \qquad (19)$$

It is prepared by crushing the raw materials with a ball mill in the presence of water to form a paste that is molded into pellets or ring shaped geometry. The material is then dried and roasted at 500 to 700°C giving it higher mechanical strength than BAV catalyst [98].

A low temperature catalyst designated SVS (sulfovanadate on silica gel) is prepared by depositing a potassium pyrosulfovanadate on silica gel. It has the following composition:

$$V_2O_5 \; 7.5-8.5\%, \quad K_2O \; 14.5-15.5\%, \quad SiO_2 \; 52-53\%, \quad SO_3 \; 25\% \qquad (20)$$

which gives a 3.6 molar K:V ratio [100].

This catalyst was tested for 2 yr in 5% SO_2 with 380 to 390°C inlet temperature. The inlet temperature had to be gradually increased to 390 to 400°C after 1 yr to maximize conversion. Thereafter it remained stable for the second year [101].

A paper described SVB catalyst promoted with phosphate that has been stabilized by zirconium [48]. Its composition is described as follows:

$$V_2O_5 \cdot 4.2K_2S_2O_7 \cdot 0.38P_2O_5 \cdot 0.75ZrO_2 \qquad (21)$$

It was noted to have a factor of 1.2 to 1.5 higher reaction rate than SVD, but the reaction rate was found to drop more severely with an increase in the degree of conversion.

Other Soviet catalysts include IK-1, which is made by impregnation of a finely divided support with a solution of a mixture of vanadium sulfate and potassium bisulfate. A high temperature (IK-2), low temperature (IK-3), and a thermally stable (TS) catalyst are also available, as well as a fluid bed type (KS). Table IV compares the reported properties of Soviet catalysts.

An article on Russian catalyst usage [102] summarized the situation in 1979 as follows: SVD and SVNT catalysts are inexpensive and provide high technological performance. They account for more than 65% of the entire output of acid in Russia. In heavy-duty systems, the use of BAV is no longer possible because of health considerations (emissions?). Thermally stable TS catalyst has hardly been developed commercially. Types SVS and IK-1–4 are expensive (a factor of 2.5 and 6 higher, respectively, than SVNT). Consequently, they are produced in very limited amounts. There has been limited experience with operation of industrial systems using SVD catalyst in combination with IK-1–4 (2 yr when the article was written), but combinations of SVD and SVNT have been used for about 10 yr.

An approach has been to use combinations of available catalyst formulations in each of the individual passes in a converter. For this purpose, catalysts of two types are proposed in the first pass: low temperature to initiate the reaction (SVNT, SVS, IK-1–4) and high temperature (SVD, BAV, TS) for regions deeper in the layer. Low temperature catalyst is charged to form the upper part of the first bed and the fourth and fifth beds; the high temperature catalyst is charged to form the lower part of the first bed. Reported bed loadings for three different Russian plants are shown in Table V. See references 19, 85, 101, 103, and 104 for additional data on these catalysts.

Workers in other countries have written computer programs that permit investigation of a specific promoter to vanadium or actives to support ratio within a formulation "map" or grid to define relationships (patent numbers* 4,285,927 and 4,294,723). Usually variables such as actives loading on support, binder level, excess alkali or acid must fall within a relatively narrow range to achieve an active formulation. A Japanese patent [59] claims there is an optimum sum of the V_2O_5 content (weight percent) and the alkali ratio expressed as mole percent K_2O plus Na_2O divided by mole percent V_2O_5 to produce maximum conversions.

Herce et al. [105] have tried to characterize catalysts with five parameters as follows: x and y, characterizing the chemical composition with $x = V_2O_5$

* For a complete citation of these patents, refer to the Appendix.

TABLE IV

Comparison of Soviet Catalyst Properties[a]

Property	BAV[b]	SVD[c]	SVS[d]	SVNT[e]	SVB[f]	MSV/D[g]	MSV/A[h]	IK-4[i]	KS[j]
V_2O_5 content (dry basis—wt %)	8.0	7.2	12.0	5.7	—	2.4	2.95	9.3	—
Molar K:V ratio	2.5	2.7 2.8	2.4–3.6	—	4.2	9.0	9.0	3.0	1.2
Bulk density (g/cc)									
Dry basis	0.48								
Calcined basis	0.65	0.57–.60							
Internal surface area (m²/g)	5–10	3–4	9.5	2.3	—	3.4	27.4	10.8	
Ave pore radius (Å)									
Low	700	2000	3000	1000	—	1000	100	1000	
High	1000	3000				10000	1000		
Pore volume (cc/g)	0.45	0.32–0.58	0.50	0.42	—	0.41	0.63	0.37	
Referenced in	87	87, 134, 135	85, 86	87	36.0	134	134	87, 135	11, 91
Catalyst intended use	Standard high temp	Improved high temp	Low temp	Low temp	Low temp	High SO_2	High SO_2	Low temp	Fluid bed

[a] Suchkova et al. [154]; Valitov et al. [100].

[b] BAV—barium, aluminum, vanadium.

[c] SVD—sulfovanadium-diatomite.

[d] SVS—sulfovanadate on silica gel.

[e] SVNT—sulfovanadate (low temperature).

[f] SVB—sulfovanadium-phosphorous.

[g] MSV/D—molten sulfovanadate on diatomite.

[h] MSV/A—molten sulfovanadate on Aerosil.

[i] IK-1—see text (limited data). IK-2—high temperature [102]. IK-3—low temperature [102]. IK-4—low temperature [102].

[j] KS—fluid bed.

TABLE V

Combinations of Soviet Catalyst Used in Fixed Bed Converters[a]

System	Bed	Catalyst	Height of the bed (mm)	Volume of catalyst (m³)	Overall gas conversion and plant conditions
1	I	IK-1-4	100	10.4	Gas conversion 98.2%
		SVD	530	55.1	Catalyst charge 300 liter/MT
	II	SVD	500	52.0	7.5% SO_2 (pyrite roast)
	III	IK-1-4	120	12.4	Capacity 1000 MT/day
		SVD	550	57.2	Single absorption
	IV	SVD	600	62.4	
	V	SVD	650	67.6	
2	I	IK-1-4	200	8.8	Gas conversion 99.6%
		SVD	200	8.8	Catalyst charge 212 liter/MT
	II	SVD	600	25.4	Sulfur burning
	III	SVD	400	17.6	Capacity 450 MT/day
	IV	IK-1-4	400	17.6	Double absorption (2 + 3)
	V	IK-1-4	400	17.6	
3	I	SVNT	125	18.1	Gas conversion 99.8%
		SVD	525	76.1	Catalyst charge 285 liter/MT
	II	SVD	440	63.8	8.5–9.0% SO_2
	III	SVD	360	52.2	Capacity 1000 MT/day
	IV	SVD	400	58.0	Double absorption
	V	SVNT	100	14.5	
		SVD	210	30.4	

[a] Maslennikov et al. [133]

concentration, y = molar ratio of K:V; δ, thickness of liquid layer corresponding to a unit mass of catalyst (mass of liquid layer/density of liquid times specific surface), which is a function of chemical composition; X, the degree of conversion, and S_{sp}, the specific surface. With all conditions equal, graphical dependencies were found for $X = f(x,y)$; $S_{sp} = g(y,x)$, $\delta = (y,x,S_{sp})$. The dependence of conversion on K:V ratio and V_2O_5 concentration at low degrees of conversion is shown by contour plots with an optimum value of X corresponding to an atomic K:V ratio of 0.73 at 9.3% V_2O_5. This K:V ratio is near unity, which makes one think that the optimum liquid layer composition is approximately $V_2O_5 \cdot K_2S_2O_7$.

However, this simple relationship may be misleading because the solubility of SO_3 in a pyrosulfate ($K_2S_2O_7$) melt is known to be a function of temperature. Most commercial catalysts have K:V ratios of about 2 to 3.5. Other vanadates such as KV_3O_8, K_3VO_4, $K_3V_2O_{14}$, and $KV_4O_{10.4}$ may be present in addition to silicates, sulfates, and aluminates [106].

A different optimum composition probably exists at different temperatures and gas concentrations. The dependence of conversion on catalyst liquid layer thickness (in angstroms) as a function of K : V atomic ratios of 0.5, 0.6, 0.73, 1.10, and 1.5 is given by Mayagoitia *et al.* [107] and is also discussed by others [108]. The optimum melt thicknesses reportedly range from 3.0 to 500 Å depending on K : V ratio.

Jensen-Holm *et al.* [109] and Grydgaard *et al.* [110] report the results of about 500 SO_2 oxidation rate measurements using catalyst made by impregnating a controlled-pore glass support (CPG). Turnover frequencies for K : V ratios of 3.5 using a 3060-Å CPG loaded with different levels of actives are shown as functions of reciprocal absolute temperature. These plots show straight line segments with various slopes (i.e., they have "breaks").

The general conclusion one draws from the extensive literature is that this type of catalyst is highly complex and that all the interactions between variables are not fully known.

VI.   Catalyst Testing

Catalyst testing serves to answer questions from users, from producers and from research workers interested in mechanisms or fundamentals. Consequently, the measurement techniques used will depend on who asks the questions or how answers are to be used. Most acid plant operators are not equipped to measure catalyst properties and instead rely on data provided by suppliers.

Standardized test methods have not been available and each major catalyst supplier developed his own. However, ASTM Committee D32 is beginning to change this situation for physical properties, but as of late 1982 had no plans to develop an activity test or chemical analysis procedures for sulfuric acid catalyst. The following tabulation of ASTM general catalyst methods appear potentially useful for sulfuric acid catalysts [111].

ASTM method number	Title
D3663-78	Surface area by multipoint N_2 adsorption
D3766-80	Definitions and nomenclature
D4058-81	Attrition–abrasion (formed catalysts)

Other test methods are being worked on by Committee D32 and should become available in 1983 and 1984.

Properties of importance are catalytic activity (and/or ignition temperature), pressure drop, hardness (attrition or crush resistance), life, chemical composition, resistance to contaminants, pellet size and shape, bulk density, porosity, surface area, and thermal stability. Discussions of major properties from a catalyst user's viewpoint are given in several articles [74, 75, 97, 112–114].

The property of greatest importance is catalytic activity because this is the "reason for being" of any catalyst. It is the property that is most difficult to measure satisfactorily, in part because the reaction rate is so rapid and highly exothermic that large temperature changes can occur in relatively small catalyst samples. Hence defining the reaction rate (or activity) as a function of temperature is very difficult. In addition, the catalyst appears to exhibit a form of reaction hysteresis, i.e., differences in reaction rate depending on the direction of approach to a specified condition. This phenomenon is particularly evident at low temperatures (below about 450 to 475°C) and may require many hours (or days) of operation before true equilibrium conditions are established [3, 115].

Activity is reported in various ways, generally as conversion for specified conditions and volume of catalyst or as a relative percentage versus a standard catalyst. For research or fundamental use, activity is frequently reported as gram moles SO_2 oxidized per second per gram atom of vanadium or gram mole of V_2O_5 (turnover number). Obviously, different values of activity are obtained at different temperatures, pressures and gas compositions, which means that tables or graphs are required for complete comparisons of different catalysts. It is not unusual to find that a catalyst that is superior to another at one operating condition is inferior at a different condition.

A measurement related to activity is ignition ("kindling") temperature. The operator of an acid plant is interested in this property because he observes that there is a critical gas temperature at the inlet of each adiabatic pass below which normal reaction gradually declines and stops. Also, highest overall conversions and heat recoveries are obtained with catalysts having relatively low ignition temperatures, which gives this property considerable economic significance. A simple definition of ignition temperature is that it represents the lowest entering gas temperature for a specified adiabatic operating situation that will sustain a reasonably close approach to equilibrium at the outlet of a particular catalyst pass.

Unfortunately, most measurements of ignition temperature are very arbitrary and reported values can vary considerably. Observed ignition temperatures in acid plants are functions of gas velocity, gas composition, reactor geometry and heat transfer characteristics, catalyst history (or extent of fouling), and temperature measurement accuracy. Even in the laboratory, this measurement is difficult to make and interpret. Usually it is carried out by measuring conversions obtained with a small (differential) sample of catalyst as temperature is gradually increased (or decreased) and then plotting the results. With this type of test, ignition temperature has been variously defined as the temperature equivalent to 1% conversion, as the temperature at which a specified, measurable temperature increase is obtained, or as the limiting lower temperature at which the reaction rate decreases to zero (as determined analytically or graphically) [116]. In a full-scale plant, observed ignition temperatures are generally slightly higher than values observed in the laboratory. Catalyst activity (conversion performance) varies inversely with ignition temperature in a general sense but the relationship usually is complex.

Measurement of activity is carried out differently by various catalyst suppliers, which makes it almost impossible to compare results directly. In some cases the measurement is made by passing gas of known or standard composition through a small bed of catalyst (± 1 or 2 in. deep) to approximate differential conditions [12, 62, 117, 118]. Occasionally, the extent of reaction and temperature rise is limited by mixing catalyst pellets with inert ceramic pieces of about the same size. Alternately, tests are sometimes made in so-called "integral" (adiabatic) reactors that use relatively deep beds of catalyst to simulate full-scale plant conditions [115], or test "sleeves" (cylindrical inserts ranging from a 2- to 12-in. diameter) are installed within a plant catalyst bed [119]. Alternately, samples can be withdrawn from the plant catalyst beds, but it is very difficult to get truly representative gas samples and temperature measurements in a full-scale plant. Larger "pilot plant" adiabatic units are occasionally used [120].

Determination of catalyst kinetics requires a gradientless or recycle reactor [39, 41, 121] allowing sufficient time for the sample to reach a stable condition. A recycle reactor resembles the differential reactor, but with internal or external gas circulation to minimize temperature and concentration gradients across the sample owing to reaction. Both differential and recycle reactors try to simulate isothermal conditions, but in actual practice, only the recycle reactor closely approaches them. Note that measurement of true kinetics requires very small catalyst particles, instead of pellets, to minimize mass transfer effects.

One of the newer reactor designs is a microreactor that contains a small catalyst sample (0.5 g). The catalyst is brought to a steady state condition and then pulsed with a specific gas mixture. The objective is to obtain information on mechanisms or rate limiting steps [122]. Other pulse techniques have been described for studying vanadium species on the catalyst surface [123]. Special tests have also been developed for fluidized bed sulfuric acid catalyst, (patent numbers* 3,793,230; 3,933,991; and Re 29,145; see also [7]).

Catalyst activity testing is dependent on the selection of analytical appa-

* For a complete citation of these patents, refer to the Appendix.

Fig. 1. General view of facilities for testing sulfuric acid catalysts.

ratus or technique as well as the choice of reactor. In today's world of microcomputers, reactor control and data precision have improved significantly as compared with earlier years. Generally, the SO_2, O_2, and N_2 content of gas streams are determined before and after the catalyst sample by gas chromatography. Direct measurement of SO_3 concentration is very difficult and is not normally used. Also, it is difficult to maintain SO_3 completely in vapor form from the sample point in a reactor system until it is processed through an analytical apparatus.

To avoid deactivation of chromatograph columns by SO_3, samples of gas generally are passed through a sulfuric acid scrubber or cold trap to remove SO_3 prior to analysis. Proper corrections are made to the SO_2 concentration for the amount of SO_3 removed plus reduction in gas volume as a function of the amount of reaction occurring. A check is made on the analysis by computing the oxygen balance on each gas sample. Figure 1 shows a general view of some Monsanto laboratory facilities for testing sulfuric acid catalysts.

A complete chromatographic separation of a gas mixture into its components and subsequent analysis typically can be completed in 6 to 7 min. An excellent review of analytical procedures for gases is reported by Wainwright and Westerman [124], and this topic is also discussed for plant conditions by Donovan *et al.* [125].

Physical property evaluation involves the tests outlined in the following paragraphs. Additional information is given in the literature (patent number* 4,284,530; see also [111, 112, 117, 126, 127]).

1. Bulk Density

Weight of whole pellets per unit volume is measured in a large container to minimize wall effects. The sample is tapped to form a packed bed or layer after first measuring bulk density in the original expanded or "loose" state. As a first approximation, plant catalyst beds can be assumed to have bulk densities intermediate between loose and compacted values.

2. Hardness

Several different hardness tests may be made depending on concern. Compression hardness is usually reported as the mean of 12 to 25 individual pellets crushed singly in a mechanical variable loading device. Ball mill attrition loss is measured by percentage weight loss of the sample that takes place during tumbling for a specified amount of time, followed by screening to separate fines. Tests for compression strength of bulk catalyst beds are sometimes used, employing a variably loaded piston at the top of a sample mass.

3. Pellet Length

Mean pellet length (and length distribution) may be determined by segregating the sample into different length fractions and then computing a mean length. Alternately, when extrudates are reasonably uniform, a relationship may be established between pellet count per unit volume and mean length.

4. Pressure Drop

Pressure drop generally requires a large sample for accuracy. Both loose and packed (vibrated) bed pressure drops have been measured in a 1-ft deep layer 4 in. in diameter, using air at room temperature. This measurement provides an excellent relative comparison of one catalyst with another but extrapolation to actual operating conditions is dangerous. Obvious sources of error are wall effects and variations in catalyst bulk density (or void fraction).

---

* For a complete citation of this patent, refer to the Appendix.

In general, pressure drop is correlated well by the Ergun equation [128] if a proper value of void fraction is selected. Unfortunately, laboratory measurements of void fraction (bulk density) give a range rather than an unequivocal value. Hence the best way of correlating plant pressure drop data is to use the Ergun equation to calculate a void fraction value that best fits the particular data.

Usually, pressure drop data produce reasonably straight line plots versus superficial gas velocity raised to the 1.5 power. Some typical pressure drop values are given in references 74, 75, 112, 119, 129.

5. Accelerated Aging

A considerable number of simulated aging tests are described in the literature, most involving heating to abnormally high temperatures. In Monsanto's test, a standard activity measurement is made on catalyst subjected to thermal cycling between an elevated temperature and a low temperature in a specified gas environment. Other analyses may also be performed on the aged samples including x-ray diffraction, surface area, etc. A detailed discussion of aging is presented in Section VII.

As a check and confirmation of laboratory accelerated aging tests it is highly desirable to expose catalyst samples to actual plant conditions for periods of about 1 to 5 yr. Typically this is done by placing samples in "sleeves" installed within plant catalyst beds.

6. Chemical Analyses

Standard wet chemical or instrumental analysis procedures are used for sulfuric acid catalyst after initial sample dissolution or treatment. Typically, bulk samples are riffled or quartered to obtain a representative portion.

The sample is then finely ground with a mortar and pestle or in a mechanical grinder or homogenizer. Depending on the particular analysis to be made, the ground sample is then either fused with a mixture of Na_2CO_3 and K_2CO_3 (equal parts) in a platinum dish or solubilized by boiling H_2SO_4 (5–25%), boiling HCl (5–10%) or an acid mixture. Silica usually is determined by treatment with 48% HF in a platinum dish to volatilize SiF_4.

Because the SO_3 content of various catalyst samples can vary widely, depending on exposure conditions and other constituents, it is desirable to standardize this factor to report weight percentages on a common basis. Normal Monsanto practice is to measure "loss on ignition" by heating samples in a Vycor® dish for at least 2 hr at 800°C. Most analyses are then made on the ignited sample. It should be noted that not all SO_3 is eliminated by heating at 800°C because some normal sulfates do not decompose at this temperature.

7. *Other Physical Property Measurements*

Pore volumes and pore size distribution are frequently measured by mercury intrusion. Only a few papers have appeared that describe some of the fundamental properties of sulfuric acid catalyst. Kadlec, Hudgins, and Silveston [130] presented data from steady state counter diffusion experiments. The results were correlated with three popular gas diffusion models and with the Bruggeman model for tortuosity. Other diffusivity, activation, tortuosity, and rate constant data have been reported as a function of liquid loading and vanadium content by Villadsen *et al.* [41, 110, 121, 131, 132]. A general correlation is presented by Chen and Rinker [133]. Thermal conductivity values are reported by Balakrishnan and Pei [134]. Note again that most measurements of this kind are made at room temperature, where the active molten salt is a solid.

VII. Catalyst Deterioration or Poisoning

A. GENERAL CONSIDERATIONS

Catalyst life can be shortened significantly whenever the catalyst is misused either during storage or in service. Moisture is the biggest concern during storage. Free SO_3 retained by the pellets after calcining or removal from a plant converter may be hydrated to sulfuric acid on moisture contact. This in turn forms brown or reddish crystals at the pellet surface that can cause local decrepitation and material losses during handling. In addition, moisture softens all catalyst pellets, which also results in handling losses. Catalyst suppliers generally use containers fabricated with an internal moisture barrier or use a plastic bag inside an outer container. In a dry condition, the catalyst can be stored for many years.

After the catalyst has been placed in service, performance gradually deteriorates in high temperature or dusty operating locations by an "aging" process. Catalyst life is generally considered long when compared with other catalysts, although there are significant differences in aging behavior and life between different commercial sulfuric acid catalysts. Some particularly stable catalysts have occasionally given relatively satisfactory service for 30 to 40 yr in converter passes 3 and 4 where temperatures are low (below 450 to 500°C) and gases are clean.

The general practice is to screen catalyst during an annual plant turnaround although longer intervals are sometimes feasible. The first pass is always screened plus, occasionally, one of the other passes on a rotating basis. Most of the dust and scale collects on the first layer of catalyst. The volume of material lost during screening is replaced with fresh catalyst. The new catalyst is always placed at the top of the layer because its ignition temperature is lowest when new. At intervals of about 5 to 10 yr it usually is desirable to completely replace the first pass.

Catalyst aging apparently is caused by several mechanisms operating simultaneously that may be classified into three groups [135]:

(1)   thermal deactivation,
(2)   mechanical deactivation,
(3)   chemical deactivation.

Some of the mechanisms are inherent in the process whereas others are preventable by using efficient gas cleaning techniques and careful operation.

B.   THERMAL DEACTIVATION

Thermal deactivation is caused by heating the catalyst to about 600 to 650°C or higher causing irreversible degradation of the support. Several papers [136–138] have reported studies of thermal aging by subjecting samples to 700°C operating conditions for up to 100 hr. According to Boreskov the observed change in porosity and surface area does not directly explain the sharp decrease in activity. However, this conclusion must be considered tentative because it does not truly relate to operating conditions where active ingredients are molten.

High temperatures always occur at the bottom of the first catalyst layer, but may occur in other locations when the inlet SO_2 concentration is incorrectly controlled during plant start-up [139]. At these high temperatures, fine globules of silica appear to be dissolved in the melt. As a result, the thickness of the melt film increases and possibly blocks part of the pore structure. The role of silicon dioxide is not fully understood because it can reportedly react with vanadium species that in turn can react with potassium sulfates or pyrosulfates with the formation of potassium sulfovanadates [80, 140].

Stability also depends on the expertise of the catalyst manufacturer and his choice of a particular support material. During operation, sintering closes the smallest pores first, resulting in reduced total surface area per unit of catalyst [139, 141]. The reduction in surface area can be closely correlated with a reduction in catalytic activity. During 2 yr of exposure at 600°C, the volume of small and medium pores decreases while the volume of large pores increases so that the overall pore volume remains practically the same [136].

High temperatures can also cause phase transitions of the silica support. These transitions may cause softening of the catalyst matrix via expansion or contraction. Putanov [137, 142] studied industrial catalyst for over 3 yr and found a decrease in specific surface area, an increase in the volume of pores, and a gradual transformation of SiO_2 to α-cristobalite. The increase in crystallinity was found to be proportional to the time of industrial use of the catalyst. That cristobalite forms at a lower temperature in the presence of vanadium actives as compared with SiO_2 alone suggests that a chemical interaction of SiO_2 occurs with the molten salt. It was reported that the crystalline transition is accelerated by higher temperature, but is reduced with increasing K:V ratio and increasing concentration of SO_3 in the gas. Results are reported for catalyst samples from Badische Aniline u Soda Fabrik, Grillo, Imperial Smelting Corporation, Girdler, and Monsanto Company for various experimental conditions. Because two of the samples contained α-cristobalite and still had high thermal stability, it was concluded that the role of phase conversion in the support may not be critical in the aging process.

Other investigators [143, 144] suggest that pore structure is altered as a result of chemical processes that can form new phases and rupture the internal pore structure. Catalysts having high $K_2O:V_2O_5$ ratios had higher thermal stabilities, although some of the potassium formed a water insoluble salt independent of the promoter ratio.

Illarionov et al. [143] investigated thermal deactivation of catalyst as a function of color of vanadium species owing to chemical changes in the active component. It was determined that thermally deactivated catalysts take on a reddish-brown or black color. Catalysts activated by sodium are apparently less thermally stable than catalysts activated by potassium. The dark phase was attributed to the formation of sodium vanado–vanadates. However, the same black color can be formed when cooling catalyst in the presence of SO_2 resulting in formation of $V_{12}O_{26}$. It was concluded that the main cause of thermal deactivation in catalysts promoted by potassium is a combination of potassium with SiO_2 and as a result, formation of an inert vanadium phase.

Thermal deactivation generally proceeds rather slowly, and it has been found by plant experience that good quality catalysts can be exposed to abnormally high temperatures (700–800°C) for short periods without causing drastic deactivation.

C. MECHANICAL DEACTIVATION

Depending on design of the acid plant (sulfur burning or metallurgical type), dust or scale coming into contact with the catalyst will tend to plug voids between the catalyst pellets to differing degrees. Sources of dust in sulfur burning plants include: contaminated sulfur, failure of molten-sulfur filters, spalling of combustion chamber brick, equipment or duct scale, and vibration of the catalyst layer itself during operation [135]. Sulfur burning plants typically do not have gas cleaning capability. Metallurgical acid plants located at copper, zinc, or lead smelters do have gas cleaning facilities, but removal of dust is never wholly complete. A small fraction of extraneous dust is carried into the reactor where it typically accumulates in the top portion of the first catalyst pass. Some individual contaminants that form submicrometer fumes may preferentially deposit at subsequent, lower temperature catalyst passes. Arsenic and lead frequently act in this way.

As previously outlined, molten salts in the catalyst migrate into dust, which causes an overall loss of vanadium salts from the catalyst.

D.   CHEMICAL DEACTIVATION

"Poisoning" of a catalyst is usually interpreted to be a drastic or sudden loss of activity caused by reaction of a particular trace contaminant with the catalyst. In this sense, vanadium sulfuric acid catalysts are not subject to poisoning. However, many substances will react with the catalyst either to reduce activity, cause gradual losses of vanadium, or alter pellet physical strength. Even the major reactant SO_2 causes some loss of activity when present in high concentrations at low temperatures (below $430-450°C$). However, this decrease in activity is limited in extent and is reversible. Full activity is restored by exposing the catalyst to relatively high oxygen (or $SO_3 + O_2$) concentrations and/or raising its temperature.

A few laboratory studies have been made to quantify the effects of various contaminants but most information on this topic has been obtained from acid plant experience. For this reason the available information is only semiquantitative in many cases. Literature information is reasonably complete for arsenic compounds, selenium, and carbon monoxide, and some work on fluorides has been reported [98, 129, 135]. Arsenic compounds and fluorides appear to be the most objectionable substances in terms of concentrations that can be tolerated.

Some materials, specifically arsenic, lead, mercury, and selenium, are undesirable contaminants in product acid even though small amounts can be tolerated to some degree by the catalyst. For this reason, particular attention is paid to removing these substances when designing SO_2 cleaning facilities. Special procedures for removing all except lead from gases are outlined in the literature, but descriptions of such procedures are outside the scope of this article.

Table VI, derived in part from Donovan and Stuber [145], Duecker and West [3] and Malin [129] outlines the effects of contaminants in feed gases on catalyst.

TABLE VI

Effects of Feed Gas Contaminants on Catalysts

Contaminant	Effect on vanadium catalysts
H_2O	Does not have an effect at temperatures above typical dew points for sulfuric acid (150–200°C). At lower temperatures, there may be degradation of the catalyst with loss of activity and mechanical strength, depending on the extent of condensation. Catalyst can usually be regenerated by careful heating.
As_2O_3	At temperatures significantly below 600°C, the catalyst is saturated with arsenic and a reduced plateau of catalytic activity is reached, which apparently does not change appreciably with further exposure. The decrease of activity appears to be connected with blocking of the catalyst surface by arsenic pentoxide (As_2O_5). At temperatures near 600°C the volatile compound $V_2O_3 \cdot As_2O_5$ can be formed, and some long-term loss of activity may be noted because of vanadium losses.
AsH_3	Because of its easy oxidizability it has the same effect as As_2O_3.
Se	Harmful effect only at temperatures below 400°C; initial activity is restored after heating.
C_nH_m (hydrocarbons)	Harmless in small concentrations. In individual cases there has been catalyst activity loss as a result of surface deposition of carbon formed by incomplete oxidation of hydrocarbons. The amount of carbon produced is dependent on properties and concentration of the hydrocarbon, the concentration of oxygen, and temperature. In large amounts, heat release from oxidation can be a serious problem.
SiF_4, HF	Sharply reduce activity, but extremely low levels act relatively slowly. HF reacts with silica supports forming volatile SiF_4, and deposition of silica gel on the catalyst surface has also been noted.
$FeSO_4$	Mechanically covers the surface of the catalyst and causes a loss of activity and pressure drop increase. When catalyst beds contaminated with iron are observed in a cold condition, hard crusts between pellets are noted. The crusts contain appreciable potassium plus vanadium and it is evident that substantial migration of molten-salt actives occurs.
S, CS_2, H_2S	Are not objectionable in small amounts if there is sufficient oxygen to permit oxidation. Large amounts can produce harmful heat release or block the catalyst surface with sulfur deposits.
H_2	May cause loss of catalyst activity by reducing vanadium pentoxide to a lower oxidation state. Heat release may be a problem.
NH_3	Harmless in reasonable quantities. Can be oxidized with objectionable heat release when present in large amounts.
NO, CO_2	Are not objectionable at reasonably low concentrations. (Note that NO may be troublesome in acid plants because it can contaminate product acids and/or cause formation of submicrometer acidic mists.)

Table VI *continues*

TABLE VI *continued*

Contaminant	Effect on vanadium catalysts
CO	According to most workers, does not harm catalyst, but in the presence of large quantities of CO it is theoretically possible for the reaction to be inhibited owing to reduction of SO_3 (SO_3 + $CO \rightleftarrows SO_2 + CO_2$). Heat release from oxidation can be troublesome and there is some evidence of vanadium reduction at low temperatures (below approximately 450 to 475°C).
Cl_2, HCl	Do not cause significant problems in low concentrations. If there is extended exposure, losses of vanadium from the catalyst occur as a result of volatile $VOCl_3$ formation.
Pb, Hg	Information is limited, but analyses of spent catalyst from a number of acid plants indicate that compounds of these elements are readily deposited from very low concentrations in gases. Significant catalyst activity loss then occurs. Where *elemental* mercury is present in small concentrations, its volatility is apparently sufficient to prevent deposition on catalyst at operating temperatures.

Analysis of catalyst samples by x-ray diffraction after more than 12 months in service revealed the formation of several double salts including $KFe(SO_4)_2$ and $KAl(SO_4)_2$. Catalyst pellets containing the aluminum double salt were found to be a factor of four harder (by compression) than the original pellets. Another species of vanadium (V_5O_9) plus α-K_2SO_4 has been reported in exhausted catalyst [146].

TABLE VII

Approximate Upper Impurity Limits in 7% SO_2[a]

Substance	Gas before cleaning[b] (after coarse dust removal)	Gas to catalyst[b] (after cleaning)
Chlorides, as Cl	125	1.2
Fluorides, as F	25	0.25
Arsenic, as As_2O_3	200	1.2[c]
Lead, as Pb	200	1.2[c]
Mercury, as Hg	2.5	0.25[c]
Selenium, as Se	100	50.0[c]
Total solids (dust)	1000	1.2
H_2SO_4 mist (as 100%)	—	50.0
Water (as H_2O)	4×10^5	—

[a] Donovan and Stuber [145].

[b] Units are mg/NM³, dry basis. Limits for uncleaned gas may require special plant design features and in some cases should be raised or lowered to fit particular plant designs.

[c] May be objectionable in product acid, with possible exception of lead for technical grade acid. Halogen acids typically have limited solubility in concentrated sulfuric acid but may cause localized corrosion.

Table VII gives approximate numerical values for tolerable amounts of common impurities in SO_2 gases from smelters.

E.   CATALYST REGENERATION OR DISPOSAL

Little information has been published regarding catalyst regeneration after service, probably because of the relatively low price of new catalyst and the irreversible changes in support and vanadium content which occur in many cases. Several processes have been described to recover the vanadium content of various types of catalyst. An exception is a paper by Hassan and Iskandar [146] in which a process is described to regenerate 10-yr old catalyst. Good results are reported, undoubtedly because the spent catalyst was not extensively degraded. Normal practice in the United States is to ship spent catalyst to companies that extract vanadium compounds from ores, slags, or residues. The material has only limited value because it is essentially a medium to low grade ore and shipping costs are significant.

VIII.   Future Trends

Future developments in any industrial process depend both on new technology and on the effects of external economic or process constraints in the areas of raw materials, product uses, emission limitations, etc. Frequently these external constraints actually are the motivating forces for technology changes.

As we see it, external constraints are promoting the following trends in the sulfuric acid industry:

(1)   evolution of the process to maximize steam and energy production;

(2)   continuing construction of large capacity new plants to replace old or obsolete small units;

(3)   a continuing effort to reduce plant capital costs by process or equipment changes, with particular emphasis on higher concentrations and/or pressures of reactants;

(4)   continued search for new or better catalysts;

(5)   continued use and availability of elemental sulfur as the major and preferred raw material for sulfuric acid plus increased emphasis on recovering and reusing spent or sludge acids.

Several symposia have addressed various facets of these issues, in particular the November, 1981 National A.I.Ch.E. Meeting [147] and a November 1981 London Conference sponsored by the British Sulphur Corporation [5].

We conclude that the present contact sulfuric acid process will continue to be the preferred process for the foreseeable future. A few acid plants now using sulfur as raw material may possibly be converted to use sulfur dioxide from nearby power plant stack scrubbing processes. Acid plants in nonferrous smelters will continue to utilize byproduct sulfur dioxide, with increases in concentration occurring as smelter processes are improved to reduce emissions and increase efficiency.

Although many novel process modifications have been suggested [148, 149] we do not see them as economically viable except under unusual conditions. The use of fluidized bed reactors, cyclic changes in gas concentrations, or cyclic processes using catalyst in its most active state followed by transfer to another reactor for reactivation appear unlikely to be adopted widely, even though they represent some ingenious thinking.

The pressure plant has some future potential but currently is, at best, a break even cost proposition versus current low pressure technology. A major factor is the relatively high cost of large gas compressors and expanders. If current work on coal gasification and liquefaction brings down the relative cost of such equipment, then a higher pressure process might well become attractive.

It is clear that increasing concentrations of sulfur dioxide (and possibly oxygen) will be used in the future. Currently, the cost of oxygen produced from liquid air is too high to make it economical as a purchased raw material, but if new, lower cost processes for producing oxygen enriched air are developed, then the use of this auxiliary raw material could become attractive.

With regard to catalyst developments, if a catalyst active at $350°C$ or lower could be developed, then it would be possible to achieve desired high conversions in a single absorption process rather than using the more complex double absorption process that is now required. However, this is a very difficult goal and the likelihood of success appears low.

The present basic type of vanadium catalyst has been in existence for well over 60 years, and there has been an extensive, worldwide research effort on it. We estimate that over 1000 papers and patents have appeared. In spite of this substantial effort, catalyst fundamentals or mechanisms are not yet clearly understood, which is an indication of their complexity as well as the challenge and opportunity for further research.

Some significant improvements in catalyst have occurred over the years, primarily in making it more active, harder and more stable (longer life) in service. In addition, changes in pellet geometry have been adopted to reduce gas pressure drop (and dust fouling effects) in catalyst beds. It is likely that further, evolutionary improvements in vanadium catalyst will be made, with particular emphasis on improving stability under projected, more intensive process conditions, on reducing susceptibility to fouling by extraneous materials and on improving activity at low temperatures (400°C ±).

In spite of reported favorable results with Cr_2O_3 catalysts [150–153], it appears unlikely that any new catalyst will be developed that does not use vanadium or a noble metal as its major active ingredient.

Appendix

The following list arranged in numerical order (or date) includes only a few selected patents dealing with oxidation of organics, H_2S or CO over vanadium catalysts, silica carriers, and catalyzed oxidation of SO_2 in water solutions. It supplements an earlier list covering 1957 and prior years [3].

Patent number	Inventors, assignee, and date
2,016,135[a]	T. Cummings (to General Chemical Co.), 10-1-1935.
2,799,560[a]	P. Davies (to Imperial Chemical Industries, Ltd.), 7-24-1957.
2,863,838	G. C. Vincent (to Imperial Chemical Industries, Ltd.), 12-9-1958.
2,941,957	P. H. Pinchbeck and F. B. Popper (to The Coal Tar Research Assoc.), 6-21-1960.
2,973,371	N. Chomitz and W. R. Rathjens (to American Cyanamid Co.), 2-28-1961.
2,977,324	D. A. Dowden and A. M. Ure (to Imperial Chemical Industries, Ltd.), 3-28-1961.
3,005,687	M. J. Udy (to Strategic Materials Corp.), 10-24-1961.
3,038,911	D. J. Berets, R. A. Hermann, and H. A. Van Brocklin, (to American Cyanamid Co.), 6-12-1962.
3,186,794	P. Davies (to Imperial Chemical Industries, Ltd.), 6-1-1965. (Corresponds to Canada 655,525, 1-8-1963.)
3,215,644	H. Kakinoki, I. Kamata, and Y. Aigami (to Dainippon Ink Seizo Kabushiki Kaisha), 11-2-1965.
3,216,953	R. Krempff (to Compagnie de Saint Gobain), 11-9-1965.
3,226,338	H. L. Riley and A. Romanski (to United Coke & Chemicals Co. Ltd.), 12-28-1965.
3,243,261	W. Dieters (to Inventa A.G. für Forschung u. Patentverwertung), 3-29-1966.
3,275,406	R. Krempff (to Compagnie de Saint Gobain), 9-27-1966.
3,282,645	B. G. Mandelik (to Pullman, Inc.), 11-1-1966.
3,300,280	R. Terminet (to Soc. Nationale Petroles d'Aquitaine), 1-24-1967.
3,318,662	E. Pauling (to Metallgesellschaft Aktiengesellschaft), 5-9-1967.
3,345,125	M. Kruel, H. Juentgen, and H. Dratwa (to Bergwerks—Verband G.m.b.H), 10-3-1967.
3,362,786	D. B. Burkhardt, 1-9-1968.
3,438,722	L. A. Heredy, D. E. McKenzie, and S. J. Yosim (to North American Rockwell Corp.), 4-15-1969.

Patent number	Inventors, assignee, and date
3,438,727	L. A. Heredy (to North American Rockwell Corp.), 4-15-1969.
3,438,728	L. R. F. Grantham (to North American Rockwell Corp.), 4-15-1969.
3,438,733	L. R. F. Grantham and C. M. Larsen (to North American Rockwell Corp.), 4-15-1969.
3,438,734	L. R. F. Grantham (to North American Rockwell Corp.), 4-15-1969.
3,448,061	G. Mika (to Farbwerke Hoechst), 6-3-1969.
3,454,356	A. K. S. Raman (to Esso Research and Engineering Co.), 7-8-1969.
3,518,206	D. M. Sowards and A. B. Stiles (to E. I. DuPont de Nemours), 6-30-1970.
3,552,912	R. F. Bartholomew and H. M. Garfinkel (to Corning Glass Works), 1-5-1971.
3,552,921	G. C. Blytas (to Shell Oil Co.), 1-5-1971.
3,574,543	L. A. Heredy (to North American Rockwell Corp.), 4-13-1971.
3,574,544	L. A. Heredy, D. E. McKenzie, and S. J. Yosim (to North American Rockwell Corp.), 4-13-1971.
3,574,545	L. F. Grantham (to North American Rockwell Corp.), 4-13-1971.
3,615,196	A. B. Welty, Jr., A. K. S. Raman, and C. M. Lathrop (to Esso Research & Engineering Co.), 10-26-1971.
3,667,910	Y. Eguchi (to Takeda Chemical Industries, Ltd.), 6-6-1972.
3,755,549	H. Guth (to Bayer Aktiengesellschaft), 8-28-1973.
3,776,854	F. M. Dantzenberg, H. W. Konwenhaven, and J. E. Naber (to Shell Oil Co.), 12-4-1973.
3,773,893	D. J. Mandelin (to Occidental Petroleum Corp.), 11-20-1973.
3,789,019	A. B. Stiles (to E. I. DuPont de Nemours), 1-29-1974.
3,793,230	L. Dorn, G. Heinze, J. Wokulat, M. Möller, and F. Rubsam (to Bayer Aktiengesellschaft), 2-19-1974.
3,795,732	J. W. Fleming (to Koppers Co., Inc.), 3-5-1974.
3,829,457	E. Berrotti and P. Koch, 8-13-1974.
3,848,056	A. G. Fonseca (to Continental Oil Co.), 11-12-1974.
3,855,386	R. H. Moore (to Battelle Memorial Institute), 12-17,1974.
3,856,927	P. L. Silveston and R. R. Hudgens (to University of Waterloo), 12-24-1974.
3,873,670	J. J. Dugan and I. S. Pasternak (to Exxon Research & Engineering Co.), 3-25-1975.
3,873,674	R. Rosenthal, R. W. Rieve, J. A. Kieras, and G. A. Bonetti (to Atlantic Richfield Co.), 3-25-1975.
3,875,294	L. Reh, K. Vydra, K-H. Doerr, H. Grimm, and K-H. Hennenberger (to Metallgesellschaft Aktiengesellschaft), 4-1-1975.
3,880,985	H. Haeseler, L. Dorn, W. Möller, J. Wokulat, F. Rubsam, and G. Heinze (to Bayer Aktiengesellschaft), 4-29-1975.
3,897,545	L. Reh, K-H. Dörr, H. Grimm, and K. Vydra (to Metallgesellschaft Aktiengesellschaft), 7-29-1975.
3,926,854	J. M. Whelan and R. J. Brook (to University of Southern California), 12-16-1975.
3,933,991	L. Dorn, G. Heinze, J. Wokulat, W. Möller, and F. Rubsam (to Bayer Aktiengesellschaft), 1-20-1976.
3,987,153	A. B. Stiles (to E. I. DuPont de Nemours), 10-19-1976.
3,997,655	L. Reh, K-H. Dörr, H. Grimm, and K. Vydra (to Metallgesellschaft Aktiengesellschaft), 12-14-1976.

Patent number	Inventors, assignee, and date
Re 29,145	L. Dorn, G. Heinze, J. Wokulat, W. Möller, and F. Rubsam (to Bayer Aktiengesellschaft), 3-1-1977.
4,042,668	J. L. Balmat (to E. I. DuPont de Nemours and Co.), 8-16-1977.
4,124,695	J. M. Whelan (to University of Southern California), 11-7-1978.
4,126,578	F. G. Sherif (to Stauffer Chemical Co.), 11-21-1978.
4,127,509	P. Leclercq (to Produits Chimiques Ugine Kuhlmann), 11-28-1978.
4,158,048	P. Leclercq (to Produits Chimiques Ugine Kuhlmann), 6-12-1979.
4,184,980	F. G. Sherif and W. H. Smith (to Stauffer Chemical Co.), 1-22-1980.
4,193,894	J. Villadsen (to Monsanto Co.), 3-18-1980.
4,206,086	F. W. Sherif (to Stauffer Chemical Co.), 6-3-1980. (Corresponds to German 2,640,169, 3-24-1977.)
4,213,882	H. Kranich (to Johns-Manville Corp.), 7-22-1980.
4,244,937	J. A. Durkin (to Texaco, Inc.), 1-13-1981.
4,284,530	F. G. Sherif (to Stauffer Chemical Co.), 8-18-1981.
4,285,927	H. Hara, T. Kanzaki, H. Motomura, and A. Adashi (to Nippon Shokubai Kagaku Kogyo Co. Ltd.), 8-25-1981. (Partly corresponds to Japan 51(76)—139,586, 11-31-1976.)
4,294,723	H. Hara, T. Kanzaki, H. Motomura, and A. Adashi (to Nippon Shokubai Kagaku Kogyo Co. Ltd.), 10-23-1981. (Partly corresponds to Japan 51(76)—139,586, 11-31-1976).

[a] Also included in previous list [3].

References

1. B. Vidon (1972). *Chem. Proc. Eng.* **53**(9), 34–35.
2. National Materials Advisory Board (1978). "Vanadium Supply and Demand Outlook" (Publication NMAB-346). National Academy of Sciences, Washington, D.C.
3. W. W. Duecker, and J. R. West (eds.) (1971). "The Manufacture of Sulfuric Acid", Reinhold, New York. (Originally printed 1959.)
4. C. N. Kenney (1975). *Cat. Rev. Sci. Eng.* **11**(2), 197–224.
5. A. I. More (ed.) (1981). "Making the Most of Sulphuric Acid," Proc. British Sulphur Corp. 5th International Conference, London, England.
6. I. P. Mukhlenov, E. I. Dobkina, S. M. Kuznetsova, Z. G. Filippova, B. A. Lipkind, G. L. Kustova, N. M. Korchagin, and T. F. Merkulova (1976). *Khim. Prom* **10**, 769–770. (English version, pp. 807–880.)
7. I. P. Mukhlenov, E. I. Dobkina, and G. M. Arnautova (1972). *Issled. Obl. Neorg. Tekhnol.* 305–309.
8. R. S. Rozentsueig, A. M. Larionov, S. M. Kuznetsova, and E. I. Dobkina, (1980). *Geterog. Katal. Protsessy* pp. 86–90.
9. I. P. Mukhlenov, I. S. Safonov, Z. G. Filippova, and E. I. Dobkina (1977). *Khim. Prom.* **1**, 35–36. (English version, pp. 37–38.)
10. B. T. Vasilev, N. P. Dobroselskaya, and B. M. Maslennikov (1979). *Khim. Promst. Rubezhom* **2**, 40–57.

11. M. Cuccetto, F. Traina, A. Cappelli, A. Collina, and M. Dente (1971). "Sulfur and SO_2 Develop" pp. 123–127. AIChE, New York.
12. R. H. Fariss (1963). Inst. Chem. Eng. London June 20–21, 51–54. (CHEME Symp. Series No. 11.)
13. M. Bodenstein, and W. Pohl (1905). Z. Elektrochem. 11 373.
14. W. H. Evans, and D. D. Wagman (1952). J. Res. Nat. Bur. Stand. 49, 141.
15. A. F. Kapustinsky, and L. A. Shamovsky (1936). Acta Physicochim. URSS 4, 791.
16. R. W. Lovejoy, J. H. Colwell, D. F. Eggers, Jr., and G. D. Halsey, Jr. (1962). J. Chem. Phys. 36(3), 612–617.
17. G. V. Mikhailov, and V. S. Vitkov (1975). Zh. Prikl. Khim. 48(10), 2292–2293. (English version, p. 2365, Part 2.)
18. L. W. Ross (1966). Sulfur 65, 37.
19. D. R. Stull, and H. Prophet (1971). "JANAF Thermochemical Tables" 2nd edition (NSRDS-NBS37). U. S. Department of Commerce, Washington, D.C.
20. G. B. Taylor and S. Lehner (1931). Z. Phys. Chem., pp. 30–43.
21. C. N. Kenney (1980). Catalysis London 3, 123–135.
22. A. Urbanek, and M. Trela (1980). Catal. Rev. Sci. Eng. 21(1), 73–133.
23. P. Mars, and J. G. Maessen (1965). Proc. 3rd Int. Cong. Catal. 1, 266.
24. A. R. Glueck, and C. N. Kenney (1968). Chem. Eng. Sci. 23(10), 1257–1265.
25. M. Dojcinovic, M. Susic, and S. Mentus (1981). J. Mol. Catal. 11(2/3), 275–282.
26. C. Vayenas (1977). "Electrochemical and Kinetic Studies of the SO_2 Oxidation over Pt and V_2O_5 Based Melts." Ph.D. Thesis, University of Rochester.
27. S. Desagher, L. T. Yu, and R. Buvet (1975). J. Chem. Phys. 72(3), 390–396.
28. D. V. Fikis, K. W. Heckley, W. J. Murphy, and R. A. Ross (1978). Can. J. Chem. 56(24), 3078–3083.
29. N. H. Hansen (1979) Ph.D. Thesis, (in Danish) Technical University of Denmark, Chemistry Department, Lyngby, Denmark.
30. F. P. B. Holroyd (1968). Ph.D. Thesis, Cambridge, England, Chapter 6, pp. 81–91 and Fig. 6.1–6.9.
31. A. Kato, K. Tomoda, I. Mochida, and T. Seiyama (1972). Bull. Chem. Soc. Jpn. 45(3), 690–695.
32. V. M. Mastikhin, O. B. Lapina, V. F. Lyakhova, and L. G. Simonova (1981). React. Kinet. Catal. Lett. 17(1–2), 109–113.
33. D. V. Tarasova, O. M. Chentsova, V. I. Zaikovskii, G. N. Kustova, and G. E. Selyutin (1979). React. Kinet. Catal. Lett. 12(1), 13–17.
34. M. M. Tomishko, and V. M. Maslennikov (1979). Khim. Promst. Ser. Promst. Miner. Udobr. Sernoi Kisloty 2, 12–14.
35. M. M. Tomishko, B. M. Maslonnikov, V. N. Nosov, and L. Ya. Kulikova (1979). Khim. Promst. Miner. Udobr. Sernoi Kisloty 11, 6–9.
36. H. Flood, and T. Forland (1947). Acta Chem. Scand 1, 592–604.
37. H. Lux (1939). Z. Elektrochem. 45, 303.
38. N. Neth, G. Kautz, H. J. Huster, and U. Wagner, (1980). Ger. Chem. Eng. 3, 44–53.
39. G. K. Boreskov, R. A. Buyanov, R. A. and A. A. Ivanov (1967). Kinet. Katal. 8(1), 126, 153–159. (English version, pp. 126–132).
40. S. Weychert, and A. Urbanek (1969). Int. Chem. Eng. 9(3), 396–403.
41. H. Livbjerg, and J. Villadsen (1972). Chem. Eng. Sci. 27(1), 21–38.
42. J. Happel (1979). Oxid. Commun. 1(1), 15–21.
43. J. Happel, M. A. Hnatow, and A. Rodriguez (1973). AIChE J. 19(5), 1075–1078.
44. J. Happel, and M. A. Hnatow (1973). EPA report 650/2-73-020.
45. L. A. Kasatkina, and V. I. Shustov. (1964). Kinetika Kataliz 5(5), 945–948. (English version, pp. 833–836.)

46. V. I. Artoshenko, V. I. Toshinski, A. I. Kharkov, and V. N. Shamraev (1977). *Tezisy Dokl. Ukr. Resp. Konf. Fiz Khim* **12**, 101.
47. J. L. Herce, J. B. Gros, and R. Bugarel (1977). *Chem. Eng. Sci.* **32**, 729–732.
48. P. K. Kokhreidze, S. V. Ivanenko, V. P. Saltanova, and N. S. Torocheshnikov (1979). *Khim. Prom.* **9**, 556–557 (English version, pp. 620–621.)
49. I. P. Mukhlenov, G. T. Slavin, V. E. Soroko, and V. A. Konovalov (1979). *Zh. Prikl. Khim.* **52**(12), 2677–2683. (English version, pp. 2534–2539, Part 1.)
50. S. Novak (1963). *Chem. Prum.* **13**, 516.
51. A. A. Ivanov, G. K. Boreskov, R. A. Buyanov, E. P. Polyakova, L. P. Davydova, and L. D. Kochkiva, (1968). *Kinet. Katal.* **9**(3), 560–564. (English version, pp. 463–466, Part 1.)
52. S. Kovenklioglu, and G. B. De Lancey (1979). *Chem. Eng. Sci.* **34**, 811.
53. C. N. Satterfield (1970). "Mass Transfer in Heterogeneous Catalysis." M.I.T. Press, Cambridge, Massachusetts.
54. P. R. Rony (1968). *Chem. Eng. Sci.* **26**, 1021.
55. P. R. Rony (1969). *J. Catal.* **14**(21), 142–147.
56. J. L. Harris, and J. R. Norman (1972). *Ind. Eng. Chem. Process Des. Dev.* **11**(4), 564–573.
57. G. Chartrand, and C. M. Crowe (1969). *Can. J. Chem. Eng.* **47**, 296.
58. M. Appl, G. Zirker, B. Triebskorn, P. R. Laurer, and H. J. Becker (1980). European Patent No. 019,174.
59. H. Hara (1954). Japanese Patent No. SHO-29-6560.
60. E. I. Dobkina, V. I. Deryuzhkina, and I. P. Mukhlenov (1965). *Kinet. Katal.* **6**(2), 352–355. (English version, pp. 304–306.)
61. V. V. Illarionov (1955). "Heterogeneous Catalysis in Chemical Industry," 393–395. Gos. Nauch. Tekh. Izdatel. Khim. Lit. Moscow. (Russian.)
62. H. F. A. Topsoe, and A. Nielson, (1947). *Trans. Dan. Acad. Tech. Sci.* **11**(1), 3–24.
63. J. Tandy (1956). *Appl. Chem.* **6**, 68–74.
64. K. I. Brodovich, N. P. Krasilnikova, L. P. Shakhunova, N. D. Zavyalova, and M. F. Shaskkina (1963). "Vanadium Catalysts for Contact Production of Sulfuric Acid", pp. 85–91. Goskhimazdat, Moscow.
65. B. S. Milisavlevich, A. A. Ivanov, G. M. Polyakova, and V. V. Serzhautova (1975). *Kinet. Katal.* **16**(1), 103–107. (English version, pp. 80–83. Pt. 1.)
66. I. P. Mukhlenov, S. A. Ikramov, and V. I. Deryuzhkina, (1974). USSR Patent No. 432,916.
67. V. I. Spitsyn, and M. A. Meerov (1952). *Zh. Obshch. Khim* **22**, 90–95. (English version, pp. 959–962).
68. I. P. Muhklenov, S. A. Ikramov, and V. I. Deryuzhkina (1976). *Khim. Prom* **3**, 226–227. (English version, pp. 217–219.)
69. I. P. Mukhlenov, N. N. Selyutina, E. A. Parkhomova, E. S. Rumyantseva, and V. I. Deryuzhkina, (1974). *Zh. Prik. Khim.* **47**(11), 2612. (English version, p. 2702, Part 2).
70. E. V. Gerburt, I. M. Gleikhenganz, B. T. Vasilev, and V. V. Illarionov, (1963). "Vanadium Catalysts For Contact Production of Sulfuric Acid", (V. V. Illarinov and E. V. Gerburt, eds.), pp. 36–42. Goskhimizdat, Moscow.
71. P. Jiru, and L. Kubica (1962). *Int. Chem. Eng.* **2**(4), 531–536.
72. A. V. Osokin, E. I. Dobkina, S. M. Kuznetsova, I. P. Mukhlenov, and S. S. Ordanyan (1980). *Geterog. Katal. Protsessy* 82–86.
73. A. Simecek (1970). *J. Catal.* **18**(1), 83–89.
74. M. Appl, and N. Neth (1979). "Fert. Acids, Proc. Brit. Sulphur Corp. Int. Conf. Fert. 3rd", 42IMA8, pp. XX1–26.
75. M. Appl, and N. Neth (1979). *Sulphur* **145**, 26–29, 38–39.
76. E. K. Dienes, and R. B. Pohlman (1974). Canadian Patent No. 947,271.

77. V. Grosser, and E. Antkowicz (1980). Czech Patent No. 208,272.
78. R. Haase, R. Weiher, H. Deparade, D. Heinz, and D. Radeck, (1970). East German Patent No. 76,216.
79. H. Hara, T. Kanzaki, H. Motomuro, and A. Adachi (1981). Japanese Patent No. Sho 56 31137.
80. A. J. Jeffries, D. E. Cozens, and J. L. Pegler (1974). U.K. Patent No. 1,358,905.
81. P. R. Laurer, and H. J. Becker, (1980). German Patent No. 2,922,116.
82. F. Wolf, and R. Haase (1965). East German Patent No. 116,593.
83. M. M. Tomishko, B. M. Maslennikov, V. N. Talanova, V. S. Lyutikov, and A. M. Bondarenko (1979). *Khim. Promst. Ser. Prom. St. Miner. Udobr. Sernoi Kisloty* **9**, 4–7.
84. G. M. Polyokova, G. K. Boreskov, A. A. Kvanov, L. P. Davydova, and G. A. Marochkina (1971). *Kinet. Katal.* **12**(3), 666–671. (English version, pp. 586–590, Pt. 1.)
85. I. V. Shvedova, I. P. Mukhlenov, and E. I. Dobkina (1978). *Geterog. Katal. Protsessy* pp. 44–46.
86. V. I. Vanchurin, A. I. Malakov, A. G. Amelin, and I. M. Semenov (1979). *Tr. Mosk. Khim Teknol. Inst. im D.I. Mendelleva* **107**, 138.
87. B. Neuman (1935). *Z. Elektrochem. Angew. Phys. Chem* **41**, 589.
88. V. I. Vanchurin, A. I. Malakhov, A. G. Amelin, V. I. Grachev, and N. A. Bukharova (1978). *Tr. Mosk. Khim. Teknol. Inst. im D.I. Mendeleeva* **99**, 83–85.
89. L. V. Bezrukov, V. V. Denisov, and V. L. Bezrukova (1980). *Geterog. Katal. Protsessy* 7–13.
90. A. A. Ivanov, G. K. Boreskov, and V. S. Beskov (1972). *In* "Porous Structure of Catalysts and Transport Processes in Heterogeneous Catalysis: Symp. III, 4th Int. Congress on Catalysis, 1968", (G. K. Boreskov, ed.), pp. 383–404. Akademiai Kiado, Budapest. (English, 1972.)
91. V. Y. Gavrilov, V. B. Fenelonov, A. A. Samakhov, A. A. Ivanov, and G. K. Boreskov (1978). *Kinet. Katal.* **19**(2), 428–434. (English version, pp. 337–342, Part 2.)
92. V. A. Kzis'Ko, V. A. Dzis'Ko, D. V. Tarasova, A. S. Tikhova, R. I. Nazarova, G. P. Vishnyakova, and N. A. Guseva (1968). *Kinet. Katal.* **9**(3), 668–675. (English version, pp. 552–558, Part 2.)
93. K. Lasiewics, A. Dziewanowska-Pudliszak, and J. Weselowski (1969). *Int. Chem. Eng.* **9**(4), 583–587.
94. D. V. Tarasova, V. A. Dzis'Ko, and V. P. Balaganskaya (1969). *Kinet. Katal.* **10**(2), 406–410. (English version, pp. 329–332, Part 2.)
95. D. V. Tarasova, G. K. Boreskov, and Z. Z. Dzis'Ko (1968). *Kinet. Katal.* **9**(6), 1347–1352. (English version, pp. 1111–1115, Part 2.)
96. S. Weychert, and E. Zbiec (1972). *Przem. Chem.* **51**(10), 657–661.
97. H. J. Becker, W. Gosele, N. Neth, and J. Adlkofer (1979). *J. Chem. Ing. Tech.* **51**(8), 789–795.
98. A. G. Amelin, (1971). "Technology of Sulfuric Acid," pp. 151–159, 189–215. Khimiya, Moscow.
99. V. V. Illarionov, and E. V. Gerburt, (eds.) (1963). "Vanadium Catalysts for Contact Production of Sulfuric Acid," pp. 7–19. Goskhimazdat, Moscow.
100. N. Kh. Valitov, F. Kh Ibragimov, G. P. Malyatova, and V. M. Lysikov (1974). *Zh. Prik. Khim.* **47**(3) 550–554. (English version pp. 552–555).
101. N. M. Petrov, V. I. Malkiman, I. A. Apakhov, L. M. Gribanova, A. L. Olesova, L. N. Manaeva, L. A. Artomasova, and A. A. Zharova (1973). *Khim Prom.* **49**(1), 850–852. (English version, pp. 714–716.)
102. V. M. Maslennikov, G. I. Petrovskaya, N. N. Vidyakin, and N. P. Dobroselskaya (1979). *Kim Promst. Ser. Promst. Miner. Udobr. Sernoi Kisloty* **9**, 1–3.

103. N. M. Petrov, V. I. Malkiman, D. I. Milman, E. A. Kravchenko, and T. Ye. Rodkina (1977). *Tr. Ural. Nauchno Issled. Khim. Inst.* **43**, 11–17.
104. A. V. Shkarin, I. A. Ovsyannikova, N. G. Tatarnikov, A. A. Samakhov and M. V. Kozlov (1976). *Kinet. Katal.* **17**(1), 250–252. (English version, pp. 218–220, Part 2.)
105. J. L. Herce, J. Besombes-Vailhe, and R. Bugarel (1973). *Kinet. Katal.* **14**(1), 118–122. (English version, pp. 90–94, Part 1.)
106. N. A. Andreeva, S. Yu. Eliseev, N. P. Matveeva, and E. G. Semin (1978). *Geterog. Katal. Protsessy uzv. Filtruys. Sloe* pp. 38–43.
107. V. Mayagoitia, J. Besombes-Vailhe, and R. Bugarel (1972). *Ind. Chim. Belg.* **37**, 647–653.
108. G. K. Boreskov, V. A. Dzis'ko, D. V. Tarasova, and G. P. Balaganskaya (1970). *Kinet. Katal.* **11**(1), 181–186. (English version, pp. 144–148, Part 2.)
109. H. Jensen-Holm, H. Livbjerg, J. Villadsen (1977). Soltoft Memorial Publication, Research Results. Department of Chemical Engineering Technical University, Copenhagen. Denmark,
110. P. Grydgaard, H. Jensen-Holm, H. Livbjerg, and J. Villadsen (1978). "Chemical Reaction Engineering-Houston", pp. 582–596, (V. W. Weekman, Jr. and D. Luss, eds.), pp. 582–596. ACS Symposium Series 65. Am. Chem. Soc., Washington, D.C.
111. American Society Testing and Materials (1981). "Annual Book of ASTM Standards" Vol 25, pp. 1108–1113. (Note: Method D4058 will appear in the 1982 edition).
112. J. R. Donovan, R. M. Smith, and J. S. Palermo (1977). *Sulphur* **131**, 46–50.
113. B. Dukenovic, and G. Valcik (1980). *Hem. Ind.* **34**(4), 105–109.
114. S. K. May (1980). *Sulphur* **146**, 30–34.
115. S. Kovenklioglu, and G. B. DeLancey (1979). *Can. J. Chem. Eng.* **57**, 165–175.
116. J. Michalek, and J. Vosolsobe (1980). *Chem. Prum.* **30**(55), 566–569.
117. G. K. Boreskov, L. G. Ritter, and E. I. Volkova (1949). *Zh. Pviv. 1. Khim* **22**(3), 250–260.
118. H. Hara (1967). *Annales Ge'nie Chim. Congre's Int. Soufre, Toulouse,* **3** 187–196.
119. P. Schoubye, and A. Albjerg (1978). *Sulphur* **138**, 34–36.
120. J. R. Donovan, and J. S. Palermo (1978). *Sulphur* **138**, 39–41.
121. H. Livbjerg, and J. Villadsen (1971). *Chem. Eng. Sci.* **26**, 1495–1503.
122. G. P. Korneichuk, N. A. Stukanovskaya, and I. T. Chashechnikova (1979). *Katal. Katal.* **17**, 28–31.
123. A. Miyamoto, Y. Yamazaki, M. Inomata, and Y. Murakami (1978). *Chem. Lett.* **12**, 1355–1358.
124. M. S. Wainwright, and D. W. B. Westerman (1977). *Chromatographia* **10**(11), 665–668.
125. J. R. Donovan, E. D. Kennedy, D. R. McAlister, and R. M. Smith (1977). *Chem. Eng. Prog.* **73**(6), 89–94.
126. J. R. Donovan, J. S. Palermo, and R. M. Smith (1978). *Chem. Eng. Prog.* **74**(9), 51–54.
127. B. Dukenovic, G. Valcik, and P. Putanov (1980). *Hem. Ind.* **34**(9), 256–259.
128. S. Ergun (1952). *Chem. Eng. Prog.* **48**, 89–94.
129. K. M. Malin, (ed.) (1971). "Sulfuric Acid Workers Handbook," pp. 496–517. Khimiya, Moscow.
130. B. Kadlec, R. R. Hudgins, and P. L. Silveston (1973). *Chem. Eng. Sci.* **28**(3), 935–945.
131. H. Livbjerg, B. Sorensen, and J. Villadsen, (1974). "Chemical Reaction Engineering II: Third Int'l. Symp." (H. M. Hulburt, ed.), pp. 242–258. In Advances in Chemistry, Vol. 133. Am. Chem. Soc., Washington, D.C.
132. J. Villadsen, and H. Livbjerg (1978). *Catal. Rev. Sci. Eng.* **17**(2), 203–272.
133. O. T. Chen, and R. G. Rinker (1978). *Chem. Eng. Sci.* **33**(9), 1201–1209.
134. A. R. Balakrishnan, and D. C. T. Pei (1974). *Ind. Eng. Chem. Process Des. Dev.* **13**(4), 141–146.

135. J. Michalek (1980). *Sb. Vys. Sk. Chem. Technol. Praze* **B25**, 57–65.
136. V. I. Malkiman, N. M. Petrov. E. A. Kravchenko, L. N. Manaeva, G. M. Kesareva, and V. N. Sabel'ev (1979). *Khim Prom.* **8**, 475–476. (English version, pp. 540–542.)
137. P. Putanov, D. Smiljanic, B. Djukanovic, N. Jovanovic, and R. Herak (1973). *Catalysis* **2**, 1061–1071.
138. V. I. Vanchurin, A. G. Amelin, A. I. Malakhov, and G. M. Semenov (1978). *Tr. Mosk. Khim. Teknol. Inst. im D.I. Mendeleeva* **99**, 86–89.
139. A. Tamura, R. R. Hudgins, P. L. Silveston (1976). *J. Catal.* **42**(1), 122–130.
140. G. V. Cherepkov, I. P. Mukhlenov, A. M. Shevyakov, and E. I. Dobkina (1976). *Kinet. Katal.* **17**(1), 204–207. (English version, pp. 172–174, Part 2.)
141. I. P. Mukhlenov, S. A. Ikranov, N. P. Matveeva, and V. I. Deryuzhkina (1972). *Zh. Prikl. Khim.* **45**(6), 1341–1344. (English version, pp. 1384–1387, Part 2.)
142. P. P. Putanov (1978). *Mekh. Katal. Prot. Sib. Chteniya Po Katal.* pp. 53–66, Novosibirsk.
143. V. V. Illarionov, D. F. Terentov, and N. I. Buryak (1963). "Vanadiam Catalysts for Contact Production of Sulfuric Acid," pp. 48–56. Goskhimazdat, Moscow.
144. E. I. Makarova, and N. I. Buryak (1963). "Vanadium Catalysts for Contact Production of Sulfuric Acid", pp. 57–61. Goskhimazdat, Moscow.
145. J. R. Donovan, and P. J. Stuber (1967). *J. Metals* **19**(11), 45–50.
146. S. A. Hassan, and F. T. Iskandar (1976). *J. Catal.* **43**,(1–3), 243–251.
147. Am. Inst. Chem. Engs. (1982). "Sulfuric/Phosphoric Acid Plant Operations" (CEP Technical Manual). New York.
148. J. R. Donovan, and D. R. McAlister (1982). "Sulfuric/Phosphoric Acid Plant Operations" (CEP Technical Manual). pp 212–215, Am. Inst. Chem. Engs. , New York.
149. R. R. Hudgins, and P. L. Silveston (1981). *Environ. Sci. Tech.* **15**(4), 419–422.
150. V. I. Artoshchenko, N. F. Kleshchev, N. E. Kucherenko, and A. N. Butenko (1977). *Tezisy Dokl. Ukr. Resp. Konf. Fiz. Khim.* **12**, 106–107.
151. G. M. Schwab (1957). "Handbook of Catalysis: Heterogeneous Catalysis," Vol 5, pp. 611–614 Springer-Verlag, Vienna.
152. L. V. Stainov, V. A. Taranushich, and K. G. Il'in (1980). *Izv. Vyssh. Uchebn. Laved Khim. Khim. Technol.* **23**(12), 1528–1531.
153. V. A. Taranushich, A. P. Savostyanov, K. G. Ilin, L. V. Stainov, and V. B. Ilin (1979). USSR Patent No. 703,132.

Index